LIFE IN EUROPE UNDER CLIMATE CHANGE

LIFE IN EUROPE UNDER CLIMATE CHANGE

JOSEPH ALCAMO

UNITED NATIONS ENVIRONMENT
PROGRAMME
NAIROBI, KENYA

JØRGEN E. OLESEN

AARHUS UNIVERSITY
DEPARTMENT OF AGROECOLOGY
TJELE, DENMARK

WILEY-BLACKWELL

A John Wiley & Sons, Ltd., Publication

This edition first published 2012
© 2012 Joseph Alcamo and Jørgen E. Olesen

Wiley-Blackwell is an imprint of John Wiley & Sons, formed by the merger of Wiley's global
Scientific, Technical and Medical business with Blackwell Publishing.

Registered Office
John Wiley & Sons Ltd, The Atrium, Southern Gate, Chichester, West Sussex, PO19 8SQ, UK

Editorial Offices
350 Main Street, Malden, MA 02148-5020, USA
9600 Garsington Road, Oxford, OX4 2DQ, UK
The Atrium, Southern Gate, Chichester, West Sussex, PO19 8SQ, UK

For details of our global editorial offices, for customer services and for information about how to
apply for permission to reuse the copyright material in this book please see our website at www.
wiley.com/wiley-blackwell.

Library of Congress Cataloging-in-Publication Data

Alcamo, Joseph.
  Life in Europe under climate change / Joseph Alcamo, Jørgen E. Olesen.
      p. cm.
  Includes bibliographical references and index.
  ISBN 978-1-4051-9619-2 (cloth) – ISBN 978-1-4051-9618-5 (pbk.)
  1. Climate change mitigation–Europe. 2. Climatic changes–Social aspects–Europe. 3. Human
beings–Effect of climate on–Europe. I. Olesen, Jørgen E. II. Title.
  QC989.A1A46 2012
  304.2'8–dc23
                                    2011045562

A catalogue record for this book is available from the British Library.

Wiley also publishes its books in a variety of electronic formats. Some content that appears in print
may not be available in electronic books.

Set in 10 on 12.5 pt Minion by Toppan Best-set Premedia Limited
Printed and bound in Malaysia by Vivar Printing Sdn Bhd

1   2012

Contents

Color plate pages fall between pp. 150 and 151

Foreword

After a cold winter, scientists fighting over some leaked documents, and a Climate Summit in Copenhagen that fell far short of delivering everything hoped for, it is tempting to relax. Maybe it wasn't that serious anyhow? Maybe scientists were mistaken? Maybe we don't have to bother about climate change?

This book shows us how dangerous this approach could be. *Life in Europe under Climate Change* demonstrates why action on climate change is the challenge of our time. And it makes us realize why continuing with business as usual isn't a possibility. Or rather it still is, but it would have dire consequences and therefore be highly irresponsible.

Even if the international community manages to keep global warming below the danger level of around 2°C above the pre-industrial temperature – equivalent to some 1.2°C above today's level – climatic changes with serious effects will be unavoidable.

We will all be affected, so we all have a direct interest in understanding what life in Europe under climate change will be like. This book paints a comprehensive picture. I particularly appreciate that it does so in plain language that is easily accessible to non-scientists, like me.

Climate change is already making itself felt and, as the central chapters of the book show, virtually no sector of the economy will be immune to its impacts. It will affect our health and that of the ecosystems on which our prosperity and well-being depend; the production of food and of other economically essential products and services; and our seas, our coasts and the security of our water supplies.

By anticipating and preparing for these impacts, however, we can strengthen our resilience and minimize our vulnerability. The earlier we act, the less this process of adaptation will cost.

Adaptation needs to be undertaken primarily at the local, regional and national levels, but a coordinated approach across the EU can support and strengthen these efforts. That is why the European Commission is developing an EU adaptation strategy.

Adaptation is of course no substitute for mitigating the greenhouse gas emissions that are causing climate change in the first place. Adapting to climate change and building a low-carbon future have to go hand in hand.

With EU's climate and energy targets for 2020, Europe has already taken a first big step on the road to building a low-carbon economy. But we must continue on this path. This book provides strong evidence of why.

Connie Hedegaard
EU Commissioner for Climate Action

Preface

Many different forces shape the European way of life. Some are geographic, some historical and still others, political. But among the most pervasive forces is the long term pattern of weather, better known as "climate". Consider how much the patterns of rain, the cycles of seasonal temperatures, the duration of cloudiness or sunshine hours influence our water supply and energy use, as well as our source of food. Think about how they determine the widths of rivers or depths of lakes, the greenness of the landscape, the height of trees, and the whole complex of ecosystems. Consider further that the scientific community is all but certain that this climate is changing. So we are left with the question, what will life be like under climate change?

It is clear that there is wide public interest in the impacts of climate change in Europe, and a vast literature on the subject. But it is also clear that most of this literature is too technical for a general audience. This is not a good situation because the impact of climate change is too critical a subject to be left to the experts. Knowledge of these impacts and their implications should be widely available to teachers, students, policymakers, and the general public. And that is the crux of this book – to review what scientists say about the possible effects of climate change on European society and nature, and to make this knowledge accessible to a wide readership.

Much of this book is based on the "Europe" chapter of the Fourth Assessment Report on climate impacts of the Intergovernmental Panel on Climate Change (IPCC). In this book we often follow the review of literature presented in that chapter, but expand and explain it in our own words. We have also added new subjects, such as the discussion of Climate Protection found in Chapter 10. Throughout the book we interpret the IPCC report in our own way, and retain responsibility for this interpretation.

The authors of this book co-wrote the Europe IPCC chapter along with many other colleagues, and we are indebted to them for the excellent compilation of knowledge and astute analysis found in that chapter: José M. Moreno, Bela Novaky, Marco Bindi, Roman Corobov, Robert Devoy, Christos Giannakopoulos, Eric Martin, Anatoly Shvidenko, M. Araújo, A. Bristow, J. de Ronde , A. Dlugolecki , S. des Cler, L.P. Graham, E. Jeppesen, A. Guisan, S. Kovats, P. Lakyda, J. Sweeny and J. van Minnen.

We are grateful to many others who have helped in provision and preparation of material for this book and to those who have read and commented upon drafts of the chapters. Many scientists and organizations have provided useful graphs that we have been allowed to reproduce here and which greatly helps in illustrating the causes and effects of climate change. The secretaries at Department of Agroecology at Aarhus University, Jytte Christensen, Julie Jensen and Birgit L. Sørensen provided invaluable assistance in preparing illustrations and reference lists and getting the manuscripts organized. The authors are grateful to Harsha Dave and Ernest Imbamba for assistance in proofing reference lists of various chapters.

In closing, the message of this book is that it is likely that climate change will influence and shape European society and its environment for decades to come. We believe that understanding these changes and acting on them will help secure the future of coming generations. But while we can provide data and analysis in this book, we all need to act on this knowledge.

Joseph Alcamo, Chief Scientist, United Nations Environment Programme,
Nairobi, Kenya; and Professor of Environmental Science and Engineering,
University of Kassel, Germany (on temporary leave)
Jørgen E. Olesen, Professor of Agricultural Sciences,
Aarhus University, Tjele, Denmark
March, 2012

1 Introduction

Observe constantly that all things take place by change, and . . . consider that the nature of the Universe loves nothing so much as to change the things which are, and to make new things like them. (Marcus Aurelius, Meditations)

What this book is about

Life in Europe will indeed go on as the climate changes, but it will not be the same. The air will be warmer, the winds stronger or in some places weaker, and the patterns of rainfall and snowfall will change. It seems likely that when it rains it will rain harder, and when it snows the whiteness will cover the hills and valleys for a briefer time than it once did. Much of the vegetation cover of Europe will be altered, and along with it the available habitat for mammals, birds, insects and other animals. Ice-locked harbors will become rare, and heatwaves and forest fires less so. Since alterations in the weather and its long-term patterns (called "climate") touch everything in nature and society, we should expect that life in Europe will also differ. These changes will be minor in some cases, profound in others, but in any case, pervasive.

Considering how all-encompassing these changes will be, it is unfortunate that a larger readership is not better informed about them. But this deficit is understandable because the latest scientific information continues to be mostly locked away in technical papers with a limited readership, in reports too voluminous to be read, and in scientific workshops too specialized for most concerned citizens and students. This brings us to the point of this book. It is the aim of the authors to mine the latest insights and information from the scientific community to

Life in Europe Under Climate Change, First Edition. Joseph Alcamo and Jørgen E. Olesen.
© 2012 Joseph Alcamo and Jørgen E. Olesen. Published 2012 by John Wiley & Sons, Ltd.

explain what life in Europe under climate change could be like, or will be like, and to bring this knowledge to an audience outside the specialized scientific community. But just anticipating future impacts of climate change is not enough, since we also need to know how to cope with these impacts. Another goal of this book is, therefore, to report how Europeans can help slow down the tempo of climate change by reducing greenhouse gas emissions, and how we can adapt to the climate change we cannot avoid.

Since much has already been written about climate change, it is fair to ask in which way this book will be different from earlier books on the subject. It is a fact that scientists have recently made available a new wealth of information that is relevant for those outside the circle of climate researchers. First, climate change is no longer considered just a phenomenon of the future because the first signs of climate change and its impacts are evident today. Recent trends in precipitation and temperature (see Chapter 2) have already stimulated an earlier budding of plants and flowers, a faster tempo in glacier melting, and perhaps more frequent heatwaves and floods. Another new aspect is that the variety and intensity of different kinds of climate impacts on particular parts of Europe are better understood. Previously a great number of scientific studies limited themselves to general conclusions about how climate change will affect crop production, forest growth, the distribution of plants, or the sea level. But now, after a decade of computer modeling, analogue experiments, and field observations, we can make more confident statements about specific impacts in particular places. It is as if an out-of-focus photo has suddenly become sharper. There is abundance of new analyses for Europe providing estimates about the future frequency of storm surges along its coastline, about the changing rate in the occurrence of floods and droughts in its river basins, about threatened ecosystems in its mountains, and about new climate-related health risks in its cities. Novel patterns are emerging in our understanding of how and where climate change will affect Europe. One of the newest insights is the presumption that climate change will play out very differently in various parts of Europe. The computer models suggest that Europe is getting warmer overall, but in the coming decades wetter north of the Alps and drier to the south. The implications of this are many as shown in Chapter 9. Finally, as knowledge about the specific impacts of climate change increases, it becomes clearer how we can best cope with the threat of climate change. Hence, there is a growing amount of information available on how Europe can adapt to climate change impacts along its coastline, on its agricultural fields, within its cities, and elsewhere. In many ways, this book will present an up-to-date view of life in Europe under climate change.

Which impacts have already been observed?

When climate impact studies first began to appear in the 1980s, scientists were faced with the dilemma of trying to anticipate climate impacts that had little precedence. But scientists no longer need to speculate about the future because we are already experiencing a foretaste of what it will be like to live under climate change. Scientists now agree that climate change is already having an impact on Europe and worldwide. As we will see, the signs are generally mild and mostly unthreatening, but they imply that Europe is on a trajectory to more significant impacts.

What are the signs of these changes, and how can we distinguish them from the strong year-to-year variability of climate? We know from everyday experience that climate has a significant seasonal and year-to-year variation and causes a similar variation in biological processes. If spring in northern Europe is particularly cold, plants tend to start budding later than in the previous year. Likewise, a year with an especially windy, dry autumn will see deciduous trees lose their leaves earlier than usual. The question is, if climate and its impacts already have a strong variability, how can we discern the effect of long-term climate change amid this variability? The answer is provided by "attribution studies" (see Box 1.1). To attribute a particular change to global warming, scientists must first establish if a long-term trend in climate can be distinguished from natural climate variability. Then they must demonstrate that the trend is related to society's impact on climate, either through its greenhouse gas emissions or through its modification of land use. Based on this procedure, scientists have attributed numerous changes in Europe's environment to climate change as shown in Table 1.1.

Even though patterns of precipitation have a profound impact on society and the rest of nature, scientists until now have found a closer connection between temperature changes and climate impacts. This is because precipitation has a much stronger variability over time and space than temperature, and this makes it harder to discern long-term trends in precipitation. One place where the effect of warmer temperatures has become apparent is in the upper reaches of Europe's mountains. Because of warming conditions, high elevations have become steadily more habitable for trees with the result that the tree line has moved higher in many parts of the continent.[21] In the mountains of sub-Arctic Sweden, the tree line has climbed about 60 meters during the 20th century.[22] In the same way that higher elevations have become more habitable for trees, global warming has also made Europe's northern latitudes more habitable for plants that would previous have been unable to survive. Indeed, weather conditions are gradually becoming less harsh at the southern boundary of Europe's tundra zone and its hardy indigenous vegetation is gradually being replaced by less resilient types of brush and trees.

If it seems that spring arrives sooner now than it did some years ago, this perception has been confirmed by science. Based on extensive fieldwork carried out worldwide, but concentrated in Europe, we now know that the northern spring has been occurring about 2.3–5.2 days earlier each decade over the last 30 years.[23] This estimate is based on *phenological studies*, the investigation of phenomena of plants and animals that recur periodically, such as leaf unfolding, flowering, leaf fall, appearance or emergence of butterflies, and bird migration.

Box 1.1 Climate change attribution studies. How science tests for the occurrence of climate change impacts

"Attribution studies" are carried out with two basic aims. First, to determine whether an observed change in society or the rest of nature, say the retreat of a glacier, has to do with natural variability of climate or with a new long-term trend in climate. Typical questions might be: has a particular glacier receded over the past decades, and if so, is this retreat consistent or not with its "expected" decade-to-decade fluctuation? If the retreat of a glacier is found to be statistically outside its normal decade-to-decade changes, then scientists declare that they have detected the impact of long-term climate change on the retreat of the glacier. The second aim is to investigate whether the observed climate changes can be clearly attributed to anthropogenic causes. This is a trickier question because, over geological time, strong fluctuations occur in climate having nothing to do with humans. An example of a climate change caused by a factor independent of humanity is the effect of a volcanic eruption. When volcanoes erupt they send large quantities of dust into the upper atmosphere; this material inhibits sunlight from penetrating the atmosphere and can have a major cooling effect on the earth's surface for several years. Hence, "detecting" the impact of long-term climate change is not the same as "attributing" it to anthropogenic causes.

Three similar, but slightly different approaches are taken to test attribution:[1]

One way is to use climate models to make different "predictions" of the climate conditions observed over recent decades. The key point here is that scientists include the effect of anthropogenic greenhouse gas emissions in one set of model runs and exclude it in another. The two sets of model runs, with and without emissions, are then compared to see which one gives a better prediction of the actual observed climate. Model experiments of this type have shown that including the effect of greenhouse gases produces a better simulation of observed climate, thereby confirming the impact of anthropogenic greenhouse gases on climate. Continuing with the example of the retreating glacier, the question here is, do the emissions theoretically make a difference in the climate computed at the location of the glacier in question? If yes, then the retreat of the glacier can be attributed to anthropogenic climate change.

Another approach is to summarize data from many different observational sites and show that changes in physical and biological systems are consistent with the hypothesis of a warming world caused by greenhouse gases emitted by society. For example, scientists have observed changes in the timing of the life cycles of plants (such as flowering and ripening of fruit) at numerous sites in Europe and have determined that these changes are, in sum, consistent with an observed anthropogenically-caused warming trend.

In a third approach, scientists have compared changes in many different aspects of nature (river run-off, plant growth characteristics, extent of glaciers) to observed regional warming trends related to global greenhouse gas emissions from society and have concluded (in the tortuous language of science) that it is extremely unlikely that observed changes are not related to anthropogenic climate change.

Table 1.1 Examples of changes in natural and managed ecosystems in Europe attributed to recent temperature and precipitation trends. Source: Review in Alcamo et al.[2]

European Region	Observed change
Coastal and marine systems	
Northeast Atlantic, North Sea	Northward movement of plankton and fish[3]
Terrestrial ecosystems	
Europe	Upward shift of the tree line[4]
Europe	Phenological changes (earlier onset of spring events and lengthening of the growing season)[5]; increasing productivity and carbon sink during 1950–1999 of forests (in 30 countries)[6]
Alps	Invasion of evergreen broad-leaved species in forests; upward shift of *Viscum album*[7]
Scandinavia	Northward range expansion of *Ilex aquifolium*[8]
Fennoscandian mountains and subartic	Disappearance of some types of wetlands (palsa mires) in Lapland; increased species richness and frequency at altitudinal margin of plant life[9]
High mountains	Change in high mountain vegetation types and new occurrence of alpine vegetation on high summits[10]
Agriculture	
Northern Europe	Increased crop stress during hotter drier summers; increased risk to crops from hail[11]
Britain, southern Scandinavia	Increased area of silage maize (more favourable conditions due to warmer summer temperatures)[12]
France	Increases in growing season of grapevine; changes in wine quality[13]
Germany	Advance in the beginning of growing season for fruit trees[14]
Cryosphere	
Russia	Decrease in thickness and areal extent of permafrost and damages to infrastructure[15]
Alps	Decrease in seasonal snow cover (at lower elevation)[16]
Europe	Decrease in glacier volume and area (except some glaciers in Norway)[17]
Health	
North, East	Movement of tick vectors northwards, and possibly to high altitudes[18]
Mediterranean, Atlantic, Central	Heat-wave mortality[19]
Atlantic, Central, East, North	Earlier onset and extension of season for allergenic pollen[20]

Since these things happen regularly and seasonally, and are closely connected to weather events, it appears that they are sensitive to long-term climate trends. In general, if temperatures get consistently higher, the occurrence of phenological events will occur earlier in the year. Indeed, many phenological changes have already been observed in Europe, for example, the overall lengthening of the growing period of plants. Since 1951 the growing period of some trees in Germany has increased by one to two days per decade.[24] Over the same period, the length of the growing season of selected types of vegetation in Switzerland have become nearly three days longer.[25] The earlier onset of spring and summer is also apparent in advances in leaf unfolding and flowering of many different plant species; data from the past 30–50 years from 21 different European countries show that these phenomena have been occurring, on the average, one to three days earlier each decade.[26]

While spring has arrived earlier in the year for many types of wild vegetation, the same applies to cultivated crops and fruit trees. The difference is that changes in crop growth are not as pronounced as for uncultivated vegetation because the growth of crops is also strongly influenced by farming practices. For instance, the effects of changing weather conditions could be easily masked by a farmer introducing a new crop variety to his fields. Nevertheless, climate change has had a discernible effect on the overall timing of crop growth in Europe. In Germany, for instance, various junctures in the growth cycle of crops have moved up in the calendar year an average of 2.1 days per decade between 1951 and 2004.[27] Wine grapes have been similarly affected. Based on data from a wide range of vineyards and grape types across Europe, it was found that the bloom dates of wine grapes advanced an average of 3 days per decade over a 30– to 50-year period preceding 2004.[28]

Not only plants, but animals also have experienced phenological changes related to climate change. The appearance of butterflies in the United Kingdom has advanced about three days per decade since the mid-1970s.[29] The spring migration of selected bird species in Europe and North America has been occurring around 1–4 days earlier each decade over the past 30–60 years.[30]

Climate change impacts have manifested themselves not only on land, but in the sea as well. For example, scientists have observed changes in the distribution of fish species along Europe's Continental Shelf, particularly a northward extension of the ranges of such fish as sardines, anchovies, red mullet and bass, and they have related these changes to regional warming.[31]

As ocean temperatures increase, water also expands and raises the average level of the seas. This effect has contributed to a sea-level rise averaging about 8–30 mm per decade since the beginning of the 20th century.[32] Many other factors, however, have contributed to this effect (see Chapter 6). As the sea has risen it has become one of the major causes of increasing erosion along Europe's coastline.[33]

Changes have also been observed in Europe's "cryosphere", the term given to its frozen realm of ice and snow. The cryosphere includes mountain glaciers, seasonal snow cover, sea ice, and land and river ice. Glaciers and permafrost are particularly useful signposts of changing climate, for two reasons. First, since they are made up of frozen water they are sensitive to temperature and other climate variables. Second, they tend to smooth out year-to-year variations in climate and are thus indicators of longer-term climate trends. For these reasons it is especially significant that the thickness and areal coverage of permafrost in Russia has been decreasing[34] and that glaciers are retreating over most of Europe.[35] It has been estimated that Alpine glaciers have lost nearly 50% of their area between 1850 and 2000.[36] The snow cover in the Alps has also been steadily declining at lower elevations[37] with obviously serious repercussions for the ski industry.

While the earlier onset of spring or disappearance of snow cover may not pose a serious risk to Europe, a more threatening trend is the apparent increase in hot summer temperatures. In 2003 an unexpectedly severe heatwave struck Europe, raising summer temperatures over most of southern and central Europe by 3–5°C above normal. Around 70,000 deaths were attributed to the heatwave[38] (see Chapter 3, Box 3.2). While climate change cannot be unequivocally blamed for this event, circumstantial evidence points in this direction.[39] Scientists have confirmed a strong, "unprecedented" trend of increasing summer temperatures since 1977 and, given scientific uncertainties, have estimated that the summer of 2003 may have been the warmest in Europe since at least 1500.[40]

Summer heat is not the only risk to health traced back to climate change. Steadily warmer spring temperatures have not only advanced the growing phases of wild and cultivated plants, but have also caused plants to release their pollen earlier in the year.[41] This, in turn, has brought forward the beginning of the hay-fever season.

The message of observed climate change

What can we conclude from the many observed impacts of climate change? First, they show that climate change already touches many different aspects of nature and society in Europe. The growing processes of plants have been altered and their spatial coverage has been changed, glaciers have shrunk, hydrology has been modified, and the frequency of heatwaves increased. Because society and ecosystems are so intertwined, these observed impacts give us a taste of the complexity and all-inclusive nature of future climate change impacts.

Second, we are seeing changes where we expect them – for example, in the earlier onset of spring and longer growing periods of plants, and in the declining areas of glaciers and permafrost – and this further confirms the theories and models used by science to anticipate future impacts.

Third, although most of the changes we have seen so far do not pose a great risk to society or the rest of nature, some occurrences such as more frequent heatwaves and forest fires are a preview of the damaging impacts that might occur over the coming decades. These events are also a warning that now is the time to begin adapting to climate change. For instance, even though an earlier onset of spring does not seem threatening, it is only one sign that the growth and life cycles of different plants and plant communities are in flux. The following chapters will show that as climate change further unfolds, the alteration in the growth characteristics of some plants and plant communities will give them a competitive advantage over others and in some cases break up important symbiotic relationships. Radical changes are likely to occur in the ranges of plants and animals, with some dying out completely. Even now, scientists believe that changing climate is reducing the habitat of plants and animals and may have caused extinctions not yet detected because studies have not been carried out on a sufficiently detailed spatial scale.[42] It seems inevitable that in the coming decades we will learn to live without some of the flora and fauna we have come to know in earlier times (see Chapter 7).

Meanwhile, the continued melting of glaciers is likely to cause larger run-off flows downstream in late winter and spring. But as the ice masses finally shrink away, they will no longer supply run-off to mountain streams and this may drastically reduce downstream water supply. Later we will see that river run-off will be even more severely curtailed south of the Alps because of declining precipitation and increasing temperatures. Many other impacts will emerge, including more frequent high summer temperatures, which will pose a serious health risk to vulnerable members of society, and which will ignite more frequent forest fires. The message we can glean from observed climate impacts is that they are the beginnings of more damaging impacts that we will experience over the coming decades.

How are future climate impacts studied?

While attribution studies can provide information about the current impact of climate change it is obvious that a different approach is needed in order to address the question of future climate impacts. The assemblage of different methods used for this task is called "climate impact and adaptation assessment". Already in the mid-1990s the Intergovernmental Panel on Climate Change outlined the basics of climate impact assessment in the seven main steps shown in Box 1.2.[43] Another useful and comprehensive guide to this sort of assessment is available from the United Nations Environment Programme.[44]

To assess the future impacts of climate change, scientists must first make assumptions about the future state of the climate and socio-economic conditions in the region or location of interest. Future climate conditions are usually taken

Box 1.2 **Climate impact and adaptation assessment.** This diagram shows the procedure for assessment of climate impacts and adaptation according to the Intergovernmental Panel on Climate Change. Source: Carter et al.[45] The approach shown here has been further elaborated since first being published in 1994.

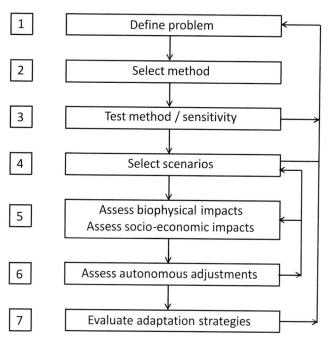

The IPCC procedure for assessing climate change impacts and adaptation involves the following seven steps as presented in the diagram above.

1 Researchers clarify the problem to be studied. They decide on the types of climate impacts to be studied (e.g. is the focus of the study on individual plants, or plant communities, or entire ecosystems at a particular location?), on its geographic coverage (e.g. field plots, a region, a country?), and on other organizational aspects of ecosystems or society.

2 Next, an appropriate methodology is selected for the study. Many options are available.[46] One alternative is to use computer models for describing socio-economic or biophysical impacts. Another is to conduct "analogue experiments", in which a micro-climate at a particular location or region is assumed to represent the future climate at that location. For example, researchers have transplanted mountain vegetation from a higher and cooler elevation to a lower and warmer site as an analogue of future climate change on that mountain (replacing "time" for "space").

3 In the next step the research methodology is tested on the small scale to make sure it is suitable for the full scale study. This can be done through feasibility or pilot studies covering a small area or a few case studies.

Continued

4 Once the methodology has been selected and tested, researchers are ready to conduct the full study. The first step here is to select the scenarios of the future that will serve as a basis for determining if impacts will occur or not. This is a key point since impacts are estimated as the difference with and without climate change of the future states of society and the environment. Hence it is necessary to first develop a baseline scenario without climate change and then to develop a picture of future climate change. Baselines are needed for climate, for the rest of the environment and for the socio-economic situation. Next, scenarios are needed for describing future climate conditions, environmental conditions, and socioeconomic conditions at the site. Not all of this information is needed for all studies. For example, socio-economic information will not always be needed in the study of the impacts of climate change on natural vegetation.

5 Using the scenarios from the previous step, researchers then assess the impact of climate change. In general, they have three options here: (i) They can either compare future changes against a fixed baseline (for example, what will be the future yield of wheat in central Germany under climate change as compared to its current average yield, i.e. assuming no change in the type of wheat or farm management?). (ii) They can compare future changes against a future baseline (what will be the future yield of wheat in central Germany under climate change and no climate change assuming that the type of wheat and farm management practices continue to evolve as they have over past decades?). (iii) They can compare future changes against a future baseline in which actions are taken to adapt to climate change (what will be the future yield of wheat in central Germany under climate change comparing the case of adaptation and no adaptation?)

6 In this step researchers evaluate the likelihood of "autonomous adjustments" of the social, economic or ecological system to climate change. "Autonomous adjustments" are spontaneous adaptations. For example, under drought conditions, plants automatically reduce their transpiration loss of water by closing their stomata. The key question here is, will autonomous adjustments prevent damage to the system under study?

7 Finally, based on the analysis of impacts and autonomous adaptation, the analyst evaluates various options for consciously adapting to climate change to minimize negative effects. Different time scales may need to be considered here, depending on the adaptation problem being studied. If the issue is how to adapt the design of a bridge or dam to climate change then the analysis needs to consider conditions several decades into the future. On the other hand, if the topic is how agriculture can adapt, then perhaps the time horizon of interest is only a few years hence.

Procedures for the systematic analysis of adaptation options are presented in Chapter 9 of this book.

To sum up, these seven steps are a very simplified representation of the complexities of climate impact assessment, and the reader should consult Carter et al.[47] for further details.

from published results of climate models. Assumptions about future socio-economic trends are based on various sources (demographic projections, economic growth forecasts) coming from studies in the social and economic sciences.

Analysts must also determine whether the subject being considered, whether it be an ecosystem or segment of the population, could automatically adjust to changes in climate. These are called "autonomous adjustments". An example of such an adjustment is the case of a wetlands ecosystem moving inland in response to rising sea level. If such an adjustment is possible, it would obviously lessen the negative impacts of climate change. If, however, the wetlands ecosystem is hemmed in by coastal settlements, then an "autonomous adjustment" is not feasible. However, society could still step in and help. In the case of the hemmed-in wetlands, society could actively promote adaptation by re-establishing a similar wetland at another location or by changing local land use ordinances so that wetlands have the capacity to move inland. The evaluation of adaptation measures along these lines is an essential part of climate impact assessment.

Although the impact assessment procedure presented in Box 1.2 has been successfully used in many studies, scientists are now trying out other approaches to make the estimation of climate impacts even more scientifically credible and policy relevant. These approaches include:[48]

- *Vulnerability-based assessment* This identifies the vulnerability of the subject being studied and actions to reduce this vulnerability. An example of such an assessment is the analysis of socio-economic factors such as a lack of social services that make a population in one region more vulnerable than another to the occurrence of drought.[49] Recently, vulnerability assessments has been linked to poverty reduction and other key social issues.
- *Adaptation-based assessment* This focuses on the capacity of the subject to adapt to climate change and includes an evaluation of measures for increasing adaptive capacity. As compared to vulnerability studies, this type of assessment places its accent on the solution side of the problem. An example of an adaptation assessment is the evaluation of various options available to a coastal population for coping with sea-level rise, coastal flooding and other negative impacts of climate change.
- *Risk-assessment* This is yet another way to look at climate impacts. Risk assessment has been used for many years in science and engineering to evaluate the threat of an undesirable, but plausible, future event. A typical application is assessing the risk of a serious accident at a proposed nuclear power plant. This particular way of looking at future risk is now being applied to the question of climate change and measures for minimizing its risks.

To a certain extent, these approaches complement each other, or overlap. Nevertheless, each addresses deficiencies in the others and so provides one part of a complete toolkit for climate impact assessment. In the coming years it is likely that aspects of one methodology will combine with another to form a new and more comprehensive approach to assessing climate impacts and adaptation. With this new approach in hand we will be able to gain an even clearer picture of life in Europe under climate change.

Box 1.3 The IPCC: assessing the state of play of climate science

Considering the wide-ranging importance of climate change to society, it is particularly important to grasp which scientific information is most robust, which has higher or lower confidence intervals, and which is most relevant to climate policy. A special organization to tackle this job was set up in 1988 by the World Meteorological Organization and the United Nations Environment Programme. The organization, named the "Intergovernmental Panel on Climate Change" (IPCC), is mandated by governments to assess the state of climate change science and to communicate policy-relevant points to decision makers and the public at large. The Panel publishes a set of very comprehensive reports at roughly five-year intervals, and additional reports at irregular intervals. In the spirit of full disclosure, the writers of this book are two of the authors of IPCC's latest assessment of climate change impacts on Europe.[50]

Leading members of the Panel are appointed by governments, and once selected, nominate other scientific experts to join the Panel. The disadvantage of the government connection is obvious, in that it takes away from the independence of the scientists working on the Panel. The advantage is that the evaluations produced by the Panel are taken very seriously by the different countries of the world since the IPCC was set up by the governments to advise the governments.

The IPCC is a new style of organization in that it is a vehicle for sorting out the facts about the transformation of global climate and getting these facts quickly to policy makers and the general public around the world. It is organized so that it can carry out this mandate (See Box 1.4). A full plenary session of the Panel is held once a year at which hundreds of delegates from member countries and participating organizations meet to decide on the organization's work plan, budget and other technical matters. Other decisions are vested in the Bureau of the Panel which is made up of scientists from the various member countries. A Secretariat located at the headquarters of the World Meteorological Organization, manages day-to-day activities of the IPCC.

Much of the work of the Panel is carried out in three "Working Groups" composed of international networks of scientists (Box 1.4). The first of these Working Groups assesses the latest knowledge about the physical changes going on in the climate system and about the different factors driving these changes. The second Working Group concerns itself with evaluating research on the impacts of climate change on society and the rest of nature. It also reviews the state of knowledge about adaptation to climate change. The focus of the third Working Group is on assessing options to mitigate climate change by preventing or reducing the build-up of greenhouse gases in the atmosphere. The third Working Group assesses the full spectrum of climate protection measures from the substitution of high-carbon fuels with lower-carbon alternatives to enhancing the uptake of carbon dioxide from the atmosphere through reforestation.

The working groups do not carry out scientific studies themselves, but are responsible for reviewing the scientific literature and distilling the most robust and important results from this literature. This is done after many hours of discussion and debate among Working Group members, and later, between scientists and government delegations. After extensive peer and governmental review, the three groups publish their latest findings in voluminous texts. The entire latest assessment reports from 2007 are fully accessible to readers via the Internet.[51] Because of the high international standing of these reports, they are a major source of information for this book. In 2007, in recognition of its value as a bridge between climate science and policy, the IPCC was the co-recipient of the Nobel Peace Prize.

Despite the IPCC's great efforts to get its facts straight before publishing, it became clear that errors, some large, some small, had crept their way into its reports. Considering the thousands of pages of assessment material and the literally thousands of scientists involved, it is perhaps not too surprising that some incorrect statements had slipped through the many levels of review already given to the text. In one of the most authoritative reviews of its performance, it was pointed out that the scope of IPCC's task had become much larger since it was set up in 1988, with the number of publications considered in its assessments growing from about 5,000 in the early 1990s to around 19,000 in the early 2000s.[52] Under this extended burden it was clear that the IPCC had to expand its operations and find some new ways of doing business. The same review recommended many steps in this direction, including an expansion of the IPCC management structure, a strengthening of the procedures it uses to review its findings, and using new ways to describe the uncertainties inherent in its reports.[53] No doubt these and other improvements will help the IPCC do its already useful job even better.

Box 1.4 **Organization of the Intergovernmental Panel on Climate Change (IPCC).** "TSU" stands for "Technical Support Unit" which are the secretariats of the various working groups

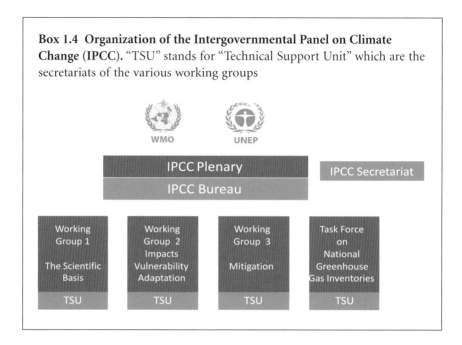

Notes

1 Rosenzweig, C., Casassa, G., Karoly, D., Imeson, A., Liu, C., Menzel, A., Rawlins, S., Root, T., Seguin, B., Tryjanowski, P., et al., 2007: Assessment of observed changes and responses in natural and managed systems. In: *Climate Change 2007: Impacts, adaptation and vulnerability. Contribution of Working Group II to the Fourth Assessment Report of the Intergovernmental Panel on Climate Change,* Parry, M.L., Canziani, O.F., Palutikof, J.P., van der Linden, P.J., Hanson, C.E. (eds). Cambridge University Press. Cambridge. UK. 79–131.

2 Alcamo, J., Moreno, J.M., Novaky, B., Bindi, M., Corobov, R., Devoy, R.J.N., Giannakopoulos, C., Martin, E., Olesen, J.E., Shvidenko, A., 2007: Europe. Chapter 12 in: Climate Change 2007: Impacts, Adaptation and Vulnerability. *Contribution of Working Group II to the Fourth Assessment Report of the Inter-governmental Panel on Climate Change (IPCC),* Parry, M.L., Canziani, O.F., Palutikof, J.P., van der Linden, P.J., Hanson, C.E., Eds., Cambridge University Press, Cambridge, UK. 541–580. www.gtp89.dial.pipex.com/12.pdf

3 Brander, K. M., Blom, G., 2003: Changes in fish distribution in the Eastern North Atlantic: Are we seeing a coherent response to changing temperature? *ICES Marine Science Symposia* 219, 261–270. Edwards, M., Richardson, A.J.,

2004: Impact of climate change on marine pelagic phenology and trophic mismatch. *Nature* 430, 881–884. Perry, A.L., Low, P.J., Ellis, J.R., Reynolds, J.D., 2005: Climate change and distribution shifts in marine fishes. *Science* 308, 1912–1915.

4   Kullman, L., 2002: Rapid recent range-margin rise of tree and shrub species in the Swedish Scandes. *Journal of Ecology* 90, 68–77. Camarero, J.J., Gutiér-rez, E., 2004: Pace and pattern of recent treeline dynamics response of eco-tones to climatic variability in the Spanish Pyrenees. *Climatic Change* 63, 181–200. Walther, G.-R., Beissner, S., Burga, C.A., 2005a: Trends in upward shift of alpine plants. *Journal of Vegetation Science* 16, 541–548. Shiyatov, S.G., Terent'ev, M.M., Fomin, V.V., 2005: Spatiotemporal dynamics of forest-tundra communities in the polar Urals. *Russian Journal of Ecology* 36, 69–75.

5   Menzel, A., Sparks, T.H., Estrella, N., Koch, E., Aasa, A., Ahas, R., Alm-Kübler, K., Bissoli, P., Braslavska, O., Briede, A., Chmielewski, F.M., Crepinsek, Z., Curnel, Y., Dalh, Å., Defila, C., Donnelly, A., Filella, Y., Jatczak, K., Måge, F., Mestre, A., Nordli, Ø., Peñuelas, J., Pirinen, P., Remišová, V., Scheifinger, H., Striz, M., Susnik, A., VanVliet, A., Wielgolaski, F.-E., Zach, S., Zust, A., 2006: European phenological response to climate change matches the warming pattern. *Global Change Biology* 12, 1969–1976.

6   Nabuurs, G.-J., Shelhaus, M.-J., Mohren, G.M.J., Field, C.B., 2003: Temporal evolution of the European forest sector carbon sink from 1950 to 1999. *Global Change Biology* 9, 152–160. Shvidenko, A., Nilsson, S., 2003: A synthe-sis of the impact of Russian forests on the global carbon budget for 1961–1998. *Tellus* 55B, 391–415. Boisvenue, C., Running, S.W., 2006: Impacts of climate change on natural forest productivity-evidence since the middle of the 20th century. *Global Change Biology* 12, 862–882.

7   Walther, G.-R., 2004: Plants in a warmer world. *Perspective in Plant Ecology, Evolution and Systematics* 6, 169–185. Dobbertin, M., Hilker, N., Rebetez, M., Zimmermann, N.E., Wohlgemuth, T., Rigling, A., 2005: The upward shift in altitude of pine mistletoe (Viscum album ssp. Austriacum) in Switzerland – the result of climate warming? *International Journal of Biometeorology* 50, 40–47.

8   Walther, G.-R., Beissner, S., Burga, C.A., 2005: Trends in upward shift of alpine plants. *Journal of Vegetation Science* 16, 541–548.

9   Luoto, M., Heikkinen, R.K., Carter, T.R., 2004: Loss of palsa mires in Europe and biological consequences. *Environmental Conservation* 31, 30–37. Kland-erud, K., Birks, H.J.B. 2003: Recent increases in species richness and shifts in altitudinal distributions of Norwegian mountain plants. *The Holocene* 13, 1–6.

10 Grabherr, G., Gottfried, M., Pauli, H., 2001: Long term monitoring of moun-tain peaks in the Alps. *Tasks for Vegetation Science* 35, 153–177. Kullman, L., 2001: 20th century climate warming and tree-limit rise in the southern Scandes of Sweden. *Ambio* 30, 72–80. Pauli, H., Gottfried, M., Grabherr, G., 2001: High summits of the Alps in a changing climate. The oldest observation

series on high mountain plant diversity in Europe. *"Fingerprints" of Climate Change – Adapted Behaviour and Shifting Species Ranges*, Walther, G.-R., Burga, C.A., Edwards, P.J. (eds), Kluwer Academic Publisher, Norwell, Massachusetts, 139–149. Klanderud, K., Birks, H.J.B., 2003: Recent increases in species richness and shifts in altitudinal distributions of Norwegian mountain plants. *The Holocene* 13, 1–6. Peñuelas, J., Boada, M., 2003: A global change-induced biome shift in the Montseny mountains (NE Spain). *Global Change Biology* 9, 131–140; Petriccione, B., 2003: Short-term changes in key plant communities of Central Apennines (Italy). *Acta Botanica Gallica* 150, 545–562. Sanz Elorza, M., Dana, E.D., 2003: Changes in the high mountain vegetation of the central Iberian Peninsula as a probable sign of global warming. *Annals of Botany* 92, 273–280. Walther, G.-R., Beissner, S., Burga, C.A., 2005: Trends in upward shift of alpine plants. *Journal of Vegetation Science* 16, 541–548.

11   Olesen, J.E., Trnka, M., Kersebaum, K.C., Skjelvåg, A.O., Seguin, B., Peltonen-Saino, P., Rossi, F., Kozyra, J., Micale, F., 2011: Impacts and adaptation of European crop production systems to climate change. *European Journal of Agronomy* 34, 96–112.

12   Olesen, J.E., Bindi, M., 2004: Agricultural impacts and adaptations to climate change in Europe. *Farm Policy Journal* 1, 36–46.

13   Duchene, E., Schneider, C., 2005: Grapevine and climatic changes: a glance at the situation in Alsace. *Agronomy for Sustainable Development* 25, 93–99. Jones, G.V., Davis, R., 2000: Climate influences on Grapewine phenology, grape composition, and wine production and quality for Bordeaux, France. *American Journal of Enology and Viticulture* 51, 249–261.

14   Chmielewski, F.-M., Müller, A., Bruns, E., 2004: Climate changes and trends in phenology of fruit trees and field crops in Germany, 1961–2000. *Agricultural and Forest Meteorology* 121, 69–78. Menzel, A., 2003: Plant phenological anomalies in Germany and their relation to air temperature and NAO. *Climatic Change* 57, 243.

15   Mazhitova, G., Karstkarel, N., Oberman, N., Romanovsky, V., Kuhty, P., 2004: Permafrost and infrastructure in the Usa Basin (Northern European Russia): Possible impacts of global warming. *Ambio* 3, 289–294; Frauenfeld, O.W., Zhang, T., Barry, R.G., Gilichinsky, D., 2004: Interdecadal changes in seasonal freeze and thaw depths in Russia. *Journal of Geophysical Research*, 109, doi:10.1029/2003JD004245.

16   Laternser, M., Schneebeli, M., 2003: Long-term snow climate trends of the Swiss Alps (1931–99). *International Journal of Climatology* 23, 733–750. Martin, E., Etchevers, P., 2005: Impact of climatic change on snow cover and snow hydrology in the French Alps. *Global Change and Mountain Regions (A State of Knowledge Overview)*. Huber, U.M., Bugmann, H., Reasoner, M.A. (eds), Springer, New York, 235–242.2005

17   Hoelzle, M., Haeberli, W., Dischl, M., Peschke, W., 2003: Secular glacier mass balances derived from cumulative glacier length changes. *Global and Planetary Change* 36, 295–306.

18   Beran, J., Asokliene, L., Lucenko, I., 2004: Tickborne encephalitis in Europe: Czech Republic, Lithuania and Latvia. Euro Surveill. Weekly Release 8. Daniel, M., Kriz, B., Danielová, V., Materna, J., Rudenko, N., Holubová, J., Schwarzová, L., Golovchenko, M., 2005: Occurrence of ticks infected by tickborne encephalitis virus and Borrelia genospecies in mountains of the Czech Republic. *Euro Surveill.* 10, E050331.1. At: www.eurosurveillance.org/ew/2005/050331.asp#1 (accessed 05.11.2006). Danielova, V., Kriz, B., Daniel, M., Benes, C., Valter, J., Kott, I., 2004: Effects of climate change on the incidence of tick-borne encephalitis in the Czech Republic in the past two decades. *Epidemiology Mikrobiology Immunology* 53, 174–181. Izmerov, N.F., Revich, B.A., Korenberg, E.I. (eds), 2004: Climate Change and Public Health in Russia in the XXI century. *Proceeding of a workshop, April 2004, Moscow, Adamant, Moscow, Russia.* Lindgren, E., Gustafson, R., 2001: Tick-borne encephalitis in Sweden and climate change. *Lancet,* 358, 16–18. Materna, J., Daniel, M., Danielova, V., 2005: Altitudinal distribution limit of the tick Ixodes ricinus shifted considerably towards higher altitudes in central Europe: results of three years monitoring in the Krkonose Mts. (Czech Republic). *Central European Journal of Public Health* 13, 24–28. Randolph, S., 2002: The changing incidence of tick-borne encephalitis in Europe. *Euro Surveill.* Weekly 6. From www.eurosurveillance (retrieved 30.10.2004). org/ew/2002/020606.asp.

19   Fischer, P.H., Brunekreef, B., Lebret, E., 2004: Air pollution related deaths during the 2003 heat wave in the Netherlands. *Atmospheric Environment* 38, 1083–1085. Kosatsky, T., Menne, B., 2005: Preparedness for extreme weather among national ministries of health of WHO's European region. *Climate Change and Adaptation Strategies for Human Health,* Menne, B., Ebi, K.L. (eds), Springer, Darmstadt, 297–329. Nogueira, P.J., 2005: Examples of heat warning systems: Lisbon's ICARO's surveillance system, summer 2003. *Extreme Weather Events and Public Health Responses,* Kirch, W., Menne, B., Bertollini, R. (eds), Springer, Heidelberg, 141–160. Pirard, P., Vandentorren, S., Pascal, M., Laaidi, K., Le Tertre, A., Cassadou, S., Ledrans, M., 2005: Summary of the mortality impact assessment of the 2003 heatwave in France. *Euro Surveill.* 10, 153–156.

20   Beggs, P.J., 2004: Impacts of climate change on aeroallergens: past and future. *Journal of the British Society for Allergy and Clinical Immunology* 34, 1507–1513. Huynen, M., Menne, B., Coord., 2003: Phenology and Human Health: Allergic Disorders. Report of a WHO meeting, Rome, Italy (16–17 January), *Health and Global Environmental Change* Series no. 1, WHO, Europe, 55 pp. van Vliet et al., 2003.

21   Camarero, J.J., Gutiérrez, E., 2004: Pace and pattern of recent treeline dynamics response of ecotones to climatic variability in the Spanish Pyrenees. *Climatic Change* 63, 181–200. Kullman, L., 2002: Rapid recent range-margin rise of tree and shrub species in the Swedish Scandes. *Journal of Ecology* 90, 68–77; Walther, G.-R., Beissner, S., Burga, CA, 2005: Trends in upward shift of alpine plants. *Journal of Vegetation Science* 16, 541–548.

22 Callaghan, T.V., Johansson, M., Heal, O.W., Sælthun, N.R., Barkved, L.J., Bayfield, N., Brandt, O., Brooker, R., Christiansen, H.H., Forchhammer, M., Høye, T.T., Humlum, O., Järvinen, A., Jonasson, C., Kohler, J., Magnusson, B., Meltofte, H., Mortensen, L., Neuvonen, S.I., Pearce, R.M., Turner, L., Hasholt, B., Huhta, E., Leskinen, E., Nielsen, N., Siikamäki, P., 2004: Environmental changes in the North Atlantic region: SCANNET as a collaborative approach for documenting, understanding and predicting changes. *Ambio Special Report* (Tundra-Taiga2 Treeline Research) 13, 39–50. Truong, G., Palmé, A.E., Felber, F., 2006: Recent invasion of the mountain birch Betula pubescens ssp. tortuosa above the treeline due to climate change: genetic and ecological study in northern Sweden. *European Society for Evolutionary Biology*, doi: 10.1111/j. 1420–0–4 9101.2006. 01190.x.

23 Rosenzweig et al., 2007.

24 Menzel, A., 2000: Trends in phenological phases in Europe between 1951 and 1996. *International Journal of Biometeorology* 44: 76. Menzel, A., 2003: Plant phenological anomalies in Germany and their relation to air temperature and NAO. *Climatic Change* 57: 243.

25 Defila, C., Clot, B., 2001: Phytophenological trends in Switzerland. *International Journal of Biometeorology* 45: 203.

26 Rosenzweig et al., 2007.

27 Menzel, A., Vopelius, J. von, Estrella, N., Schleip, C., Dose, V., 2006: Farmers' annual activities are not tracking speed of climate change. *Climate Research* 32: 201–207.

28 Jones, G.V., Duchene, E., Tomasi, D., Yuste, J., Braslavksa, O., Schultz, H., Martinez, C., Boso, S., Langellier, F., Perruchot, C., Guimberteau, G., 2005: *Changes in European winegrape phenology and relationships with climate*, GESCO 2005.

29 Roy, D.B., Sparks, T.H., 2000: Phenology of British butterflies and climate change. *Global Change Biology* 6(4): 407.

30 Rosenzweig et al., 2007.

31 Brander, K., Blom, G., Borges, M.F., Erzini, K., Henderson, G., Mackenzie, B.R., Mendes, H., Ribeiro, J., Santos, A.M.P., Toresen, R., 2003: Changes in fish distribution in the eastern North Atlantic: are we seeing a coherent response to changing temperature? *ICES Marine Science Symposium*, 219, 261–270. Beare, D., Burns, F., Jones, E., Peach, K., Portilla, E., Greig, T., McKenzie, E., Reid, D., 2004: An increase in the abundance of anchovies and sardines in the north-western North Sea since 1995. *Global Change Biology*, 10, 1209–1213. Genner, M.J., Sims, D.W., Wearmouth, V.J., Southall, E.J., Southward, A.J., Henderson, P.A., Hawkins, S.J., 2004: Regional climatic warming drives long-term community changes of British marine fish. *Proceedings of the Royal Society Biological Sciences* 271, 655–661. Perry, A.L., Low, P.J., Ellis, J.R., Reynolds, J.D., 2005: Climate change and distribution shifts in marine fishes. *Science* 308, 1912–1915.

32 Liebsch, G., Novotny, K., Dietrich, R., 2002: Untersuchung von Pegelreihen zur Bestimmung der Änderung des mittleren Meeresspiegels an den

europäischen Küsten, Technische Universität Dresden (TUD), Germany. As quoted in: EEA (European Environment Agency). 2004. Impacts of Europe's changing climate. An indicator-based assessment. EEA Report No 2/2004. 101 pp.

33 Rosenzweig et al., 2007.

34 Mazhitova, G., Karstkarel, N., Oberman, N., Romanovsky, V., Kuhty, P., 2004: Permafrost and infrastructure in the Usa Basin (Northern European Russia): Possible impacts of global warming. *Ambio* 3, 289–294. Frauenfeld, O.W., Zhang, T., Barry, R.G., Gilichinsky, D., 2004: Interdecadal changes in seasonal freeze and thaw depths in Russia. *Journal of Geophysical Research Atmospheres* 109, doi:10.1029/2003JD004245.

35 Hoelzle et al., 2003.

36 Zemp, M., Haeberli, W., Hoelzle, M., Paul, F., 2006: Alpine glaciers to disappear within decades. *Geophysical Research Letters* 33, L13504, doi:10.1029/2006GL026319.

37 Laternser, M., Schneebeli, M., 2003: Long-term snow climate trends of the Swiss Alps (1931–99). *International Journal of Climatology* 23, 733–750. Martin, E., Etchevers, P., 2005: Impact of climatic change on snow cover and snow hydrology in the French Alps. *Global Change and Mountain Regions (A State of Knowledge Overview)*, Huber, U.M., Bugmann, H., Reasoner, M.A. (eds), Springer, New York, 235–242.

38 Kosatsky, T., 2005: The 2003 European heatwave. *EuroSurveillance* 10, 148–149.

39 Confalonieri, U., Menne, B., Akhtar, R., Ebi, K.L., Hauengue, M., Kovats, R.S., Revich, B., Woodward, A., 2007: Human health. *Climate Change 2007: Impacts, Adaptation and Vulnerability. Contribution of Working Group II to the Fourth Assessment Report of the Intergovernmental Panel on Climate Change*, Parry, M.L., Canziani, O.F., Palutikof, J.P., van der Linden, P.J., Hanson, C.E. (eds), Cambridge University Press, Cambridge, UK, 391–431.

40 Luterbacher, J., Dietrich, D., Xoplaki, E., Grosjean, M., H., Wanner, 2004: European seasonal and annual temperature variability, trends, and extremes since 1500. *Science* 303 (5663): 1499–1503.

41 Bortenschlager, S., Bortenschlager, I., 2005: Altering airborne pollen concentrations due to the global warming. A comparative analysis of airborne pollen records from Innsbruck and Obergurgl (Austria) for the period 1980–2001. *Grana* 44, 172–180. Emberlin, J., Detandt, M., Gehrig, R., Jaeger, S., Nolard, N., Rantio-Lehtimaki, A., 2002: Responses in the start of Betula (birch) pollen seasons to recent changes in spring temperatures across Europe. *International Journal of Biometeorology* 46, 159–170.

42 Thomas, C.D., Franco, A.M.A., Hill, J.K., 2006: Range retractions and extinction in the face of climate warming. *Trends in Ecology & Evolution* 21(8): 415–416.

43 Carter, T.R., Parry, M.L., Harasawa, H. et al., 1994: *IPCC technical guidelines for assessing climate change impacts and adaptations.* Geneva, Intergovernmental Panel on Climate Change.

44 Feenstra, J., Burton, I., Smith, J., Tol, R. (eds), 1998: *Handbook on methods for climate change impact assessment and adaptation strategies.* United Nations Environment Programme and Free University of Amsterdam. www.cordelim.net/extra/cd%20forestal/Adaptaci%F3n%20al%20CC%20y%20MNR/Literatura/APF/HMC.pdf. Retrieved. November, 2008.

45 Carter et al., 1994.

46 Carter, T.R., Jones, R.N., Lu, X., Bhadwal, S., Conde, C., Mearns, L.O., O'Neill, B.C., Rounsevell, M.D.A., Zurek, M.B., 2007: New assessment methods and the characterization of future conditions. Climate Change 2007: Impacts, Adaptation and Vulnerability. Contribution of Working Group II to the Fourth Assessment Report of the Intergovernmental Panel on Climate Change, Parry, M.L., Canziani, O.F., Palutikof, J.P., van der Linden, P.J., Hanson, C.E., Eds., Cambridge University Press, Cambridge, UK, 133–171. Carter et al., 1994.

47 Carter et al., 2007. Carter et al., 1994.

48 Carter et al., 2007.

49 Alcamo, J., Klein, R., Carius, A., Acosta-Michlik, L., Krömker, D., Tänzler, D., Eierdanz, F., 2008: A new approach to quantifying and comparing vulnerability to drought. *Regional Environmental Change* 8:137–149. DOI 10.1007/s10113–3–008–8–0065–5–5.

50 Alcamo et al., 2007.

51 www.ipcc.ch/ipccreports/assessments-reports.htm

52 InterAcademy Council. 2010. InterAcademy Council Review of IPCC. http://reviewipcc.interacademycouncil.net/report/Climate%20Change%20Assessments,%20Review%20of%20the%20Processes%20&%20Procedures%20of%20the%20IPCC.pdf

53 InterAcademy Council, 2010.

2 Climate and climate change

Introduction

The earth's climate is changing, and further changes are foreseen having wide-ranging impacts on ecosystems and the human societies. According to the IPCC,[1]

> warming of the climate system is unequivocal, as is now evident from observations of increases in global average air and ocean temperatures, widespread melting of snow and ice, and rising global average sea level.

Climate change is without question one of the biggest challenges that Europe and the rest of humankind has ever faced. This is, not least, due to the enormous

Life in Europe Under Climate Change, First Edition. Joseph Alcamo and Jørgen E. Olesen.
© 2012 Joseph Alcamo and Jørgen E. Olesen. Published 2012 by John Wiley & Sons, Ltd.

consequences that climate change will have for ecosystems and human society. Unfortunately, climate change also poses a very difficult problem for politicians to deal with. The core of the problem is that most people experience very little relationship between greenhouse gas emissions, climate change and their everyday life. There is both a temporal and spatial separation between emissions and impacts of climate change. The industrialized countries, which have emitted most of the greenhouse gases, are in general the least vulnerable to climate change effects. In addition, many of the detrimental effects of climate change will happen far later (decades to centuries) than the greenhouse gas emissions. It is therefore difficult to achieve substantial popular support for necessary and effective measures to mitigate climate change. A first step to achieve this is to increase the awareness of how human activities are currently changing the climate, how the future climate of Europe might look like, and how this impacts life in Europe.

Scientific understanding of anthropogenic global climate change has advanced considerably in recent years, and this has led to intensive studies on how climate may change in the future, and what can be done to avoid or at least mitigate the most serious climate change scenarios. It has also led to numerous studies on impacts of climate change and how ecosystems and human life can adapt to the changing environment. Scientists from Europe have been leading in many of these studies, and several large-scale European research projects such as PRUDENCE[2] and ENSEMBLES[3] have paved the way for a far better understanding of what climate and climate impacts in Europe may look like in the future.

Climate, defined as the long-term weather (see Box 2.1), affects every aspect of human, animal, and plant life. It forces humans and all other life forms to continuously adapt in order to survive most efficiently within the climate type of a given region. On a timescale of thousands or millions of years, the earth has experienced much warmer and much cooler climates than today. Humans, however, are influenced by climate changes occurring over much shorter timescales. It is therefore critically important not only to know how climate will change, but also how fast such changes will occur. Successive mean changes are of course important, but many aspects of climate change will also be felt as new weather (i.e. the day-to-day variation in temperature, rainfall, windiness etc.). Some extreme events such as storms, droughts or floods have particularly serious consequences, and they have therefore received considerable recent scientific attention, also in terms of evaluating changes in the frequency and magnitude of such events.

Europe's climate

Europe stretches from the Arctic, over the temperate zone, and just into the subtropical climatic zone of the northern hemisphere (Figure 2.1). The continent covers about 3,900 km from north to south and more than 5,000 km from west to east, not including the smaller Atlantic islands.

Box 2.1 Defining climate

Often there is a considerable gap between common perception of what we mean by 'climate' and a more scientific definition. This mismatch often gives rise to considerable misleading statements and assessments of the role of humankind as a driver of observed changes. Climate is basically the statistical properties of weather in the atmosphere, but also (e.g. the "weather" in the oceans), over a long time period. The simplest and best understood indicators of weather are mean temperature and mean precipitation amounts, whether on a monthly, seasonal or annual basis. Climate can, in principle be defined with different time frames. This is one reason for the gap between common perceptions and scientific definitions. When weather changes from one year to another, it is not in agreement with a simple definition of climate as being constant or invariant in time. In practice, long time periods are used in defining climate in such a way that in the absence of trends, climate is close to invariant. This means that yearly and even decadal variations should show up only marginally. In practice, climatologists in the first part of the 20th century decided to use 30-year periods as a compromise between the wish for reasonably short periods and the need for invariance in the conditions from one period to another. This led to the definition of 30-year climate norms, which started with the period 1901–30. The latest norm is for the period 1961–90. This period is also sometimes called the "climate normal period" in this book and elsewhere. Therefore, many climate variables (i.e. annual temperature and precipitation) are compared with respect to this reference "climate normal" period. With changing climate, one can question the applicability of 30-year periods for defining climate. In some countries, use is made of "running" 30-year periods, such as 1971–2000 and later 1981–2010 to continue from 1961–90, instead of waiting for the whole 1991–2020 to come about. Another way is to focus more on trends and tendencies in climate than some fixed period.

Parts of Europe have been by and large shaped by the movement of ice during the Ice Ages when glaciers covered much of the continent. It is highly indented and has a long and complex coastline, with many islands and peninsulas of all sizes. This means that there is a large oceanic influence from the surrounding seas in north, western and southern Europe. The seas are also very different, ranging from enclosed to totally open seas (Figure 2.1).

Europe generally has a milder climate than parts of Asia and North America at the same latitude. Europe's mild climate is caused by effective northward transport of heat from the Equatorial regions by warm ocean currents and atmospheric circulation over the Atlantic Ocean. This is compounded by predominantly westerly winds that bring relatively warm air in to Europe from the Atlantic Ocean.

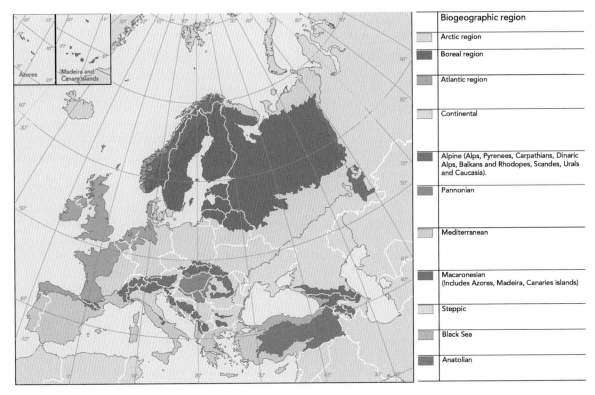

Biogeographic region
Arctic region
Boreal region
Atlantic region
Continental
Alpine (Alps, Pyrenees, Carpathians, Dinaric Alps, Balkans and Rhodopes, Scandes, Urals and Caucasia).
Pannonian
Mediterranean
Macaronesian (Includes Azores, Madeira, Canaries islands)
Steppic
Black Sea
Anatolian

Figure 2.1 Biogeographic regions of Europe. Source: EEA.[4] Reprinted with permission from European Environmental Agency, Copenhagen. (See color plate.)

The ocean circulation mentioned above is conventionally described as driven by the sinking of cold and salty – and consequently dense – surface waters in the Arctic Ocean. This is part of the so-called thermohaline (or Meridional Overturning) circulation, which transports water, heat and salinity. The ocean circulation is also affected by the surface wind regime, and constrained by the placement of the continents and the structure of the sea floor. One of the most spectacular impacts of these warm currents is the surface Gulf Stream current and its extension to the north along the Norwegian coast: Much of Norway's coastline lies within the Arctic region, but almost all of it remains free of ice and snow throughout the winter.

In general, northern Europe has longer, colder winters and shorter, cooler summers than southern Europe. In addition, winters are longer and colder, and summers shorter and hotter, in the east, which is more continental than the more maritime west. Most of Europe receives from 500 mm to 1500 mm of precipitation each year. The greatest annual precipitation occurs in areas just west of mountains. Such regions include parts of western Britain and western Norway. The continent's smallest annual precipitation (less than 500 mm) occurs in three types of areas: (1) east of the various mountain chains; (2) far inland from the Atlantic

Ocean; and (3) along the Arctic coast. Such regions include central and southeastern Spain, northern Scandinavia, northern and southeastern parts of European Russia, and western Kazakhstan.

Climates of the past

The climate has varied considerably more over longer timescales (from thousands to hundreds of thousands of years and longer) than over the past 150 years for which we have a continuous instrumental record of temperature and precipitation with global coverage. Assessment of climate change over longer timescales depends on indirect measurements of temperature and precipitation (Box 2.2). Translating such measurements (e.g. different oxygen isotopes in ice cores or sediments), into a consistent record of global or regional temperature involves considerable uncertainties. Even so, the IPCC estimates that the mean temperature over the Northern Hemisphere (such records are less well established on a global scale) over the recent 50 years is probably higher than for the past 1300 years.[5]

During the past 800,000 years climate on earth has been oscillating about every 100,000 years between glacial periods, during which the global mean temperature was a few degrees Celsius lower than today, and inter-glacial periods, during which temperature was about comparable to that of today. The transition between these two states was primarily driven by variations in the earth's orbit around the sun (Box 2.3), which led to a pattern of changes in the seasonal amount and distribution of solar radiation on earth, affecting global energy balances and heat transport (as well as the possibility for snow to accumulate over the continents), and thus climate. Herein it needs to be recalled that the actual variations in solar radiation during summer are quite small. What seems to come into play are some so-called feedback effects and especially increasing (decreasing) albedo with accumulating (decreasing) snow and ice cover and changes in the carbon cycle, leading to change in the natural greenhouse effect.

During glacial periods large amounts of water are accumulated in glaciers, which cause the sea level to decline. The lower temperature reduces evaporation

---

Box 2.2 Measuring temperature

The thermometer was originally invented by Galileo Galilei in the early 17th century, but regular temperature measurements on a global scale have only been conducted since the mid-19th century. Standard temperature measurements are made at a height of 2 meters in a screen that provides shielding and ventilation to ensure that measurements reflect the air temperature and not that of solar radiation on the thermometer. Temperature measurements are affected by urbanization, and it has often been debated whether such effects have affected the global and regional temperature records. It is well known that temperatures in a city tends to be considerably higher than in adjacent rural areas because of the "urban heat island effect". There are many causes for this effect, including the fact that concrete and other urban surfaces absorb and re-radiate solar radiation more effectively than areas with heavy foliage, and because cities are continuously warmed by waste heat from their concentrated use of energy. On the other hand, the majority of climate stations are located at a distance from cities. Recent studies show that the urban heat island effect adds only a small uncertainty (0.06°C over 100 years) to long-term temperature trends, which is far less than the observed temperature increase.

Satellites have recently provided new opportunities for measuring the distribution of temperature over the earth. However, such measurements primarily reflect changes in temperature over relatively thick atmospheric layers. These measurements show that the temperature in the upper part of the atmosphere has been declining over time, which accords with the enhanced greenhouse effect giving an increased thermal irradiance from the upper part of the atmosphere. To reconstruct temperature over longer timescales requires the use of indirect methods. Geological sediments give information about past plant and animal life, and the composition of ecosystems can be translated into information about temperature and precipitation when these organisms lived. Similar information can be derived, for example, from the thickness of tree rings.

For temperature studies over very long timescales, science relies very much on studies of ice cores in Greenland and Antarctica, from which it is possible to indirectly estimate variations in temperature over past eons. After all, the ice has accumulated over long periods of time, carrying the history of past times into the depths of the ice sheets. One of the most important methods for reconstructing temperature records is based on the study of oxygen in the ice itself in these ice cores. The bubbles in these cores provide a glimpse of the atmosphere of the earth hundreds of thousands of years ago. It turns out that the relative amount of the typical oxygen isotope (^{16}O) versus the heavier and rarer isotope (^{18}O) provides information about temperature at the time these molecules were trapped as part of snowfall in the ice core. This is because oxygen constitutes the heaviest part of a water molecule (H_2O), and because the lighter ^{16}O evaporates more easily than the heavier ^{18}O; hence, there will be a higher proportion of ^{18}O in the atmosphere when air temperature is warm than when it is cold. Armed with this knowledge, scientists can roughly estimate the temperature at the time the snowfall became part of the ice core and hence hundreds of thousands of years back in time. Strictly speaking, the derived temperature only describes the conditions of the surrounding ocean areas from which the snowfall has originated via evaporation.

Box 2.3 Milankovitch and the Ice Ages

The earth's orbit about the sun is not constant, but varies over time owing to attraction from other planets. In addition, the axial tilt and the direction of the tilt (precession) vary over time. These variations affect the total solar radiation and its distribution over earth. The Serbian mathematician Milutin Milankovitch (1879–1958) was the first to calculate these effects for the past million years by hand, and thus demonstrate that such variations are the primary drivers affecting glaciation and the shifts between glacial and interglacial periods. The main characteristics of earth's orbit that change are:

Orbital shape (eccentricity): The earth's orbit changes shape from almost circular to slightly elliptic and back again over a period of about 100,000 years. This gives changes in distance from the earth to the sun, resulting in small changes in total solar radiation received on earth.

Axial tilt (obliquity): The angle between earth's rotational axis and an angle perpendicular to the plane of the orbit varies between 22.1° and 24.5° and back again over a period of about 41,000 years. Changes in axial tilt changes the distribution of solar radiation, but not the total amount received on earth. An increased tilt causes larger differences between summer and winter at high latitudes. Cooler summers are believed to encourage the start of an ice age by melting less of the previous winter's ice and snow.

Precession (wobble): The direction of the earth's rotational axis in space changes over a period of about 21,000 years. Currently it is summer on the northern hemisphere when earth is closest to the sun. In about 10,500 years it will be summer on the southern hemisphere, when earth is closest to the sun. This does not affect the total radiation received on earth, but its seasonal distribution on the hemispheres.

from the oceans leading to less precipitation. Since a large proportion of the precipitation falls as snow, glacial periods are often accompanied by marked dry periods over the ice-free land surfaces. Glacial periods therefore mean a global climate that is tougher for life on earth than the current climate, in particular in Europe.

Over the past one million years or so, the duration of a glacial period has been about 100,000 years, and inter-glacial periods usually 10,000–30,000 years. We are currently within an inter-glacial period – Holocene – that has some resemblance to the quite long one, roughly 400,000 years ago. The Holocene has lasted about

11,700 years. This, however, does not mean that a transition to a new glacial period is imminent. According to the IPCC, it is unlikely that the earth over the next 30,000 years for natural reasons will go into a glacial period. The reason is that the next sufficiently large reduction in summer radiation on the northern hemisphere will not happen for the next 30,000 years or so.

The indirect temperature records for the Northern Hemisphere over the past 1300 years suggest that mean temperatures have varied within a range of 0.5°C. Variability within this range can be explained by drivers other than greenhouse gases (e.g. solar radiation, volcanic eruptions leading to large amount of particles in the middle atmosphere, etc.). It has often been argued that a medieval warming (medieval climate optimum) existed around 950 to 1200 and that the rise and fall of the Viking civilization in Greenland and Iceland was directly linked to climate changes. The currently available information indicates that mean temperatures over the Northern Hemisphere during medieval times (950–1200) were indeed relatively warm, and considerably warmer than the widespread cooler conditions during the 17th century (often termed the Little Ice Age). However, there is no evidence to show that hemispheric and, even less so, global mean temperatures were as warm, or the extent of warm regions as large, as those during the latter years of the 20th century.

Recent climate change

The warming trend throughout Europe of 0.9°C over the period 1901–2005 is well established.[6] This is slightly above the global average. The recent period shows a trend considerably higher than the mean trend, about 0.4°C per decade for the period 1979–2005. Eight of the 12 years between 1996 and 2007 rank among the 12 warmest years since regular measurements were started in 1850 in Europe. Seasonally, Europe has warmed mostly in spring and summer, but with large differences between regions (Figure 2.2). The warming over the past 30 years has thus been strongest in Scandinavia during the winter, whereas the Iberian Peninsula has warmed during the summer. An increase of daily temperature variability is observed during the period 1977–2000 due to an increase in warm extremes, rather than a decrease in cold extremes.[7] The average length of summer heatwaves over western Europe has doubled since the 1880s, and the frequency of hot days has almost tripled.

Precipitation in Europe has generally increased over the 20th century, however, with large spatial variations. Mean winter precipitation is increasing over most of the Atlantic and parts of northern Europe (Figure 2.3), most likely in part due to stronger advection of wet Atlantic air masses over this part of the continent. Drying has been observed in the Mediterranean area and parts of eastern Europe, whereas no clear trends have been seen in western Europe. The proportion of Europe experiencing meteorological droughts has not changed significantly over the 20th century. An increase in mean precipitation per wet day has been observed in

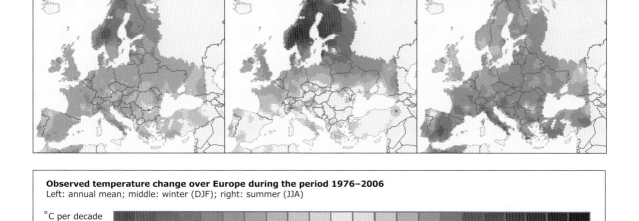

Observed temperature change over Europe during the period 1976–2006
Left: annual mean; middle: winter (DJF); right: summer (JJA)

°C per decade

-2 -1.8 -1.6 -1.4 -1.2 -1 -0.8 -0.6 -0.4 -0.2 0 0.2 0.4 0.6 0.8 1 1.2 1.4 1.6 1.8 2

Figure 2.2 Observed temperature change over Europe during the period 1976–2006. Left: annual mean; middle: winter (December to January); right: summer (June to August). Source: EEA.[8] Reprinted with permission from European Environmental Agency, Copenhagen, and KNMI, Netherlands. (See color plate.)

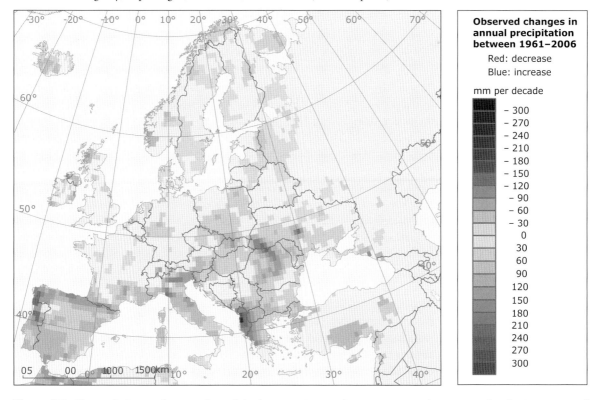

Observed changes in annual precipitation between 1961–2006

Red: decrease
Blue: increase

mm per decade

- 300
- 270
- 240
- 210
- 180
- 150
- 120
- 90
- 60
- 30
0
30
60
90
120
150
180
210
240
270
300

Figure 2.3 Observed changes in annual precipitation over Europe for 1961 to 2006 in mm per decade. Source: EEA.[9] Reprinted with permission from European Environmental Agency, Copenhagen, and KNMI, Netherlands. (See color plate.)

most parts of the continent, even in some areas which are, on average, getting drier.

Wind speeds and storminess in Europe have shown considerable variation over the past century, with no clear long-term trend. However, some places have seen decreases and other increases. For example, there has been a decrease in wind storms over the Netherlands over the past 40 years.[10]

How do greenhouse gases work?

The radiation balance

Climate on earth, as described above, has varied considerably more than observed over the past century. The temperature on earth is determined by the balance between the energy gained through solar radiation and the energy lost in the form of thermal radiation. Two factors have the principal control over this radiation balance: (1) the amount of solar radiation absorbed at the earth's surface; and (2) the strength of the greenhouse effect.

The earth receives about $340 \, W/m^2$ solar radiation as a global average over the entire year. This is equal to sufficient energy to power more than three 100-Watt light bulbs day and night on one square meter of surface. About a third of this radiation is reflected by clouds, particles in the atmosphere and by the earth surface. This is called the planetary albedo. The remaining incident radiation (around $240 \, W/m^2$) warms the surface and the atmosphere.

The surface of the earth as well as the atmosphere releases the heat through thermal radiation, which is proportional to the fourth power of the absolute temperature (Stefan Boltzmann law). Over a long averaging period, the outgoing thermal radiation will balance incoming solar radiation ($240 \, W/m^2$). Using the Stefan Boltzmann law, this gives an average temperature for the globe of -18 to $-19°C$. This is some $33°C$ lower than the mean global temperature of 14–$15°C$. So another mechanism must be responsible for warming the earth. This is called the "the greenhouse effect".

Particular gases, including water vapor and carbon dioxide, are called "greenhouse gases", because they are rather transparent to solar radiation but absorb much of the outgoing long-wave thermal radiation. After absorbing thermal radiation they re-radiate to the atmosphere, thereby warming the air. The warming occurs in the lower part of the atmosphere. The earth as seen from space will still have a temperature of -18 to $-19°C$, corresponding to the effective outgoing radiation as given from the Stefan-Boltzmann law. Temperature declines with altitude and this temperature is currently obtained at a height of about 5 km. An elevated concentration of greenhouse gases in the atmosphere will effectively lead to a warming of the lower part of the atmosphere and raise the altitude in the atmosphere of the effective radiation.

Table 2.1 Global radiative forcing estimates and uncertainty ranges for 2005 from a range of factors. The radiative forcing is the change in net irradiance at the tropopause since pre-industrial times. Best estimates and uncertainty ranges cannot be obtained by adding the individual terms, since the uncertainty ranges are not symmetric for some factors. Source: Solomon et al.[12] Reprinted with permission from Intergovernmental Panel on Climate Change, Geneva.

Driver	Type	Factor	Radiative forcing (W/m²)	
			Best estimate	Range
Antropogenic	Long-lived greenhouse gases	Carbon dioxide (CO_2)	1.66	(1.49–1.83)
		Methane (CH_4)	0.48	(0.43–0.53)
		Nitrous oxide (N_2O)	0.16	(0.14–0.18)
		Halocarbons (CFC)	0.34	(0.31–0.37)
	Ozone	Stratospheric decrease	−0.05	(−0.15–0.05)
		Tropospheric increase	0.35	(0.25–0.65)
	Stratospheric water from CH_4		0.07	(0.02–0.12)
	Surface albedo	Land use change	−0.20	(−0.4–0.0)
		Black carbon on snow	0.10	(0.0–0.2)
	Aerosols (particles)	Direct effect	−0.50	(−0.9–−0.1)
		Cloud albedo	−0.70	(−1.8–−0.3)
	Contrails (from aircraft)	Linear contrails	0.01	(0.003–0.03)
Natural	Solar radiation	Irradiance	0.12	(0.06–0.30)
Total anthropogenic			1.60	(0.6–2.4)

Greenhouse gases

The most important greenhouse gases are water vapor (H_2O), carbon dioxide (CO_2), methane (CH_4), nitrous oxide (N_2O), a long list of halocarbon gases and ozone (O_3) (Table 2.1). The relative importance of the major contributors, water vapor, clouds and carbon dioxide can be roughly estimated as 2–1–1.

The anthropogenic sources of CO_2 are the burning of fossil fuels (coal, oil and natural gas) and changes in land use, particularly deforestation. The emissions of greenhouse gases have particularly increased during the past 50 years or so. The current (2007) atmospheric CO_2 concentration of 384 ppm now far exceeds the natural level of the last 650,000 years (ranging from 140 ppm to 300 ppm). The concentrations of methane and nitrous oxide have also been increasing. Methane is produced by microbial degradation of organic matter under anoxic conditions (i.e. in the absence of oxygen). Major anthropogenic sources include the rumen of cattle and sheep, in manure storages, landfills and rice paddies. Methane has a global warming potential[11] that is 23 times larger than CO_2.

Nitrous oxide is an even more potent greenhouse gas with a global warming potential 298 times larger than CO_2. Nitrous oxide originates from microbial transformations of nitrogen in soil, and the increasing fertilizer use in agriculture is the primary driver of increasing nitrous oxide emissions. Halocarbons from refrigerators, freezers, air conditioners, fire extinguishers, and so on, also contribute to global warming. Ozone, in particular from the lower part of the atmosphere (troposphere), also contributes to global warming. Ozone is formed when sunlight splits nitrogen oxides (NO_x) and hydrocarbons (e.g. from car exhaust fumes). There is thus a wide range of pollution sources that contribute to global warming.

Burning of coal and oil also emit large quantities of small particles into the atmosphere which have a cooling effect on the atmosphere because they increase the reflection of sunlight (albedo). Burning, including of biomass, also emits black carbon (soot) which absorbs sunlight. Changes in land use have also increased the albedo, but both effects are associated with significant uncertainties (Table 2.1). In addition to the anthropogenic effects on the radiation balance, there has also been relatively large, but short-lived, cooling contributions from some major volcanic eruptions. This is due to the building of small particles from the volcanic gases, and thus increases in albedo. Finally, variations in solar radiation have most likely led to some increased warming during the first part of the 20th century.

Climate sensitivity

The degree of warming that originates from increases in greenhouse gas concentrations (or from some of the other contributing factors in Table 2.1) is sometimes described in terms of the so-called "equilibrium climate sensitivity", which denotes how much the global temperature increases for an increase in radiative forcing[13] of 1 W/m². This increase in radiative forcing could come from an increase in solar radiation entering the earth's atmosphere or from a higher concentration of greenhouse gases, which would give rise to a similar change in net radiation to and from the troposphere. IPCC uses the term "climate sensitivity" for the equilibrium change in global mean surface temperature following a doubling of the atmospheric CO_2 concentration. However, here we use the term to refer only to an increase in radiative forcing of 1 W/m².

The concept of climate sensitivity makes it possible to calculate the temperature increase from different factors that directly or indirectly affect the radiation balance of the earth. If the change in radiative forcing is known (e.g. from Table 2.1), the climate sensitivity makes it possible to estimate the temperature change under equilibrium conditions. Unfortunately, the exact value of the climate sensitivity is still not known.[14] If the climate sensitivity were to be taken only from the direct radiative forcing of individual greenhouse gases, we would get a sensitivity of $0.269°C/(W/m^2)$. In reality there are a number of feedback mechanisms which enhance the temperature changes (Box 2.4). One of the most important feedbacks is the effect of temperature on the water vapor content of the atmosphere. Water vapor is a greenhouse gas, and since a warm atmosphere can contain more water vapor than a cold atmosphere, warming itself leads to an enhanced greenhouse effect through a higher concentration of water vapor in the atmosphere.

There are many of these feedback mechanisms, and the climate sensitivity is determined by the sum of these. In order to calculate the total effect of changes in greenhouse gases on the climate, models of the entire climate system are used (Box 2.6). These models represent physical processes in the atmosphere, oceans and on land. Observations during the last 150 years and reconstructions of climate over the past half million years have been used to estimate climate sensitivity. Using these methods, the value of the sensitivity centres around $0.75°C/(W/m^2)$. The feedback mechanisms thus almost triple the climate sensitivity compared to a system without feedbacks. This means that the warming effect of enhanced CO_2 will be tripled, owing to feedbacks in the climate systems. There is still considerable uncertainty about climate sensitivity and estimates range from about 0.5 to more than $2°C/(W/m^2)$.

Table 2.1 shows that the total radiative forcing from anthropogenic emissions of greenhouse gases is about $3.0 \, W/m^2$. Since the net cooling effect of anthropogenic particles and albedo changes is around $-1.4 \, W/m^2$, this means that the net effect of anthropogenic emissions to the atmosphere is about $1.6 \, W/m^2$. With a climate sensitivity of $0.75°C/(W/m^2)$ this gives a temperature increase of 1.2°C. This is more than the observed global temperature increase of about 0.8°C. This additional warming stems from a lag effect in the global climate system whereby climate continues to respond slowly for a period of time after the changes in radiative forcing occur. This inertia is primarily associated with the slow warming of the oceans.

Box 2.4 Feedback mechanisms

The climate system is made up of the atmosphere, land, water, ice, soil and vegetation, which interact. Changes in the respective climate system component cause changes in the other components, which in turn may feed back to the primary change. These feedback mechanisms can either enhance or dampen the overall changes. The feedback mechanisms come into play both when the primary effect is one of cooling and when it is one of warming. There are many feedback mechanisms, and one of the most

Continued

important objectives of climate models is to quantify these mechanisms. The most important feedback mechanisms include:

Water vapor, which is also a greenhouse gas. The amount of water vapor in the atmosphere is strongly dependent on increasing temperature by almost 7% per degree Celsius increase in air temperature. Hence, the warmer the atmosphere, the higher the vapor content, and the stronger the effect of water vapor on warming of the atmosphere (positive feedback).

Lapse rate, which describes the decline in temperature with altitude. The vertical variations of the temperature change also have a climatic effect. If the warming is relatively greater in the upper part of the lower atmosphere then this leads to a negative feedback and conversely if the warming is relatively smaller here. Climate models predict negative feedbacks in the tropics and positive ones in the mid to high latitudes. For the earth as a whole the negative feedback tends to dominate.

Ice and snow. Warming leads to melting of ice and snow and therefore to a reduction in the extent of continental snow cover and oceanic ice. This reduces the amount of solar radiation reflected to space leading to further warming (positive feedback). This is the main reason for higher rates of global warming at high latitudes.

Cloud. Clouds have a different effect on the radiation balance of the atmosphere, depending on their elevation in the atmosphere. They can slow down the atmosphere's loss of thermal radiation, thereby having a warming effect, or enhance the reflection of sunlight and have a cooling effect. High clouds will generally enhance warming, whereas low clouds will primarily cool the atmosphere. The relative importance of the two effects is not equal, however. The total effect of clouds is estimated to be one of cooling and about -20 W/m^2. This is large compared with the estimated net anthropogenic radiative forcing of 1.6 W/m^2 (Table 2.1). Therefore, even small changes in cloudiness and cloud structure may significantly affect global warming trends.

Carbon dioxide. Human-related climate changes are primarily caused by voluminous emissions of CO_2 from burning fossil fuels and from deforestation. Over many decades, CO_2 also plays a role in several climate feedback processes. For example, the solubility of CO_2 in sea water will be reduced at higher temperatures, which enhances the greenhouse effect of CO_2 (positive feedback). Another feedback effect occurs in permafrost regions, where very large amounts of carbon are stored in the soil. As permafrost melts, it releases these stocks as CO_2 to the atmosphere, which further enhances the greenhouse effect (positive feedback).

Vegetation. The vegetation on land surfaces affects the reflection of sunlight and evaporation of water. As the vegetation is affected by changes in temperatures and rainfall, several feedbacks are possible, and these can be quite difficult to predict.

Scenarios of climate change

Projecting climate change involves two steps. First, researchers must create scenarios of how the drivers of climate change will develop into the future (i.e. they must project the trends in greenhouse gas emissions and the determinants of these trends such as changes in demography, economic activity and other socio-economic factors). Second, they must translate projections of greenhouse gas emissions into atmospheric concentrations of greenhouse gases and thereafter calculate changes in climate such as in temperature, precipitation and sea level. This requires a complicated computational set-up, and there are many possible combinations of possible futures for the world and also a somewhat different formulation of models for the climate system. However, in practice, researchers work with a range of possible emission scenarios (Box 2.5) and these are applied to a range of climate models, which are of course essentially based on the same physical processes (Box 2.6).

A critical link in this chain of calculations is the specification of future socio-economic trends and their effect on greenhouse gas emissions. To capture the uncertainty of these factors, IPCC has developed different socio-economic and emission scenarios (Box 2.5). These scenarios assume a range of possible futures from those in which sustainability is a high priority for society, to those with an even stronger accent on consumerism than today. Although the projections of greenhouse gas emissions vary little between scenarios in 2030, they begin to diverge very strongly by the end of the 21st century (Figure 2.4). None of the IPCC scenarios and very few of the more recent emission scenarios have emissions in 2030 that are lower than those in 2000. The reason is, of course, that reducing emissions requires substantial changes in both energy technologies and in management in agriculture and forestry, which are the major sources of greenhouse

Box 2.5 IPCC emission scenarios

The future development of human societies leads to continued emissions of greenhouse gases at least in the short term and therefore also to increased concentrations in the atmosphere causing a further enhanced greenhouse gas effect. The main drivers of greenhouse gas emissions are population growth, economic growth, and expanded use of technologies that rely on the use of fossil fuels (coal, oil and gas). Changes in these drivers and development of technologies that rely on renewable energy sources may reduce these emissions compared to business-as-usual type developments.

It is of course very difficult to project how factors affecting emissions may develop over the next hundred years. The world might globalize to

Continued

an even greater degree or it may shift towards more regionalized markets and power centers. Society may lean towards less material consumption and thus lower greenhouse gas emissions, or it may develop towards higher energy and resource use leading to higher emissions. Another important factor is the future shape of economies in the developing world since they could become huge emitters of greenhouse gases. The climate issue is therefore deeply embedded in general issues of economic development. Considering the difficulty of estimating the socio-economic trends that will determine future emissions, a sensible approach is to develop scenarios of these trends, which describe *possible* future developments of the global society. IPCC published a comprehensive report in 2000 describing a set of scenarios, which have come to be known as the "SRES" scenarios (an abbreviation of Special Report on Emission Scenarios).[15] The IPCC scenarios cover a range of possible socio-economic developments during the 21st century. To avoid associating "values" with the scenarios, they were given the neutral and uncreative names of "A1", "A2", "B1" and "B2".

The A1 scenarios describe a world with very rapid economic growth. World population peaks during the middle of the century, and new and more energy efficient technologies are quickly introduced. The A1 family contains three sub-families, based either on fossil fuels (A1FI), non-fossil fuels (A1T) or a balanced mixture of all energy sources (A1B).

The A2 scenario depicts a more heterogeneous world with continued increase in global population and slower technological development. Power centers in this world are more decentralized than in A1 or B1.

The B1 scenario is a world that, in many respects, resembles A1, except that it puts much greater emphasis on a service- and information-based economy and on sustainable technologies. Society is much more ecologically oriented than in A1 or A2.

The B2 scenario depicts a world with continued strong population growth, although somewhat slower than in A2. This world has a slower tempo and more heterogeneous development of technology than in A1 or B2. Power centers here are more decentralized than in B1.

Taken together, the SRES scenarios make a wide range of assumptions for key socio-economic values. For example, assumed global population in 2100 ranges from 7 to 15 billion; for various scenarios, global GDP increases between 1995 and 2100 by a factor of 11 to 26. Even though some scenarios show an eventual decline in global CO_2 emissions, they all indicate an increase in CO_2 concentration from its current concentration of about 384 ppm (parts per million) to between 500 and 1000 ppm by 2100. Since the publication of the SRES scenarios in 2000, several additional emissions scenarios have been published in literature, many of them focusing on how greenhouse gas concentrations in the atmosphere might be stabilized at various specific concentrations.

Box 2.6 Global and regional climate models

Climate models describe the atmosphere as a physical system based on mathematical equations that represent the laws of physics relevant to the climate system.[18] The most important laws are:

- Newton's laws of motion;
- mass and energy conservation;
- the equation of state for ideal gases;
- radiation equations describing how solar and thermal radiation is emitted, transmitted, absorbed and re-radiated in the atmosphere and the earth's surface.

Putting it very simply, one could say that climate models distribute and redistribute heat, water and momentum within the climate system, accounting for incoming and outgoing radiative energy on the earth, and within such constraints as atmospheric composition, topography (land–sea distribution, orography (terrain relief) and bathymetry (seafloor relief)), as well as the fact that the earth rotates. The latter has some effect on circulations within the system.

Climate models can be thought of as depicting the atmosphere as a series of boxes covering the entire globe, with about 200 km horizontal distance between the centers of each box. The atmosphere is typically divided into 30–40 vertical layers, and the ocean into 20–30 vertical layers. Calculations are then repeated and repeated to simulate the evolution of temperature, moisture, clouds and winds within and in between the boxes.

Some physical processes of the climate system are represented by empirical equations rather than equations based on physical laws. Such empirical relationships are based on physical understanding of the underlying phenomena, but often also contain parameters that have been "tuned" by comparing model simulations to observations. In such cases there often is lack of precise data on these parameters, the observations offering some range of possible values. Cloud formation is one example of such a process that is addressed using both physical laws and parameterization. Since the cloud formation occurs at a scale much finer than that represented in global climate models, it is in practice not feasible to simulate these processes with equations made of the basic physical processes. Hence empirical equations based on observations are used instead.

One disadvantage of using empirical equations is that they cause uncertainty in the model calculations. One way of dealing with this problem is to increase the spatial and temporal resolutions of the models, but this greatly enhances the requirement for computer power, which is already huge. Indeed, in order to avoid the need to parameterize, for example,

Continued

convection, the spatial resolution would need to be around 1 km. Furthermore, the higher the spatial resolution, the shorter the time increments (temporal resolution) must be. Therefore, calculations with high spatial resolution have only been performed with regional climate models (RCM), which cover only a small part of the earth (e.g. Europe).

A typical RCM gets its boundary information (temperature, air pressure, water vapor etc.) from runs of global climate models. The use of RCMs in Europe has recently greatly increased the information on regional projections of climate change.[19]

gases. Alternatively it would require major changes in lifestyle, in particular for the wealthier segment of the population. Such changes take time and commitment, and are more likely to occur towards the end of the century than during the next few decades.

As described above, weather and climate are consequences of complicated interactions between physical, chemical and biological processes. To better understand this complex system and to make realistic projections of climate change, climate researchers make use of mathematical models of the climate

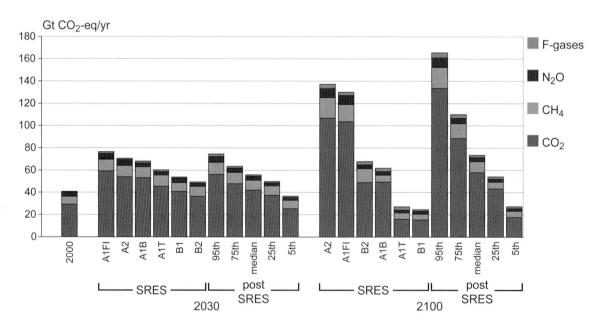

Figure 2.4 Annual global greenhouse gas emissions, expressed in Gt CO_2 equivalents, for 2000 and projected emissions for 2030 and 2100. Shown are six IPCC-SRES scenarios (Box 2.5) and a frequency distribution of emissions in post-SRES scenarios. The "F-gases" depict halocarbons. Source: Metz et al.[16] Reprinted with permission from Intergovernmental Panel on Climate Change, Geneva. (See color plate.)

system (Box 2.6). These models are computationally very demanding and require some of the world's fastest computers.

An important criterion for evaluating the credibility of climate models is to compare model simulations of current climate with observed climate. The climate models are continuously being improved, and there are currently about 20 such models worldwide. The deviations between simulated and observed temperature distribution on earth are in general quite small. These climate models are therefore considered to be a good basis for making projections of climate change under higher greenhouse gas concentrations.

Most of the present global climate model simulations are based on coupled atmosphere-ocean models (AO-GCM). Even though these models reproduce many of the observed features of the climate system (including studies made of past climates), they also reveal various contrasting results, not the least, some aspects of regional-scale climate change. The overall climate sensitivity of the global models also varies from relatively small to relatively large values. These uncertainties are largely a function of the relatively coarse resolution of global climate models and the different schemes employed to represent the processes in the atmosphere, biosphere and ocean. One example of this is the description of clouds. There are now also increasing efforts to downscale the coarser GCM results with regional climate models to give spatial resolutions of 50 km or finer.[17] This has led to improved detail in projections of regional climate changes in Europe.

How will climate change?

The modeled change in global mean temperature until 2100 is shown in Figure 2.5 and Table 2.2 for the IPCC emission scenarios. In 2100 global temperature will have further increased by 1.1–2.9°C if developments follow the B1 scenario, compared to around 1990, and by 2.0–5.4°C if the world develops according to the A2 scenario. In Chapter 10 we discuss the significance of holding the increase in global temperature to less than 2°C relative to pre-industrial conditions.

It is interesting to note that there is little difference between the projected global warming for the different emission scenarios until after 2050; in fact, they only start to differ considerably towards the end of the century. This is caused by inertia in both the climate and socio-economic systems.

A result of the inertia of the climate system is illustrated in the lower curve of Figure 2.5, which shows the change in global mean temperature if concentrations of greenhouse gases are maintained at their year 2000 level. Climate model simulations indicate that temperature will increase by some 0.3–0.9°C even if concentrations are held constant. This shows that the climate system, and in particular the temperature of the oceans, is not yet in equilibrium with current greenhouse gas concentrations. These results imply that the atmosphere will continue to warm

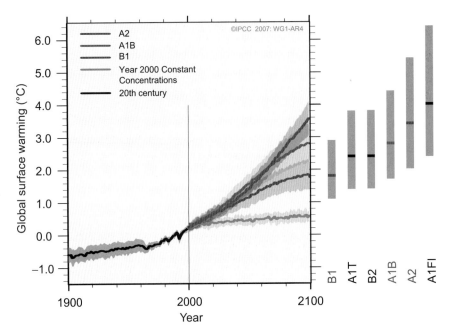

Figure 2.5 Global temperature increases from 1900 to 2100 for three IPCC-SRES emission scenarios (see Box 2.5) as estimated by climate models. The lowest line after 2000 shows results for the situation in which concentrations of greenhouse gases in the atmosphere are held constant at their year 2000 level. All temperatures are shown relative to the mean temperature for 1980–99. Source: Solomon et al.[20] Reprinted with permission from Intergovernmental Panel on Climate Change, Geneva. (See color plate.)

Table 2.2 Model projections of global warming and sea-level rise during the 21st century (from 1980–89 to 2090–99). Source: Solomon et al.[21] Reprinted with permission from Intergovernmental Panel on Climate Change, Geneva.

Scenario	Temperature rise (°C)		Sea-level rise (m)
	Best estimate	Probable interval	Excluding possible accelerating ice sheet melting
Constant 2000 concentration	0.6	0.3–0.9	
B1	1.8	1.1–2.9	0.18–0.38
A1T	2.4	1.4–3.8	0.20–0.45
B2	2.4	1.4–3.8	0.20–0.43
A1B	2.8	1.7–4.4	0.21–0.48
A2	3.4	2.0–5.4	0.23–0.51
A1FI	4.0	2.4–6.4	0.26–0.59

even if greenhouse gas emissions were reduced so as to give constant atmospheric concentrations, at today's levels.

Temperature

Not only will global average temperature change over the coming decades, but also regional patterns of temperature, rainfall, storminess and other climatic variables. Climate model results typically show a larger warming effect over land than water, and this holds generally for Europe (Figure 2.6). The temperature increase

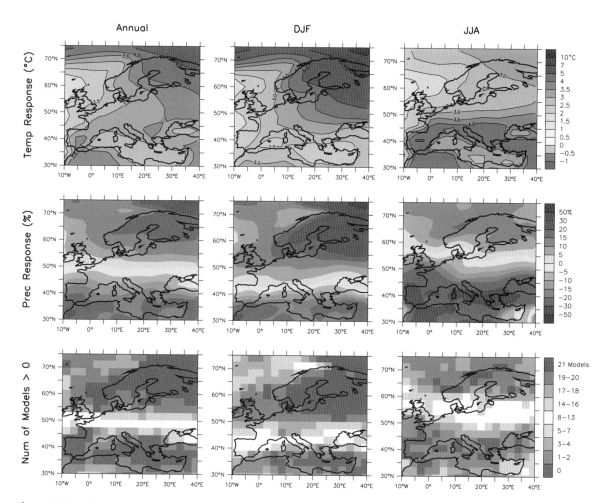

Figure 2.6 Modeled change in annual, winter and summer mean temperature (upper maps, from left to right) and precipitation (middle row maps) over Europe from 1980–1999 and 2080–2099 averaged for 21 global climate models, all using the A1B emission scenario. The bottom row maps show the number of models that out of 21 project increases in precipitation. Source: Christensen et al.[22] Reprinted with permission from Intergovernmental Panel on Climate Change, Geneva. (See color plate.)

is also much higher during winter over the Arctic than in the global mean, primarily because warming reduces the amount of snow and ice, which gives regional feedback (Box 2.4).

The annual average temperature for Europe is projected to increase by 1.0–5.5°C up to the period 2080 to 2100, as compared to 1961 to 1990. This range is built from simulations using two of the IPCC scenarios (A1B and A2) and comparing with the full range of IPCC scenarios. The warming is projected to be greatest over eastern Europe, Scandinavia and the Arctic in winter, and over the southwest, the Mediterranean region and the Balkans during summer (Figure 2.6). Over a part of France and the Iberian Peninsula the average increases in temperature may exceed 6°C during the summer, while mean increase in winter temperature in the Arctic could exceed 8°C.

Precipitation

There will also be marked changes in the magnitude and distribution of precipitation. The overall pattern is for dry areas to become even drier, and for areas with plentiful rainfall to become even wetter. Decreasing precipitation will occur primarily over the currently dry tropics and subtropics (e.g. the Mediterranean region), whereas increases in precipitation will be seen largely at higher latitudes in cool climates.

Climate models project changes in precipitation that vary considerably from season to season and across regions in Europe. Geographically, projections indicate a general precipitation increase in northern Europe and a decrease in southern Europe (Figure 2.6). The change in annual mean between 1980–89 and 2080–99 for the intermediate A1B scenario varies from 5% to 20% in northern Europe and from −5% to −30% in southern Europe and the Mediterranean region. Projections of changes under the A2 scenario follow the same pattern, but are mostly larger in magnitude.

Wind speed

Changes in atmospheric circulation will cause storm tracks at middle latitudes to move pole-wards. Climate model simulations indicate that the number of storms in the North Atlantic might decrease, but that the strength of the heaviest storms might, at the same time, increase. These projections are still very uncertain and model-dependent.

Projections of mean wind speed are highly sensitive to projected changes in large-scale atmospheric circulation. Since this varies between different climate models, projections of wind speed are particularly uncertain.[23] Some regional climate model simulations for Europe under the A2 scenario show that mean annual wind speed over northern Europe might increase by about 8%, and decrease in the Mediterranean region. The increase for northern Europe is greatest in winter and early spring. This could generate more intense North Sea storms leading to increases in storm surges along the North Sea coast, especially in the Netherlands, Germany and Denmark.

Climate extremes

An enhanced greenhouse effect will not only lead to a generally warmer climate, but also to changes in the frequency, intensity and duration of extreme weather events. Climate models show that there will be more frequent and longer lasting heatwaves, but also more intense rainfall. These are in accordance with observations over the past 50 years, and coupled with an intensification of the hydrological cycle (see Chapter 5). The result is a higher risk of both floods and droughts in many regions of the world. This accords with the observed increase in droughts in southern Europe.

Heatwaves in Europe are projected to become more frequent, intense and of longer duration through the 21st century. Likewise, night temperatures are expected to increase considerably. Geographically, the maximum temperatures during summer are projected to increase far more in southern and central Europe than in northern Europe, whereas the largest reduction in the occurrence of cold extremes is projected for northern Europe.

The combined effects of warmer temperatures and reduced mean summer precipitation would enhance the occurrence of heatwaves and droughts. The future summer climate over much of Europe would also see a pronounced increase in year-to-year variability and thus a higher incidence of heatwaves and droughts.[24] Countries in central Europe would experience the same number of hot days as currently occur in southern Europe, and in the Mediterranean region droughts would start earlier in the year and last longer. The regions most affected would be the southern Iberian Peninsula, the Alps, the eastern Adriatic seaboard, and southern Greece. In winter, the greatest effects can be expected in northern and eastern Europe, with considerably milder conditions, reduced daily temperature variability and also changes in snow cover.

Several recent modeling studies have found a substantial increase in the intensity of daily precipitation events. This holds even for areas with a decrease in mean precipitation, such as central Europe and the Mediterranean. In summer, the frequency of wet days is projected to decrease, but the intensity of extreme rain showers may increase. In addition, the frequency of several-day precipitation episodes is projected to increase, which may increase the risk of flooding.

Sea-level rise

Two factors contribute to sea-level rise under global warming. First, as water gets warmer it expands and causes a physical expansion of the oceans. Second, melting glaciers and ice sheets in a warmer environment add freshwater to the oceans.

The most likely range of global sea-level rise during the 21st century according to IPCC is 18–59 cm, depending on the emission scenario (Table 2.2). Thermal expansion is responsible for about 70–75% of this increase. As there was insufficient evidence on how the rate of melting of the Greenland and West Antarctica ice sheet might come to accelerate, possible accelerations in ice sheet melting and flotation were not factored in. Some new evidence shows that this might happen and thus adds to the estimates giving projections of sea-level rise of 1 meter or more by 2100.[25] However, since these latest estimates have not yet been confirmed they are not included in the estimates in Table 2.2.

How certain are we about climate change?

There is little doubt that the climate on earth is changing towards a warmer, and in some ways a more extreme, climate. There is also little doubt that the observed climate changes during the latter part of the 20th century can be largely attributed to anthropogenic emissions of greenhouse gases. However, there is still uncertainty about how exactly climate change will play out. This uncertainty is not so much related to the intrinsically chaotic nature of weather systems or the natural variation in climate caused by changes in solar radiation or volcanic eruptions. It is more related to our lack of complete understanding about feedback mechanisms in the climate system, which may make climate changes either more or less intense than the current range of projections.

When it comes to impacts of climate change on society, it is very important to have detailed projections of, not least, temperature and precipitation at the regional and even local level. For many applications, information on the changes in frequency of extreme weather events, such as storms, heavy rainfall and droughts, are also crucial. Here climate models have improved considerably recently, although there is plenty of scope for further improvements.

For projections of climate change that span the entire 21st century, uncertainties that relate to future greenhouse gas emissions are comparable to uncertainties related to the climate sensitivity. Hence there is a strong case for policies to reduce emissions of CO_2 and other greenhouse gases. The inertia of the climate system, the already occurring changes and the commitment to further changes, all emphasize that adaptation to climate change is important and inevitable.

One of the more notable results from climate modeling has been indications that under high emission scenarios, thresholds in the climate system might be reached, leading to rapid changes. Such non-linear or abrupt climate changes imply transitions of parts of the climate system into a new state. This would have considerable consequences for the world, and even more so in the most affected regions. Such abrupt changes will largely be determined by how the greenhouse gas forcing affects the dynamics of the global climate system, including large changes in the hydrological cycle and feedbacks through the carbon cycling in the biosphere. Even a relatively modest pace of change might reach a threshold that, when exceeded, leads to rapid transition to some unknown new state. A few of these abrupt changes could have potentially serious consequences for Europe, including: (1) a large-scale melting of the ice sheets on West Antarctica and Greenland leading to very large sea-level rises; (2) large emissions of CO_2 from melting permafrost areas and more frequent forest and wildfires; and (3) a stronger than expected slowdown, or a rearrangement, of the Atlantic overturning circulation. The latter could counteract warming trends in Europe. Present climate scenarios typically show some weakening of this ocean circulation, which is thus accounted for in climate scenarios. The understanding of these processes is as yet limited and the chance of major implications in the current century is generally considered to be low. We return to this topic in Chapter 9.

Notes

1   Solomon, S., Qin, D., Manning, M., Chen, Z., Marquis, M., Averyt, K.B., Tignor, M., Miller, H.L. (eds) 2007. *Climate Change 2007: The Physical Science Basis. Contribution of Working Group I to the Fourth Assessment Report of the Intergovernmental Panel on Climate Change*. Cambridge University Press, Cambridge, UK.

2   PRUDENCE project. www.prudence.dmi.dk

3   ENSEMBLES project. ensembles-eu.metoffice.com

4   EEA, 2003: *Europe's environment: the third assessment. Environmental assessment report No. 10*. European Environmental Agency, Copenhagen.

5   Jansen, E., Overpeck, J., Briffa, K.R., Duplessy, J.-C., Joos, F., Masson-Delmotte, V., Olago, D., Otto-Bliesner, B., Peltier, W.R., Rahmstorf, S., Ramesh, R., Raynaud, D., Rind, D., Solomina, O., Villalba, R., Zhang, D., 2007. Paleoclimate. In: *Climate Change 2007: The Physical Science Basis. Contribution of Working Group I to the Fourth Assessment Report of the Intergovernmental Panel on Climate Change*, Solomon, S., Qin, D., Manning, M., Chen, Z., Marquis, M., Averyt, K.B., Tignor, M., Miller, H.L. (eds). Cambridge University Press, Cambridge, United Kingdom and New York, NY, USA.

6   Alcamo, J., Moreno, J.M., Novaky, B., Bindi, M., Corobov, R., Devoy, R.J.N., Giannakopoulos, C., Martin, E., Olesen, J.E., Shvidenko, A., 2007: Europe. Chapter 12 in: Climate Change 2007: Impacts, Adaptation and Vulnerability. *Contribution of Working Group II to the Fourth Assessment Report of the Intergovernmental Panel on Climate Change (IPCC)*, Parry, M.L., Canziani, O.F., Palutikof, J.P., van der Linden, P.J., Hanson C.E. (eds), Cambridge University Press, Cambridge, UK.

7   Klein Tank, A.M.G., Können, G.P., 2003: Trends in indices of daily temperature and precipitation extremes in Europe. *Journal of Climate* 16, 3665–3680.

8   EEA, 2008: *Impacts of Europe's changing climate – 2008 indicator-based assessment*. EEA Report 2008/4. European Environmental Agency, Copenhagen.

9   EEA, 2008.

10   Smits, A., Klein Tank, A.M.G., Können, G.P., 2005: Trends in storminess over the Netherlands, 1962–2002. *International Journal of Climatology* 25, 1331–1344.

11   Global warming potential (GWP) is a measure of how much a given mass of greenhouse gas will contribute to global warming compared with an equal mass of CO_2. Since different greenhouse gases have different lifetimes in the atmosphere, GWP is usually estimated for a 100–year time horizon.

12   Solomon, S. et al., 2007.

13   Radiative forcing is the change in the difference between incoming and outgoing radiation energy to the troposphere (lower part of the atmosphere). IPCC has defined it as the measured difference relative to pre-industrial levels in 1750.

14   Lenton, T.M., 2006: Climate change to the end of the millennium. *Climatic Change* 76, 7–29.

15   Nakicenovic, N., Alcamo, J., Davis, G., de Vries, B., Fenhann, J., Gaffin, S., Gregory, K., Grübler, A., Jung, T.Y., Kram, T., Emilio la Rovere, E., Michaelis, L., Mori, S., Morita, T., Pepper, W., Pitcher, H., Price, L., Riahi, K., Roehrl, A., Rogner, H.-H., Sankovski, A., Schlesinger, M.E., Shukla, P.R., Smith, S., Swart, R.J., van Rooyen, S., Victor, N., Dadi, Z., 2000: *Special Report on Emissions Scenarios*. Cambridge University Press, Cambridge.

16   Metz, B., Davidson, O.R., Bosch, P.R., Dave, R., Meyer, L.A. (eds), 2007. *Climate Change 2007: Mitigation. Contribution of Working Group III to the Fourth Assessment Report of the Intergovernmental Panel on Climate Change*, Cambridge University Press, Cambridge, United Kingdom and New York, NY, USA.

17   Christensen, J.H., Christensen, O.B., 2007: A summary of PRUDENCE model projections of changes in European climate by the end of this century. *Climatic Change* 81, Supplement 1, 7–30.

18   Houghton, J., 2004: *Global warming. The complete briefing*, third edn. Cambridge University Press, Cambridge, UK.

19   Christensen, J.H., Christensen, O.B., 2007.

20   Solomon, S. et al., 2007.

21   Solomon, S. et al., 2007.

22   Christensen, J.H., Hewitson, B., Busuloc, A., Chen, A., Gao, X., Heid, I., Jones, R., Kolli, R.K., Kown, W.-T., Laprise, R., Rueda, V.M., Mearns, L., Menéndez, C.G., Räisänen, J., Rinke, A., Sarr, A., Whetton, P., 2007: Regional climate projections. In: *Climate change 2007: The physical science basis. Contribution of Working Group I to the Fourth Assessment Report of the Intergovernmental Panel on Climate Change*, Solomon, S., Qin, D., Manning, M., Chen, Z., Marquis, M., Averyt, K.B., Tigor, M., Miller, H.L. (eds). Cambridge University Press, Cambridge, UK and New York, NY, USA. pp. 847–940.

23 Räisänen, J., Hansson, U., Ullerstig, A., Döscher, R., Graham, L.P., Jones, C., Meier, M., Samuelsson, P., Willén, U., 2004: European climate in the late 21st century: regional simulations with two driving global models and two forcing scenarios. *Climate Dynamics* 22, 13–31.

24 Schär, C., Vidale, P.L., Lüthi, D., Frei, C., Häberli, C., Liniger, M.A., Appenzeller, C., 2004: The role of increasing temperature variability in European summer heatwaves. *Nature* 427, 332–336.

25 Shepherd, A., Wingham, D., 2007: Recent sea-level contributions of the Antarctic and Greenland ice sheets. *Science* 315, 1529–1532.

3 Human health

Introduction

Health has been defined by the World Health Organization as "a resource for everyday life" rather than the avoidance of sickness. It is a positive concept emphasizing "social and personal resources, as well as physical capacities". From this perspective, human health is much more complicated than we usually assume, and it can be affected in many different ways by a changing climate and environment. The health of people depends on a high quality natural environment; conversely, a deteriorated environment will lead to poor health. This is clear from the effects of atmospheric pollution, contaminated or inadequate water supplies, and degraded soils (leading to poor crop yields and food shortages). All such

Life in Europe Under Climate Change, First Edition. Joseph Alcamo and Jørgen E. Olesen.
© 2012 Joseph Alcamo and Jørgen E. Olesen. Published 2012 by John Wiley & Sons, Ltd.

factors may threaten health and assist in spreading disease, and many of these will also tend to be exacerbated by climate change. The greater the likelihood of extreme climate events, such as droughts, floods and heatwaves, the greater the risk to the health of individuals and to the systems of disease prevention and health care.

People in Europe are exposed to climate change directly through the effects of changing weather patterns (for example, to more intense and frequent heatwaves), and indirectly through changes in water, air, food quality and quantity, ecosystems, agriculture, livelihoods and infrastructure. Studies and assessments in Europe show that climate change is likely to affect human health through:[1]

- increased risk of heatwave–related health impacts;
- continued cold related health effects, in particular where access to energy is scarce;
- increased flood related impacts;
- increased malnutrition in areas already affected;
- changes in food-borne disease patterns;
- changes in distribution of infectious diseases;
- increases in waterborne diseases, in particular where water, sanitation and hygiene standards are low;
- increases in frequency of respiratory disease from atmospheric pollution and allergenic pollen.

Such health effects will affect the population differently in different countries and regions of Europe. Effects depend on the vulnerability and adaptive capacity of the health systems and the access of such systems to the population. Some measures that are sufficient to cope with current climatic variability may need to be revised or strengthened under a warmer climate with more extremes.

Mitigation of climate change, through adoption of renewable energy sources such as solar, wind, hydro and geothermal, will have the positive side effect of reducing air pollution and improving the health of people living in cities across Europe. Conversely, higher temperatures may increase the negative health impacts of air pollution, which is a serious health problem in Europe.[2] The main problems are associated with fine particles and ozone in urban areas; this pollution is estimated to cause hundreds of thousands of premature deaths every year in Europe. Its main source is the use of fossil fuels for transport, heat and electricity production.

Direct health effects of temperature

Human populations typically have an "optimal" temperature at which the (daily or weekly) death rate is lowest. Mortality rates rise when average temperatures are outside this zone. There is typically a U-shaped relation between mortality and temperature, where the trough represents the "comfort" zone (Figure 3.1). Studies of mortality in European cities have shown that the mortality rate increases more steeply with increasing temperatures than for an increase in low temperatures.[3] A

Figure 3.1 Schematic representation of how an increase in average annual temperature would affect annual total of temperature-related deaths, by shifting distribution of daily temperatures to the right under climate change as indicated for 2050 compared with 2005. Some acclimatization occurs to the new climate regime resulting in a change in curve shape. Additional heat-related deaths in summer would outweigh the extra winter deaths averted (as may happen in some northern European countries). Average daily temperature range in temperate countries would be about 5–30°C. Source: McMichael et al.[4] Reprinted with permission from Elsevier Limited, Oxford. (See color plate.)

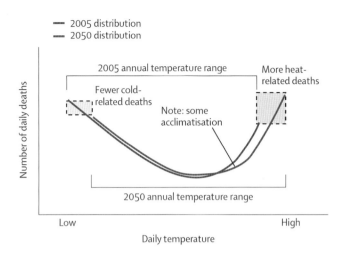

climatic warming will therefore lead to a higher number of deaths from increased frequency of heatwaves than the reduction in deaths from reductions in cold spells.

The relation between temperature and mortality varies greatly between climatic zones. People in hotter cities are more affected by colder temperatures, and people in colder cities are more affected by warmer temperatures. These differences result from differences in lifestyles and housing patterns. Acclimation and adaptation to a warmer climate will therefore shift the mortality curve towards higher temperature reducing the number of excess deaths from heatwaves (Figure 3.1).

In temperate regions, there is a seasonal variation in mortality, with higher mortality in winter than in summer. People with cardiovascular diseases are more at risk in winter, because of the cold-induced tendency of blood to clot. Accidental cold exposure occurs mainly outdoors, among socially deprived people (including alcoholics and the homeless), workers, and elderly people in temperate and cold climates. Cold also affects health by increasing some infectious agents (e.g. influenza among elderly people). Regions where housing is inadequate to protect people from cold have higher excess winter mortality.

Heatwaves have caused significant mortality in Europe in recent decades. Heat directly affects the human physiology, and thermo-regulation during heat stress requires a healthy cardiovascular system (Box 3.1). Increases in the body core temperature can lead to heat illness, or death from heat stroke, heart failure and a range of other ills. Most heatwave fatalities (heart attack and stroke) occur in people with pre-existing cardiovascular disease or chronic respiratory diseases. Children, people with chronic diseases, and those confined to bed, need particular care during extremely hot weather.

Episodes of extremely high temperatures present a challenge for health systems. For example, the heatwave in Western Europe in the summer of 2003 is estimated to have caused more than 70,000 excess deaths (Box 3.2). Overall, for populations in Europe, mortality is estimated to increase 1–4% for each one-degree increase above a cut-off point.[6]

Box 3.1 The human heat balance

The normal range of human body temperature (36.1–37.8°C) is maintained by the hypothalamus, which constantly regulates the production and release of heat.[5] Heat is lost to the environment by: (1) radiative cooling, whereby the freely exposed warm skin cools by radiating heat; (2) convection through the circulation of air around the skin; (3) conduction through direct contact of cooler objects with the skin; and (4) evaporative cooling of sweat. Conduction, radiation and convection require a temperature gradient between the skin and its surroundings, and evaporation depends on a water vapour pressure gradient. Exposure to excessive heat greatly stresses the body, particularly the cardiovascular system. When environmental heat overwhelms the body's heat-dissipating mechanisms, its core temperature rises.

There are several factors that affect the thermo-regulation of the human body:

- factors affecting behaviour that cause people not to seek proper shelter or wear proper clothing, including physical or cognitive impairment and psychiatric illness;
- increased heart rate, including exercise, outdoor activity and medications;
- factors influencing cardiac output, including cardiovascular diseases and medications;
- factors reducing blood plasma volume, including diarrhoea, renal or metabolic diseases, and medications;
- factors affecting sweating, including dehydration, ageing, diabetes, scleroderma, cystic fibrosis, and medications.

An increase of less than 1°C in body temperature is immediately detected by thermo-receptors distributed throughout the skin, deep tissues and organs. The thermo-receptors convey the information to the hypothalamic thermo-regulatory centre, which triggers two responses to increased dissipation of heat: an active increase in skin blood flow and initiation of sweating. An increase in the blood flow to the skin will cool the body when the external temperature is lower than the skin temperature. When the outdoor temperature is higher than the skin temperature, the only heat loss mechanism available is evaporation (sweating). Therefore, any factor that hampers evaporation, such as high ambient humidity, reduced air currents (no breeze, tight fitting clothes) or some drugs, will result in a rise of body temperature. This can culminate in life-threatening heatstroke or aggravate chronic medical conditions in vulnerable individuals. Mild and moderate heat-related health problems include heat rash, heat oedema, heat syncope (mild heatstroke), heat cramps and heat exhaustion.

Box 3.2 The 2003 heatwave

A severe heatwave over large parts of Europe in 2003 extended from June to mid-August, raising summer temperatures by 3–5°C in most of southern and central Europe (Figure 3.2a). The heat in June lasted throughout the entire month with monthly mean temperatures up to 6–7°C above average, but July was only slightly warmer than average (1–3°C). The highest temperatures were reached between 1 and 13 August (+7°C).[11] Maximum temperatures of 35–40°C were repeatedly recorded and peak temperatures climbed well above 40°C.

Average summer (June–August) temperatures were far above the long-term mean (Figure 3.2b), implying that this was an extremely unlikely event under current climatic conditions. However, it is consistent with the situation in which both the mean and variability of temperature increase. As such, the 2003 heatwave resembles simulations by regional climate models of summer temperatures in the latter part of the 21st century under the A2 scenario (Figure 3.2c; see also Chapter 2).[12] Hence, climate change may have already increased the risk of heatwaves like the one experienced in 2003.

The heatwave was accompanied by annual precipitation deficits up to 300 mm, and this drought had serious consequences for ecosystems (Chapter 7) and agriculture (Chapter 4). The hot and dry conditions led to many very large wildfires, particularly in Portugal. Many major rivers (e.g. Po, Rhine, Loire and Danube) were at record low levels, resulting in disruption of inland navigation, irrigation and power plant cooling (Chapter 8).

The heatwave also led to excess deaths of elderly people in many of the affected European cities (Figure 3.3). The excess deaths due to the extremely high temperatures during the period June–August are reported to have exceeded 70,000 compared to averages for the five previous years,[13] in particular among elderly people. The extreme weather also had other harmful side-effects such as an increase in air pollution caused by transportation and industry (tropospheric ozone and particulate matter) and forest fires (particulate matter).

The heatwave in 2003 has led to the implementation of "heat danger protection plans" in several European countries to avoid the consequences of a recurrence of the 2003 heatwave.[14] This includes heat danger warning systems, health and environmental surveillance, re-evaluation of care for the elderly, and structural improvements of residential institutions to maintain cooler temperatures. In 2007 a hot summer hit southeast Europe and the Balkans resulting in somewhat similar effects as for 2003 in central and western Europe.

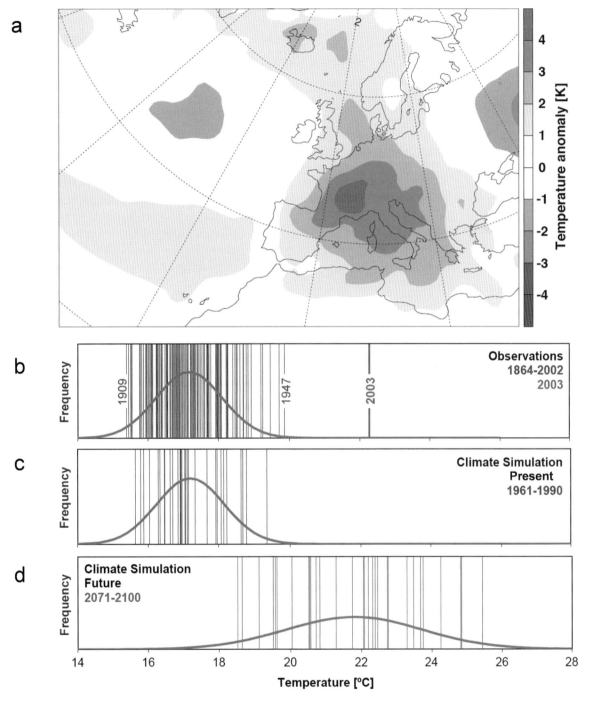

Figure 3.2 Characteristics of the summer 2003 heat wave in terms of summer (June to August) temperatures: (a) Temperature anomaly with respect to 1961–90 (b) Temperatures for Switzerland observed during 1864–2003 (c) Simulated using a regional climate model for the period 1961–1990 (d) Simulated for 2071–2100 under the A2 emission scenario. Source: Schär et al.[7] Reprinted with permission from Macmillian Publishers Limited: Nature. (See color plate.)

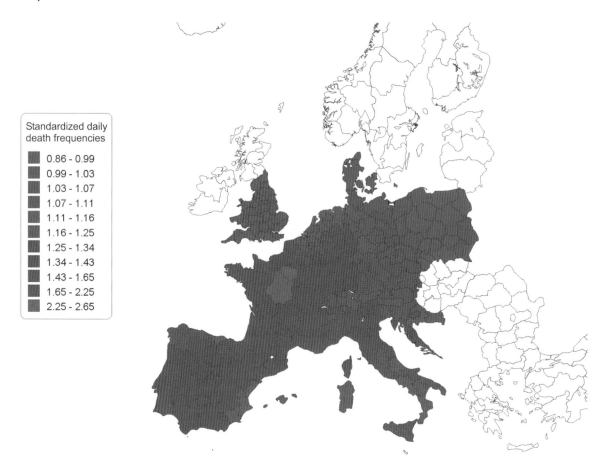

Figure 3.3 Standardized daily death frequencies during the summer heatwave from 3 to 16 August 2003 in 16 European countries (1 means equal to normal and 2 means twice the normal death rate). Source: Robine et al.[8] Reprinted with permission from Elsevier Masson SAS. (See color plate.)

People living in urban environments are at greater health risk from heatwaves than those in rural areas. Thermally inefficient housing and the so-called urban heat island effect (whereby inner urban environments, with high thermal mass and low ventilation, absorb and retain heat) amplify and extend the rise in temperatures, especially overnight. In 2003 in Paris many nursing homes and other assisted-living and retirement communities were not air-conditioned, and elderly residents might not have been promptly moved to air-conditioned shelters and rehydrated with fluids. During the summer of 2007, a heatwave emergency was declared in the Former Yugoslav Republic of Macedonia. The main problem was the very high temperatures in hospitals and emergency care units.

Heat-related morbidity and mortality is projected to increase. One projection for the EU-27 indicates 86,000 net additional deaths per year for the projected climate during the period 2071–100 as compared to the reference period 1961–90.

This projection was made under the assumption that emissions would increase according to the A2 scenario (see Chapter 2), and that no additional effort would be made to adapt or acclimatize to warmer temperatures.[9] Hence, this projection is likely (we hope) to overestimate the additional fatalities to be expected under this scenario.[10]

Winds, storms and floods

Floods are severe events that can overwhelm physical infrastructure, human resilience and social structures. They result from the interaction of heavy rainfall, surface run-off, wind, sea level and local topography. The risk of floods is substantially affected by water management practices, urbanization and land use (Chapter 5). Floods are often associated with windstorms, and it is the most common type of natural disaster that affects life and health in Europe.

Under climate change, winter floods are likely to increase throughout Europe (Chapter 5). Coastal flooding related to increased storminess and sea-level rise will also affect an increasing number of people in Europe (Chapter 6). This entails a range of health effects,[15] including:

- direct health effects: drownings, injuries, diarrhoeal diseases, vector-borne disease (e.g., from rodents), respiratory infections, skin and eye infections, and mental health problems;
- indirect health effects: damage to infrastructure for health care and water and sanitation, crops and property, disruption of livelihood and housing.

The risk of infectious disease following flooding is generally low in wealthy countries, although increases in respiratory and diarrhoeal diseases are often observed after floods. Flooding may lead to contamination of waters with dangerous chemicals, heavy metals or other hazardous substances from storage facilities or from chemicals already in the environment. This may in some cases lead to long-term contamination of soils. Increases in population density and industrial development in vulnerable areas increases the risks of exposure to hazardous materials released during storms and floods.

There is increasing evidence that mental disorders are a major impact of disasters.[16] This can result in considerable and prolonged impairment through anxiety and depression. For young children this may have medium to long-term impacts.

Nutrition, and food and waterborne diseases

Climate change will have a profound effect on food production (see Chapter 4). Increasing climate variability and the frequency of severe climate events are expected to have significant negative impacts on food production, and hence also on food security. More frequent droughts, floods and risks of fires, pests and disease outbreaks will affect food production differently in different regions thereby affecting food markets and prices. However, major impacts on food supply and human nutrition in Europe are less likely, as high income levels and good infrastructure act as a buffer to interrupted food supply.[17]

Diarrhoeal diseases from food- and waterborne infections are among the most important causes of ill health in Europe. Approximately 20% of the population of western Europe is affected by episodes of diarrhoea each year. These diseases are highly sensitive to climate and show strong seasonal variations in many countries. There are four main issues related to climate change and health aspects from food and water safety issues:[18] (1) links between poor water availability and quality and the occurrence of diarrhoeal diseases; (2) the role of intense rainfall or droughts in stimulating waterborne disease outbreaks; (3) effects of temperature and run-off on microbial and chemical contamination of coastal, recreational and surface waters; and (4) direct effects of temperature on the incidence of diarrhoeal diseases.

Climate change is likely to affect water quantity and quality in Europe (Chapter 5), and hence the risk of contamination of public and private water supplies. Both extreme rainfall and droughts can increase the total microbial loads in freshwater and have implications for disease outbreaks and water quality monitoring. Climate change will also affect the quality of coastal waters, by changing natural ecosystems or the quality of waters draining into the coastal zone. This poses risks for the recreational use of bathing waters, particular for tourists coming from other parts of Europe who may not have the built-in resistance to endemic water-related diseases. Harmful algal blooms produce toxins that can cause human illness, primarily via contaminated shellfish. The higher frequency of algal bloom incidents expected under climate change (Chapter 7) may thus increase the cases of human shellfish poisoning.[19]

Weather can also affect food consumption and preparation practices, which in turn increases the risk of food-borne diseases. Refrigeration failure is more likely to occur during hot weather. During the heatwave in 2003, interruption of refrigeration led to serious food-safety problems.[20] Warmer weather also leads to the proliferation of pests, which can affect food safety, if proper food storage conditions are not provided.

Some of the most serious diarrhoeal diseases in Europe are caused by salmonella and campylobacter infections. Several studies have documented higher incidence of salmonellosis under higher temperatures (Figure 3.4). Salmonella infections are expected to increase by 5–10% for each degree C increase in weekly temperature above 6°C. Depending on the climate scenario and other factors, this might lead to an extra 20,000 cases of salmonellosis in 2030 and 25,000–40,000 by 2080.[21] Temperature is much less important for the transmission of campylobacter.

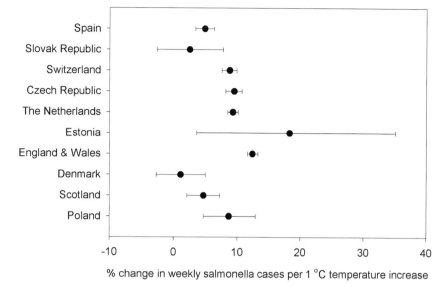

Figure 3.4 Change of weekly salmonella cases in 12 European countries for a 1°C temperature increase above a country-specific threshold, which is 6°C on average. The bars show the 95% confidence interval. Source: Kovats et al.[22] Adapted with permission from Cambridge University Press.

Vector-borne diseases

Climate change affects the occurrence of vector-borne diseases by changing the conditions which enable the spreading of vectors in a given region. It is primarily insect-borne diseases that are affected by climate change since many insect species are highly dependent on particular climatic conditions (Chapter 7). The principal disease-transmitting insects are the mosquito, tick and sandfly groups. Diseases associated with rodents are also known to be sensitive to climate change, but no assessments on impacts of climate change are available for Europe.[23]

The patterns of the vector-borne infectious diseases in Europe are affected by many factors, including movement of people and goods, changes in hosts and pathogens, land use and environmental factors. Climate change could lead to new infectious diseases being established in Europe, and previously-existing diseases could re-emerge. Whether this happens depends very much on the effectiveness of international, European and national surveillance systems for early detection and response. In many cases, it will be necessary to revise current vector-control measures, increase international collaboration between veterinary and public health services, and inform people on potential risks.

Mosquito-borne

Mosquitoes are vectors of several important diseases, including malaria, dengue, chikungunya, and West Nile Virus (WNV). These diseases are transmitted by different mosquito species, and all of these are likely to spread northwards into and across Europe under climate change. However, this does not mean that the diseases transmitted by these vectors will necessarily spread to Europe because other factors will intervene, as explained in the following paragraphs.

Malaria is an infectious disease caused by protozoan parasites that are transmitted by *Anopheles* mosquitoes. Globally, this is one of the most serious infectious diseases affecting about 500 million people every year, and killing between one and three million people, in particular poor people in Sub-Saharan Africa. The parasites multiply within the red blood cells causing symptoms of anaemia along with general symptoms of fever, chills and flu-like symptoms, and in severe cases coma and death. The *Anopheles* mosquitoes have long been present in Europe, and conditions in Europe for transmission of malaria are in principle favorable as repeated, but rare, observations of local transmission of tropical malaria show. At present, 6 of 53 European countries have indigenous malaria transmission (Azerbaijan, Georgia, Kyrgyzstan, Tajikistan, Turkey and Uzbekistan).[24] However, Europe in general has an effective health care system and measures to control mosquito populations are also in place. The re-emergence of endemic malaria in Europe due to climate change is therefore very unlikely. An increased risk of local outbreaks due to climate change is possible, but only if suitable vectors are present in sufficient numbers, and this depends on whether effective control measures are taken. The risk is probably greatest in eastern European countries, where per capita health expenditure is lowest and environmental measures to control mosquito populations are less well developed. Increases in malaria outside Europe may affect the risk of imported cases.

Dengue fever is an acute disease caused by virus that is transmitted by the *Aedes aegypti* mosquito. The fever is accompanied by severe headache, muscle and joint pains. There may also be gastritis with some abdominal pain, nausea and diarrhoea. The range of *Aedes aegypti* closely follows the 10°C winter isotherm and it is currently extending its range. This mosquito is currently absent in Europe, but in 1927–28 a dengue outbreak spread by *Aedes aegypti* in Greece affected one million people and caused 100 deaths.[25] Dengue frequently occurs in Europe, but only by travelers returning from dengue-endemic countries in tropical regions. It has been estimated that by 2080 the risk of dengue would extend to the Mediterranean region. This would require reintroduction of *Aedes aegypti* into Europe. Whether this happens and whether this leads to disease depends on the control measures taken, but under the present socio-economic conditions the risk can be considered low.

Chikungunya is a virus disease with symptoms similar to dengue fever. It causes an acute fever lasting two to five days, but this may be followed by long-term effects on the joints of arms and legs. It is transmitted by the tiger mosquito (*Aedes albopictus*). This mosquito has extended its range in Europe substantially in recent years, and has become established in at least 12 countries in southern Europe since the 1970s and in southern France since 2004.[26] Chikungunya was first observed in Europe in 2007, in the Emilia-Romagna region of Italy. This is the first example in continental Europe of an important human disease that subsequently is transmitted locally by mosquitoes.[27] The range of *Aedes albopictus* is projected to extend further to cover large parts of southern and western Europe under climate change. However, whether this will give rise to outbreaks of chikungunya depends on introduction and control of the virus in the susceptible regions. The generally high standards of the health and pest control systems at least in western Europe make major outbreaks unlikely.

West Nile Virus (WNV) is a viral disease transmitted by mosquitoes of the *Culex* family. This virus mainly affects birds, but can also be transmitted to mammals, including humans. WNV manifests itself in humans as a mild febrile syndrome which lasts 7–10 days. In some cases more dangerous encephalitis develops. The mortality rate is less than 4% of the infected. Higher transmission of WNV has been observed along major bird flyways. Human cases of WNV in Europe are rare and occur mainly in wetland and urban areas. Climate change affects the spread of this virus by migration and reproduction patterns of several bird species. When both birds and the relevant mosquitoes migrate to Europe, this creates a risk for higher frequency of occurrence in Europe. However, this depends on complex interactions among the life cycles of birds and of the mosquitoes.

Tick-borne

Two main diseases are transmitted by ticks: tick-borne encephalitis (TBE) and Lyme disease (or borreliosis). TBE is a tick-borne viral infection of the central nervous system, which can affect the brain (encephalitis), or the membrane that surrounds the brain or spinal cord (meningitis). There are about 10,000 TBE cases in Europe annually, and the disease is lethal in about 1% of the cases and leaves 10–15% of the survivors with permanent neurological damage. Lyme disease is an infection caused by bacteria belonging to the *Borrelia* group. It is the most common tick-borne disease in Europe. The infection initially causes fever, headache, fatigue, depression and skin rash. If left untreated, it may seriously affect the heart, joints and the nervous system.

Climate change can increase tick survival and thus tick density, prolong the season of tick and host activity, and shift ticks towards higher altitudes and northern latitude. Especially milder winters may enable expansion of TBE and Lyme disease to higher latitudes and altitudes. In contrast, droughts and floods may negatively affect the distribution.

Changes in tick distribution consistent with climatic warming have been reported in several European regions. The effect of climate variability on TBE or Lyme disease is still unclear. Future changes in tick-host habitats[28] and human-tick contacts may be more important for disease transmission than changes in climate.[29]

Sandfly-borne

Leishmaniasis is a disease caused by protozoan parasites and transmitted by sandflies. It is the second largest parasitic killer in the world (after malaria), and is estimated to be responsible for about half a million deaths every year globally. The parasite migrates to internal organs such as liver, spleen, and bone marrow and if left untreated will almost always result in death. Visceral leishmaniasis is present in the Mediterranean region and climate change may extend the range of the disease northwards.[30] New endemic areas have already been detected in northern Croatia, northern Italy, Germany and Switzerland. Sandfly vectors already have a wider range than the parasite, and imported dogs infected with it are common in central and northern Europe. Once the sandflies become established further north, the imported dogs could act as a source for new endemic foci. To minimize risks, it is therefore very important in future to focus on control of sandflies and the transmission via dogs.

Respiratory diseases

Diseases affecting the respiratory system are common causes of illness and death. The chronic forms of respiratory diseases include asthma and chronic obstructive pulmonary disease (COPD), which is narrowing of the airways, and include chronic bronchitis. In contrast to asthma, the limitation of airflow with COPD is difficult to treat and gets progressively worse over time. There is no current cure for COPD, whereas asthma can be treated through medication that controls the inflammation of the airways. The susceptibility to respiratory diseases are affected by: (1) the genetic background of individuals; (2) living conditions and lifestyle, which also affects the exposure to environmental effects; and (3) exposure to environmental factors, including a variety of substances present in air, water, food, soil and consumer products, as well as weather conditions and damp housing.

Climate change will affect respiratory diseases by changing the exposure to the agents affecting these diseases. This involves air pollutants such as ozone and fine particles in the air as well as allergenic pollen. All of these disease agents will be affected by climate change; in particular in urban areas climate change is expected to enhance the duration and severity of periods of poor air quality, with severe effects on respiratory diseases. Several of these disease agents interact with each other, often aggravating their individual effects. Air pollution and pollen allergy is therefore likely be more severe in a future warmer Europe.

Air quality

Air pollution has many consequences for human health, affecting average life expectancy, premature deaths and hospital admissions, use of medication and restricting activity. On the basis of the anthropogenic emissions in 2000, emissions of small particles (in particular $PM_{2.5}$) have been estimated to cause 348,000 premature deaths per year in Europe (Box 2.3). The most affected areas are Belgium, the Netherlands, northern Italy, and parts of Poland and Hungary, where the average reduction of life expectancy may reach two years.[34]

Exposure to ground-level ozone has several effects on health, primarily on the respiratory system.[35] Excess concentrations of ozone are estimated to lead to premature deaths of up to 20,000 people in the EU each year.[2] It also affects the respiratory conditions of more than 30 million person-days per year, giving rise to additional medication.

Important climate change effects on air quality are likely in Europe. Climatic warming may increase summer episodes of photochemical smog (Box 3.3) owing to increased temperatures, and decrease episodes of poor air quality during winter associated with air stagnation. Based on a limited number of modeling studies, climate change is likely to increase ozone concentrations in European urban areas, when precursor concentrations are held constant.[36] There is less certainty about the possible impacts of climate change on concentrations of fine particles.[37]

Box 3.3 Air pollution

Two air pollutants, fine particulate matter (PM) and ground-level ozone, are now generally recognized as having the greatest impacts on human health in Europe. Long-term and peak exposure can lead to a variety of health effects, ranging from minor impacts on the respiratory system to premature mortality. The negative health effects of both types of air pollution are likely to be enhanced in many European regions (in particular in urban areas) under climate change. Since 1997, up to 45% of Europe's urban population may have been exposed to ambient concentrations of PM above the EU limit set to protect human health; and up to 60% may have been exposed to levels of ozone that exceed the EU target value. It has been estimated that $PM_{2.5}$ (fine particles with sizes less than 2.5 µm) in air have reduced statistical life expectancy in the EU by nine months.[31]

Particulate matter (PM) is a collective name for fine solid and liquid particles added to the atmosphere by processes at the earth's surface. Some particulates occur naturally, originating from volcanoes, dust storms, forest and grassland fires, living vegetation, and sea spray. Significant amounts of fine particles are also produced by human activities such as the burning of fossil fuels in vehicles, power plants and various industrial processes. Some fine particles are formed by reactions of hydrocarbons and other chemicals in the atmosphere originating mostly from vehicle and industrial emissions. Increased levels of fine particles in the air are linked to health hazards such as heart disease, altered lung function and lung cancer.

Ozone is an important component of the outer atmosphere (stratosphere), where it forms the layer that protects against ultraviolet radiation from the sun. However, ozone is also formed at ground level, where it impacts health by affecting the respiratory system, reducing lung function,

(Continued)

aggravating asthma, and increasing sensitivity to allergens. Ground-level ozone is formed when nitrogen oxides (NO_x), carbon monooxides (CO) and volatile organic compounds (VOCs) react in the atmosphere in the presence of sunlight. These compounds are called ozone precursors. Motor vehicle exhaust, industrial emissions, and chemical solvents are the major anthropogenic sources of such chemicals.

Sources of air pollution are varied and may be anthropogenic (man-made) or natural. The main anthropogenic sources are: burning of fossil fuels in electricity generation, transport, industry and households; industrial processes and solvent use (e.g., chemical and mineral industries), agriculture, and waste treatment. The most important sources are energy production and road transport, which account for 49 and 17% of PM precursor emissions, respectively, and for 26 and 34% of ozone precursor emissions, respectively.[32]

In Europe, emissions of many air pollutants have fallen substantially since 1990, resulting in improved air quality. However, since 1997, measured concentrations of PM and ozone in the air have not shown any significant improvement despite the decrease in emissions. A significant proportion of Europe's urban population still lives in cities where certain EU air quality limits (set for the protection of human health) are exceeded.[33]

Pollen allergies

Hay fever (or allergenic rhinitis) is caused by the pollen of specific plants. It is characterized by sneezing, runny nose and itching eyes; these symptoms are usually seasonal and associated with the presence of pollen from specific species of groups of plants. The pollens that cause hay fever vary from person to person and from region to region; generally speaking, the pollens of wind-pollinated plants are the predominant cause of hay fever. These pollens stem from trees, grasses and weeds. Among the trees, birch pollen dominates at northern latitudes whereas olive pollen predominates in the Mediterranean region. Several grass species cause hay fever, and it is estimated that 90% of persons suffering from hay fever are sensitive to grass pollen. Several weed species are serious causes of pollen allergy, including mugwort (*Artemisia*), plantain (*Plantago*) and ragweed (*Ambrosia*) (Box 3.4). Hay fever may also be associated with other diseases, including asthma.

The burden of allergenic diseases is related to the length and intensity of the pollen season, and the frequency and height of the pollen peaks. Pollen allergy is exacerbated by pollution with fine particles and ground-level ozone, which interact with the pollen grains and also increases sensitivity to allergens. The pollen season is expanding in Europe. It has increased by 10–11 days on average over the last 30 years, primarily due to climatic warming.[41]

Climate change affects the amount, allergenicity and distribution of pollen. Experiments have shown increases in pollen production as a response to the

Box 3.4 Ragweed – a major source of allergy

Common ragweed (*Ambrosia artemisiifolia*) is an annual plant that originates from North America. It was first recorded in Europe in the mid 1800s,
but it did not start spreading until after World War I. It infests, in particular,
the Rhône valley (France), Northern Italy, and most of all, the Carpathian
Basin (Hungary). However, it has also been observed in many other European countries.[38] Ragweed prefers a dry climate and an open environment.
It thrives on sandy soils, and in France it was observed to be one of the few
plants that tolerated the heatwave of 2003. But it does not compete well with
cultivated crops, and it is therefore primarily found on land that has been
cleared and left uncultivated, including building sites. The warming of
Europe has accelerated its spread over the last few decades,[39] in particular
on fallow land set aside as part of the EU Common Agricultural Policy and
by the abandonment of the former communist-style collective agriculture
in eastern Europe.

Ragweed produces a large quantity of highly allergenic pollen that is
wind-spread. It blooms over a rather long period from early July to mid
August. The period of pollination has advanced in recent years, most likely
owing to a warmer climate. Ragweed causes severe hay fever, and in areas
infested with ragweed, this plant is responsible for about 50% of the asthma
cases. Hungary is the most severely affected country in the world. About
30% of the Hungarian population has some type of allergy, 65% of them
have pollen sensitivity, and at least 60% of this pollen sensitivity is caused
by ragweed. Here, the amount of ragweed pollen shows, with fluctuations,
a significant increasing trend. There is growing evidence that climate change
will significantly increase the problems with ragweed pollen allergy in
Europe. The plant will spread northwards in Europe as the climate warms
and it becomes possible for the plant to produce mature seeds in northern
Europe. In central and southern Europe, it is also likely to increase its dominance owing to drier summers and the abandonment of some agricultural
lands under climate change.[40]

enhanced photosynthesis that results from exposure to increased CO_2 concentration. The duration of the pollen season has been found to be extended under
climatic warming, especially in summer, and in the late flowering species. An
earlier start and peak of the pollen season is more pronounced in species that start
flowering earlier in the year. There is some evidence of stronger allergenicity in
pollen from trees grown at higher temperatures. In total, climate change is
expected to expand and worsen the situation for persons susceptible to pollen
allergy in Europe. This may also lead to more cases of asthma, in particular where
pollen interacts with other sources of air pollution.[42]

Benefits of climate change mitigation on public health

Efforts to reduce greenhouse gas emissions, in particular by burning fewer fossil fuels, will have the important side effect of reducing overall air pollution.[43] However, this benefit will not be realized if the fossil fuels are replaced with bioenergy and biofuels, which also produce various air pollutants. For Europe in particular, it is estimated that actions taken to meet the EU objective of limiting global warming to 2°C will have a considerable effect on air pollution by 2030.[44] Under this scenario EU greenhouse gas emissions are assumed to be reduced by 16–25% in 2030 compared to 2000. To achieve this, major changes would have to be made in the energy system, which would have the spin-off effect of reducing emissions of volatile organic compounds and nitrogen oxides which, in turn, would hinder the formation of ground-level ozone. Under this scenario, the costs of controlling air pollution would be 10 billion euros per year lower, and 20,000 air pollution-related "premature deaths" per year would be avoided. In the longer term, further reductions in fossil fuel use will be necessary, and this will be one of the major contributors to reducing air pollution and improving health in urban areas.

Some of the lifestyle diseases linked with living in modern wealthy societies is associated with too little physical activity, and the result has been a great increase not only in obesity, but also in diabetes and coronary artery disease. One of the reasons for this is the widespread use of cheap fossil fuels for transport, which for many Europeans has meant that cars are the favorite way of transportation, not only for work, but also for shopping and even for transporting children to school. The resulting lack of physical activity has become a serious health issue.[45] It would therefore make sense, if future policies to reduce greenhouse gas emissions would also promote an increase in physical activity (e.g. by encouraging the use of bicycling, or walking to school, work or shopping). This would probably require a considerable rethinking about traffic infrastructure and urban planning, but it would benefit both health and climate.

It has been estimated that 18% of global greenhouse gas emissions are associated with livestock production.[46] The consumption of livestock products (meat, milk, cheese, eggs, etc.) is associated with a wealthy lifestyle, and Europeans in general have a high consumption of meat and dairy products. This means that European food consumption is among the primary contributors to climate change. Particular attention should be paid to the health risks associated with the growth in meat consumption, which not only threatens climate, but is also a serious health concern, contributing to diseases such as cancer, obesity and coronary diseases.[47] To prevent increased greenhouse gas emissions from the livestock production sector, both the average worldwide consumption level of animal products and the intensity of emissions from livestock production must be reduced. The current global average meat consumption is 100 g per person per day, which should ideally be reduced 90 g per day with not more than 50 g per day coming from red meat from ruminants (i.e., cattle, sheep and goats). This would have a beneficial effect for both the environment and for human health.[48]

Ensuring good human health under climate change

Climate change will affect everybody's health, but not everyone is equally sensitive to these changes. The vulnerability of individuals depends on personal characteristics (e.g., age, income, education and health status), social and environmental contexts, access to resources (e.g., health services) and level of exposure to climate change (see Box 9.2 in Chapter 9). The populations considered to be most at risk are those living in large cities, in areas of water shortage, or along the coastline. Children are particularly vulnerable because of their physiological and cognitive immaturity and potential for long-term exposure.[49] Heat and heatwaves primarily affect old and young people; laborers in hot working environments also need special protection.

Potential options for adaptations to reduce health impacts include strengthening public health programs, including disease surveillance systems and vaccination programs for diseases such as tick-borne encephalitis.[50] Most current surveillance systems for infection have been designed to detect particular causes (e.g., food-borne disease) and individual risk factors (e.g., overseas travel or immune deficiency). Climate change requires a different perspective. The challenge under climate change is to take a more holistic approach to the causes of infection and disease, which involves the influence of climate both on the environmental sources of pathogens and disease agents, and on human behavior.

Adapting to extremes

Climatic extremes such as heatwaves, droughts, floods and storms will become more frequent in a future Europe under climate change (see Chapter 2). However, health risks need not rise, if we only can increase our preparedness for such effects. This involves changes not only in health care systems, but more importantly in infrastructure and in disaster emergency response. It is important to recognize that the variability in temperatures, rainfall and storms will increase in a future warmer Europe. Therefore plans for health protection need to target the changes in climate extremes rather than just the changes in climatic means.

There are a number of measures that can be taken to prevent heat-related illnesses and deaths (Table 3.1). In the long term, the most important actions will be improving urban planning and architecture, and changing energy and transport policies. However, such initiatives have long implementation times so they should be initiated as soon as possible.

There are many ways to reduce the vulnerability of people to the health hazards of floods. For example, new infrastructure and settlements can be located so as to minimize flood risks (see Chapter 5), and disaster relief plans can be prepared in advance of possible flooding events. Disaster relief concerns not only the operation of the primary health care sector, such as hospitals, ambulance stations, retirement homes and so on, but also the distribution of clean water and safe

Table 3.1 Adaptation measures to reduce heat-related illnesses and deaths. Source: Matthies et al.[51]

Category	Components
Action plans	Accurate, timely weather related heat alerts
	Strategies to reduce individual and community exposure to heat
	Plans for the provision of health care, social services and infrastructure
	Heat-related health information strategies
	Real-time surveillance, evaluation and monitoring
Health services	Health facility infrastructure: shading, cooling, drinking water, adapted menus
	Appropriate staff scheduling and working arrangements
	Special care for patients and residents: identification of risk, adjustment of drugs and treatment, organization of home care
	Staff training in identification and treatment of heat-related health problems

foods, and the cleaning and drying of damaged property to prevent long-term contamination. Following floods there is also the need for counseling to avoid anxiety and depression.

Food and water safety

Contamination of food products usually results from improper practices at some point from farm to fork. Therefore, providing education and timely information on how to handle food along with effective legislation and control procedures are essential; such measures will be even more important in a future, warmer climate. Food-borne disease outbreaks can be prevented by using safe water and raw materials, keeping food clean and at safe temperatures, cooking food thoroughly, and keeping raw and cooked food separately. In addition, it is essential that water-safety plans are revised to consider risks associated with climate change, so that safe drinking water is available to all households.

Vector-borne diseases

For protecting against new vector-borne diseases under climatic warming, it becomes very important to detect any changes in both disease vectors and diseases as soon as possible. Therefore, improved public health surveillance and response strategies are essential in adapting to climate change.

Control of the disease vectors (mosquitoes, ticks and sandflies) is essential to protect against the vector-borne diseases. These control measures need to be implemented at local scale, and include environmental management to reduce habitats for vector propagation (e.g., by improved operation and management of water resources) and changes in strategies for personal protection (e.g., protecting houses from mosquitoes). For control of sandflies, use of insecticide-impregnated dog collars is probably one of the measures needed to control the spread in a warmer climate.

There is also a need for health systems to be better prepared for vector-borne diseases under climate change. This involves the ability to identify and respond to new disease outbreaks, adequate logistics and sufficient supplies (e.g., pharmaceuticals, vaccines). It is also important to increase the advice for citizens on how they can protect themselves from disease, including awareness on avoiding disease transmission to those regions that are not currently affected.

Respiratory diseases

Respiratory diseases are primarily controlled by reducing the exposure of people to hazardous air pollution and allergenic pollen, by warning people about potential hazardous events and preparing health services to deal with such events. There are several international agreements that apply to control of air pollution in Europe, including the Convention on Long-range Transboundary Air Pollution. However, these conventions and agreements need to be strengthened, in particular, because climate change will worsen the effects of air pollution in Europe.

The preparedness of health systems, as well as citizens, to deal with climate-sensitive respiratory diseases can be strengthened through heatwave and pollen warning systems. This would alert vulnerable populations to reduce, or refrain from, vigorous exercise or other outdoor activity during period of high exposure, especially during the middle of the day when ozone levels are highest.

Developing healthier communities

One of the most important factors for increasing the resilience of the European population to climate change is to improve water and air quality. This needs to be promoted through effective legislation and through community action. Much of the air pollution in urban areas is caused by traffic, in particular by petrol-fueled cars. Replacing some of the car transport by walking and cycling will not only reduce air pollution, but the increased exercise will also reduce the risk of cardiovascular diseases and diabetes, as well as colon and breast cancer. Likewise, healthier diets, involving lower consumption of meat and dairy products and a higher proportion of vegetables, may also reduce greenhouse gas emissions and other pollutants.[52] Improving health under climate change is thus not only restricted to the health system, but extends to most of society and includes the actions of individual citizens.

More than 70% of Europeans live in cities, and these citizens are particularly vulnerable to climate change. One reason is that the heat-island effect aggravates the effects of global warming. Cities are therefore areas of primary concern for the health effect of global warming.

Since urban areas and their inhabitants are major consumers of non-renewable resources, they also become central to climate policy. For both adaptation and mitigation purposes, it is essential to construct new buildings that require as little heating and cooling as possible which, at the same time, provide shelter under all future climate conditions and mitigate climatic extremes. Building such systems requires long-term planning and an integrated approach to all of the services that cities provide (e.g., housing, work, transport, shopping, leisure, and education). This can probably only be developed through a positive interaction between government, local authorities and organizations, with the support of city inhabitants.

Notes

1   Menne, B., Apfel, F., Kovats, R.S., Racioppi, F., 2008: *Protecting health in Europe from climate change*. World Health Organization.
2   EEA, 2007: *Europe's environment. The fourth assessment*. European Environment Agency, Copenhagen.
3   Baccini, M., Biggeri, A., Accenta, G., Kosatsky, T., Katsouyanni, K., Analitis, A., Anderson, H.R., Bisanti, L., D'Ippoliti, D., Danova, J., Forsberg, B., Medina, S., Paldy, A., Rabczenko, D., Schindler, C., Michelozzi, P., 2008: Effects of apparent temperature on summer mortality in 15 European cities: results of the PHEWE projects. *Epidemiology* 19, 711–719.
4   McMichael, A.J., Woodruff, R.E., Hales, S., 2006: Climate change and human health: present and future risks. *The Lancet* 367, 859–869.
5   Matthies, F., Bickler, G., Marin, N.C., Hales, S., 2008: *Heat-health action plans. A guidance document*. World Health Organization.
6   Menne. B. et al., 2008.
7   Schär, C., Vidale, P.L., Lüthi, D., Frei, C., Häberli, C., Liniger, M.A., Appenzeller, C., 2004: The role of increasing temperature variability in European summer heatwaves. *Nature* 427, 332–336.
8   Robine, J.M., Cheung, S.L., Le Roy, S., Van Oyen, H., Griffiths, C., Michel, J.P., Herrmann, F.R., 2008: Death toll exceeded 70,000 in Europe during the summer of 2003. *Comptes Rendus Biologies* 331, 171–178.
9   EEA, 2008: *Impacts of Europe's changing climate – 2008 indicator-based assessment*. EEA Report 2008/4. European Environmental Agency, Copenhagen.
10 Gosling, S.N., Lowe, J.A., McGregor, G.R, Pelling, M., Malamud, B.D., 2009: Associations between elevated atmospheric temperature and human mortality: a critical review of the literature. *Climatic Change* 92, 299–341.

11 Fink, A.H., Brücher, T., Krüger, A., Leckebusch, G.C., Pinto, J.G., Ulbrich, U., 2004: The 2003 European summer heat waves and drought – Synoptic diagnosis and impact. *Weather* 59, 209–216.

12 Beniston, M., 2004: The 2003 heat wave in Europe: a shape of things to come? An analysis based on Swiss climatological data and model simulations. *Geophysical Research Letters* 31, L02202 doi:10.1029/2003GL018857.

13 Robine, J.M. et al., 2008.

14 Confalonieri, U., Menne, B., Akhtar, R., Ebi, K.L., Hauengue, R.S., Kovats, R.S., Revich, B., Woodward, A., 2007: Human health. *Chapter 8 in: Climate Change 2007: Impacts, Adaptation and Vulnerability. Contribution of Working Group II to the Fourth Assessment Report of the Intergovernmental Panel on Climate Change (IPCC)*, Parry, M.L., Canziani, O.F., Palutikof, J.P., van der Linden, P.J., Hanson, C.E., Eds., Cambridge University Press, Cambridge, UK, 391–431.

15 Kirch, W., Menne, B., Bertolinni, R., 2005: *Extreme weather events and public health responses*. Springer Verlag, Heidelberg.

16 Ahern, M., Kovats, R.S., Wilkinson, P., Few, R., Matthies, F., 2005: Global health impacts of floods: Epidemiological evidence. *Epidemiological Reviews* 27, 36–46.

17 Schmidhuber, J., Tubiello, F.N., 2007: Global food security under climate change. *Proceedings of the National Academy of Science* 104, 19703–19708.

18 EEA, 2008.

19 Hunter, P.R., 2003: Climate change and waterborne and vectorborne disease. *Journal of Applied Microbiology* 94, 37–46.

20 Menne, B. et al., 2008.

21 Kovats, R.S., Edwards, S.J., Hajat, S., Armstrong, B.G., Ebi, K.L., Menne, B., 2004: The effect of temperature on food poisoning: a time-series analysis of salmonellosis in ten European countries. *Epidemiology and Infection* 132, 443–453.

22 Kovats, R.S. et al., 2004.

23 Alcamo, J., Moreno, J.M., Novaky, B., Bindi, M., Corobov, R., Devoy, R.J.N., Giannakopoulos, C., Martin, E., Olesen, J.E., Shvidenko, A., 2007: Europe. *Chapter 12 in: Climate Change 2007: Impacts, Adaptation and Vulnerability. Contribution of Working Group II to the Fourth Assessment Report of the Intergovernmental Panel on Climate Change (IPCC)*, Parry, M.L., Canziani, O.F., Palutikof, J.P., van der Linden, P.J., Hanson, C.E. (eds), Cambridge University Press, Cambridge, UK.

24 Menne. B. et al., 2008.

25 Senior, K., 2008: Vector-borne diseases threaten Europe. *The Lancet Infectious Diseases* 8, 531–532.

26 Vazeille, M., Jeannin, C., Martin, E., Schaffner, F., Failloux, A.B., 2008: Chikungunya: A risk for Mediterranean countries. *Acta Tropica* 105, 200–202.

27 Menne, B. et al., 2008.

28   Ticks are particularly abundant in properties that border woodlots and where there are tall grasses and weeds.

29   Randolph, S.E., 2004: Evidence that climate change has caused "emergence" of tick borne diseases in Europe. *International Journal of Medical Microbiology* 293, 5–15.

30   Kuhn, K.G., Campbell-Lendrum, D.H., Davies, C.R., 2004: Tropical diseases in Europe? How can we learn from the past to predict the future. *EpiNorth* 5, 6–13.

31   EEA, 2007: Air pollution in Europe 1990–2004. EEA Report No. 2/2007. European Environmental Agency, Copenhagen.

32   EEA, 2007. Europe's environment.

33   EEA, 2007. Air pollution in Europe.

34   Amann, M., Berton, I., Cofala, J., Gyarfas, Heyes, C., Klimont, Z., Schöpp, W., Winiwarter, W., 2005: *Baseline scenarios for the Clean Air for Europe (CAFÉ) Programme*. Final Report. International Institute for Applied Systems Analysis.

35   WHO, 2003: *Health aspects of air pollution of air pollution with particulate matter, ozone and nitrogen dioxide*. World Health Organization, Regional Office for Europe, Bonn, Germany.

36   Ebi, K.L., McGregor, G., 2008: Climate change, tropospheric ozone and particulate matter, and health impacts. *Environmental Health Perspectives* 116, 1449–1455.

37   OECD, 2008: *OECD Environmental outlook to 2030*. Paris.

38   D'Amato, G., Cecchi, L., Bonini, S., Nunes, C., Annesi-Maesano, I., Behrendt, H., Liccardi, G., Popov, T., van Cauwenberge, P., 2007: Allergenic pollen and pollen allergy in Europe. *Allergy* 62, 976–990.

39   Vogl, G., Smolik, M., Stadler, L.M., Leitner, M., Essl, F., Dullinger, S., Kleinbauer, I., Peterseil, J., 2008: Modelling the spread of ragweed: Effects of habitat, climate change and diffusion. *European Physical Journal – Special Topics* 161, 167–173.

40   Dahl, A., Strandhede, S.O., Wihl, J.A., 1999: Ragweed – an allergy risk in Sweden? *Aerobiologia* 15, 293–297.

41   D'Amato, G. et al., 2007.

42   Ziska, L.H., Gebhard, D.E., Frenz, D.A., Faulkner, S., Singer, B.D., Straka, J.G., 2003: Cities as harbingers of climate change: common ragweed, urbanization, and public health. *Journal of Allergy Clinical Immunology* 11, 290–295.

43   Bell, M.L., Davis, D.L., Cifuentes, L.A., Krupnick, A.J., Morgenstern, R.D., Thurston, G.D., 2008: Ancillary human benefits of improved air quality resulting from climate change mitigation. *Environmental Health* 7, 41 doi:10.1186/1476–069X-7–41.

44   EEA, 2006: Air quality and ancillary benefits of climate change policies. EEA Technical Report No. 4/2006. European Environmental Agency.

45   Faergeman, O., 2007: Climate change and preventive medicine. *European Journal of Cardiovascular Prevention & Rehabilitation* 14, 726–729.

46 Steinfeld, H., Gerber, P., Wassenaar, T., Castel, V., Rosales, M., de Haan, C., 2006: *Livestock's long shadow. Environmental issues and options.* FAO, Rome.

47 McMichael, A.J., Powles, J.W., Butler, C.D., Uauy, R., 2007: Food, livestock production, energy, climate change and health. *The Lancet* 370, 1253–1263.

48 Stehfest, E., Bouwman, L., van Vuuren, D.P., den Elzen, M.G.J., Eickhout, B., Kabat, P., 2009: Climate benefits of changing diet. *Climatic Change* 95, 83–102.

49 Menne, B. et al., 2008.

50 Kovats, R.S, Haines, A., Stanwell-Smith, R., Martens, P., Menne, B., Bertollini, R., 2008: Climate change and human health in Europe. *British Medical Journal* 318, 1682–1685.

51 Matthies, F., et al., 2008.

52 Steinfeld, H., et al., 2006.

4 Food production

Introduction

Europe is a very culturally diverse continent. This is not the least reflected in the traditional diet and food habits of Europeans in different regions. Traditionally, dairy and livestock products play a large role in the diet in northern Europe, whereas fresh vegetables, grains and legumes, in combination with a wide range of fish and livestock products, form the basis of the Mediterranean kitchen. In central and eastern Europe soups are an essential part of the diet. All of this reflects

Life in Europe Under Climate Change, First Edition. Joseph Alcamo and Jørgen E. Olesen.
© 2012 Joseph Alcamo and Jørgen E. Olesen. Published 2012 by John Wiley & Sons, Ltd.

the differences in culture and climate throughout Europe; and even widespread globalization has not been able to erase these cultural differences. The diet of all European countries has been converging over the past few decades, and the average calorie intake is now 20% above the recommended level.[1] This is largely a result of technological improvements in livestock production, which provides animal protein and fat at very low costs. In contrast, some of the traditional European foods are more expensive to produce, and it is probably these local and often protected foods that will primarily be affected by climate change.

The countryside of Europe will also look different in the future. As the climate becomes wetter and warmer north of the Alps, fields of sunflower and maize will begin appearing where earlier they could not flourish. Even vineyards will become successful far further north than anyone can remember. Meanwhile, after decades of consistently drier and hotter conditions, many farmers south of the Alps and in the Balkans will either adapt by adopting a much less productive agriculture, or they may simply give up, and abandoned agricultural land could then become a prominent feature of the landscape.

But returning to the present, Europe is now one of the world's largest and most productive suppliers of food and fiber. In 2006 it accounted for 19% of global meat production and 18% of global cereal production. About 82% of the European meat production and 67% of the cereal production occurred in the EU27 countries. The large livestock production has caused a high dependency on the import of protein feed, in particular of soybeans from South America. The productivity of European agriculture is generally high, in particular in western Europe, and average cereal yields in the EU countries are more than 60% higher than the world average.

Europeans consume about 14% of global fish production, and the average European consumes about 20 kg fish per year, which is slightly above the world average of 16.5 kg per year. Inland fisheries in Europe have declined by 30% since 1999, which can be partly attributed to falling economic viability of this activity. Aquaculture is growing considerably on the global scale, but Europe only contributes about 4% of global aquaculture production. The major supply of fish for the European market still comes from marine fisheries, and overall Europe is a net importer of fish products. Despite policies to protect fish, overfishing has put many fish stocks in European waters outside sustainable limits (62–92% of commercial fish stocks in the northeastern Atlantic, 100% in West Ireland Sea, 75% in the Baltic Sea, and 65–70% in the Mediterranean).[2]

Agriculture and forestry, as major land users, play a key role in determining the health of the rural economy as well as the rural landscape. Even with a reduced share of economic activities, agriculture still has a valuable contribution to make to the socio-economic development of rural areas and full realization of their growth potential. The EU's common agricultural policy (CAP) has been developed in recognition of this (Box 4.1).

Over recent decades it has been increasingly recognized that agriculture contributes to some of the environmental problems in Europe, including eutrophication of ecosystems, air pollution, loss of biodiversity, pesticides in groundwater

Box 4.1 EU Common Agricultural Policy (CAP)

With the gradual expansion of the European Union, agricultural policy in Europe is now largely determined by the EU Common Agricultural Policy (CAP). The CAP dates from the early 1960s, and was based on the Treaty of Rome. Its emphasis was on encouraging better productivity in the food chain, largely for food security reasons, but also to ensure that the EU had a viable agricultural sector and that consumers had a stable supply of affordable food. This reflected the situation after World War II, where food supply could not be guaranteed in Europe and where starvation was widespread. The CAP offered subsidies and guaranteed prices to farmers, providing incentives for them to produce.

From the mid-1960s and throughout the 1970s, the CAP developed, and financial assistance was provided for the restructuring of farming, for example by aiding farm investment, aiming to ensure that farms developed in size and in management and technology skills so that they were adapted to the economic and social climate of the day. The CAP was very successful and, by the 1980s, the EU had to contend with almost permanent surpluses of the major farm commodities. This led to high cost of the CAP, and by 2008 it still consumed about 40% of the total EU budget.

During the 1990s there was a shift in the agenda, which involved reducing support prices and putting more emphasis on environmentally sound farming. In 2003 it was decided to decouple the agricultural subsidies from production. The future payment to farmers will be linked to their respect for environmental, food safety, animal and plant health and animal welfare standards, as well as the requirement to keep all farmland in good agricultural and environmental condition. This is not expected to greatly affect agricultural production outputs. However, the reform is expected to enhance the current process of structural adjustment leading to larger and fewer farms. The revised CAP has a strengthened rural development policy with increased emphasis on environmental and employment issues.

The emphasis on rural development and support for environmental issues is gradually being enhanced and the 2008 CAP agreement has for the first time included the contribution of agriculture to climate as one of the targets under the rural development pillar of the CAP. This primarily involves reductions in greenhouse gases from agriculture, but future revisions of the CAP will need to support the adaptation of European agriculture to climate change by encouraging the flexibility of land use, crop production, farming systems etc.[2] In so doing, it will be necessary to consider the multifunctional role of agriculture, and to strike a balance between economic, environmental and economic functions, which may vary between different European regions.

and greenhouse gas emissions.[4] A range of EU Directives have therefore been implemented targeting these issues, including the Nitrate Directive, Habitat Directive, Water Framework Directive, and Pesticide Directive. Many of these issues are influenced by climate, both directly and indirectly. The future implementation of both agricultural policies and of environmental targets will therefore depend on how climate change will affect European agriculture and how farmers and society will adapt to these changes.

How does climate affect crops?

Biophysical processes of agroecosystems are strongly affected by environmental conditions. The projected increase in greenhouse gases will affect agroecosystems either directly (primarily by increasing photosynthesis at higher CO_2)[5] or indirectly via effects on climate (e.g. temperature and rainfall affecting several aspects of ecosystem functioning) (Table 4.1). The exact responses depend on the sensitivity of the particular ecosystem and on the relative changes in the controlling factors.

Increasing atmospheric CO_2 concentration stimulates the yield of most European crops. These are plants that have the so-called C_3-photosynthesis pathway, and these plants include most of the crops grown in Europe. A doubling of atmospheric CO_2 concentration will in most of these crops lead to yield increases of 20–40% (Figure 4.1). The response is considerably smaller for plants that have the C_4-photosynthesis pathway,[6] which include warm-season plants like maize, sorghum, sugar cane, miscanthus, amaranth and millet. Higher CO_2 concentration not only affects photosynthesis, but also the water consumption of the plants. With higher CO_2 concentration, the amount and the openness of stomata (pores on the leaf surface that allow for gas exchange in and out of the leaf) will be reduced for both C_3 and C_4 plants. This leads to reduced transpiration and to higher water use efficiencies, resulting in higher yields under dry or drought conditions.

Higher CO_2 concentrations also affect the quality of the plant biomass, because plants accumulate more sugar leading to higher carbon contents of leaves, stems and reproductive organs. This has consequences for the quality of the food and feed, which in some cases are negative: As a result, some changes in farm management might be needed, such as changes in the way cattle are fed and in the timing of grape harvest to get the optimal quality. The attraction of plants for pests and diseases will also change, which in some cases could make the plants more resistant to attack. However, weeds will also benefit from increased CO_2, which in some cases will make these weeds a greater nuisance.[8]

Increasing temperature affects crops primarily via plant development. With warming, the start of active growth is advanced, plants develop faster, and the potential growing season is extended. This may have the greatest effect in cooler regions, and may be beneficial for perennial crops or crops, which remain in their

Table 4.1 Influence of CO_2, temperature, rainfall and wind on various components of the agroecosystem.

Component	Influence of factor		
	CO_2	**Temperature**	**Rain/wind**
Plants	Higher CO_2 leads to increased dry matter growth and decrease in water use.	Increase of temperature boosts yield up to a threshold beyond which yield declines,	Decreasing precipitation or increasing wind decreases dry matter growth,
Animals	Higher CO_2 increases yield but may reduce quality of fodder for animals.	High temperatures reduce animal growth and reproduction	Prolonged dry, wet or windy conditions negatively affect animal health,
Water	Higher CO_2 conserves soil moisture by reducing transpiration.	Higher temperatures increases evaporation leading to higher irrigation demands and, in dry environments, to salinization.	Higher rainfall will increase groundwater supply and in some areas increase groundwater levels,
Soil	Higher carbon concentrations of plant residues under higher CO_2 will lead to higher soil carbon contents.	Higher temperatures boost soil organic matter turnover leading to reduced soil carbon content but temporarily higher nutrient supply for plants.	Drier and more windy environments may lead to enhanced wind erosion, whereas more intense rainfall will enhance water erosion.
Pests/ diseases	Higher CO_2 reduces the quality of plant biomass for pests and diseases leading to fewer pests.	Higher temperatures reduces the generation time of pests and diseases and causes attacks to occur earlier in the year making pests and diseases more problematic.	Some diseases are spread by wind or rainfall. Therefore, more rainy and windy conditions will favor some diseases.
Weeds	Enhanced CO_2 concentrations will differentially favor crop and weed species. This may make some weeds more problematic. Higher CO_2 will also reduce the efficacy of some herbicides.	Higher temperatures will lead to invasive weed species in some regions. Higher temperatures will also affect the efficacy of herbicides.	More rainy conditions may make some weed species more difficult to control through herbicides.

vegetative phase, for example, sugar beets. However, increased temperature reduces crop duration for many annual crops. In wheat, an increase of 1°C during grain filling reduces the length of this phase by 5%, and yield declines by a similar amount.[9]

Compared to temperate crops such as wheat, barley and potato, sensitivity to warming may be even greater in tropical crops such as maize, soybean and cotton, in particular when they are grown at the edge of their natural range. This is illustrated by the development in yields of grain maize over the period 1961–2009 in central and southern Europe (Figure 4.2). Maize yield has been increasing in Belgium and Germany, even in recent years, while wheat yield has been levelling off. This has also resulted in a steadily increasing grain maize area in these countries and a northward shift of the grain maize cultivation in Germany. The yield

Figure 4.1 Effects of CO_2 concentration on wheat yield in experiments. Ambient CO_2 is set to 1. Open symbols represent data from field experiments, filled symbols from pot or glasshouse experiments. The solid line shows the mean estimated response. Estimates consistently show that higher CO_2 concentrations in the atmosphere will stimulate yield of wheat and other crops. In general similar effects are obtained in both pot and field experiments. Source: Olesen and Bindi.[7] Reprinted with permission from Elsevier Limited, Oxford.

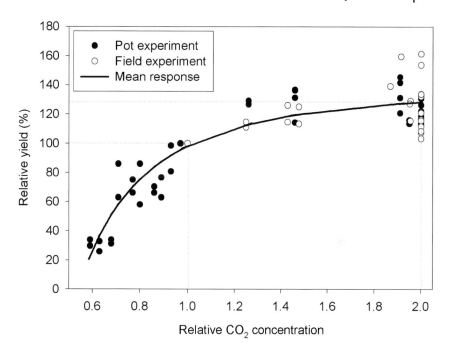

Figure 4.2 National yield and area of grain maize in four countries in northern and southern Europe for the period 1961 to 2009 (FAOSTAT database). Dots show grain yield per ha, whereas the lines show development in area cropped with maize in proportion to total cereal area. Both grain yield and area cropped with grain maize have recently increased in northern Europe. Source: Olesen et al.[10] Reprinted with permission from Elsevier Limited, Oxford.

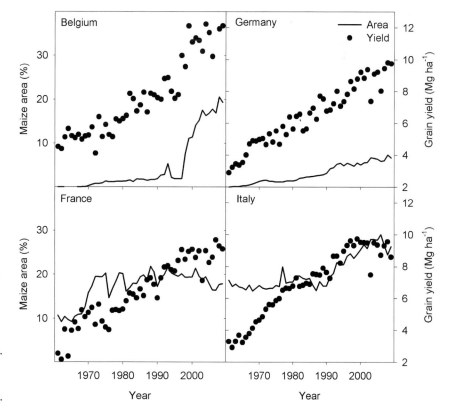

of grain maize in France and Italy has not increased in recent years. This is most likely due to warmer climate and a higher frequency of droughts, which reduces the water available for irrigation, Since maize is predominantly an irrigated crop in these countries, this has an impact on both maize yields and the area cropped with maize.

The most important effect of rainfall on crops is that it ensures a sufficient supply of water to cover the water lost through evapotranspiration during the growing season. Besides the rainfall during the growing season, crop water supply depends critically on soil waterholding capacity and plant root development. Agricultural droughts occur when the crop water demand cannot be met by either rainfall, soil water supply or through irrigation. Such droughts have occurred more frequently in recent years, particularly in southern Europe (Chapter 2). This has led to both increases and reductions in the use of irrigation in different regions depending on water availability.[11]

Climate change will also indirectly affect crop production through its impact on soil fertility, weeds, pests and disease and nutrient retention on agricultural fields, and these impacts in turn can lead to nitrate contamination, the release of greenhouse gases, and other secondary effects. Hence, climate change will have many different impacts on the ecosystem services provided by agricultural systems in Europe, and many of these are still poorly understood. A major challenge is to better understand the effects of climate change on the interactions between the various components of European landscapes, of which agriculture is only one.

What determines crop production in Europe?

In northern countries the duration of the growing season, late spring and early autumn frosts and solar radiation availability are typical climatic constraints. In these environments the duration of the growing season (frost or snow-free period) limits the productivity of crops. For example in Germany the growing season is 1–3 months longer than in Scandinavian countries, but it also varies greatly with altitude, with differences of up to three months in Austria. The short growing season is the main reason for the low crop yields in the Nordic countries. Moreover, night frosts in late spring or early autumn increase the agricultural risk in these environments. Moist conditions along the Atlantic coast and in mountainous regions limit the yield and quality of many crops, and also affect soil workability and reduce the number of machinery work-days. This is the main reason why only small areas are cultivated for cereals in the British Isles and Alpine countries compared to other regions. Permanent pastures dominate these regions instead.

In Mediterranean countries cereal yields are limited by water availability, heat stress and the short duration of the grain-filling period. Permanent crops (olive, grapevine, fruit trees, etc.) are therefore more important in this region.[12] These crops are affected by extreme weather events (such as hail and storms), which can reduce or completely destroy the harvest. Irrigation is important for crop production in many Mediterranean countries owing to high evapotranspiration and restricted rainfall. The continental climate of eastern Europe, causing drier conditions and greater amplitude of the annual temperature cycle, limits the range of crops that can be grown and the overall productivity. The most productive regions in Europe in terms of climate and soils are located in the great European plain stretching from Southeast England through France, Benelux and Germany into Poland. There are additional lowland regions (e.g. the Hungarian plains), where equally favourable soil conditions prevail, but the climate conditions (cold winters and dry summers) to some extent limit crop production.

How will individual crops be affected?

Arable crops

Global warming will expand the area of cereal cultivation northwards. Some crops that currently grow mostly in southern Europe (e.g. maize, sunflower and soybeans) will become more suitable further north or in higher altitude areas in the south.[13] The projections for a range of climate change scenarios show a 30–50% increase in suitable areas for grain maize production in Europe by the end of the 21st century, including Ireland, Scotland, Southern Sweden and Finland (Figure 4.3).

For wheat, a rise in temperatures will lead to a small yield reduction, which will often be more than counterbalanced by the effect of increased CO_2 on crop photosynthesis. The combination of both effects will, for a moderate climate change, lead to moderate to large yield increases in comparison with yields for the present situation.[15] Drier conditions and increasing temperatures in the Mediterranean region and parts of Eastern Europe may lead to lower yields there and the adoption of new varieties and cultivation methods. Such yield reductions have been estimated for Eastern Europe, and the yield variability may increase, especially in the steppe regions.

Potatoes, as well as other root and tuber crops, have shown a notable response to rising atmospheric CO_2.[17] On the other hand warming may reduce the growing season in some species and increase water requirements with consequences for yield. Climate change scenario studies performed using crop models show no consistent change in mean potato yield. For sugar beet yield an increasing occurrence in summer droughts may severely increase yield variability. However, a longer growing season will increase the yield potential for this crop in northern Europe, where water supply is adequate.

Climate-related increases in crop yields are only expected in northern Europe, while the largest reductions are expected around the Mediterranean and in the southwest Balkans and in the south of European Russia.[18] In southern Europe, particularly large decreases in yield are expected for spring-sown crops (e.g. maize,

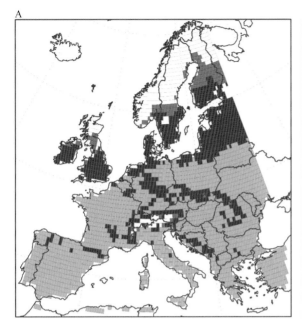

Figure 4.3 The territory where climate is suitable for cultivating grain maize. The grey area in both panels shows the suitable area under average climate conditions during the period 1961–1990. (A) The black area in the left panel shows the suitable areas under average future climate conditions during the period 2071 to 2100 as computed by 7 regional climate models under the IPCC A2 emissions scenario. (B) The black area in the right panel shows the same information for 24 climate scenarios from 6 global climate models reviewed by the IPCC A1FI, A2, B1 and B2 emissions scenarios. Medium grey areas in both panels show the uncertainty range of the respective scenario groups. Source: Olesen et al.[14] Reprinted with permission from Springer Science + Business Media.

sunflower and soybeans), while, for autumn-sown crops (e.g. winter and spring wheat) the impact is more geographically variable, and yield is expected to strongly decrease in the most southern areas and increase in the northern or cooler areas (e.g. northern parts of Portugal and Spain).[19]

Perennial crops

Many fruit trees are susceptible to spring frosts during flowering. A climatic warming will advance both the date of the last spring frosts and the dates of flowering, and the risk of damage to flower buds caused by late frost are likely to remain largely unchanged. Additionally the risk of damage to fruit trees caused by early autumn frosts is likely to decrease. However, there may very well be increased problems with pests and diseases.

Grapevine is a woody perennial plant, which requires relatively high temperatures. A climatic warming will therefore expand the suitable areas northwards and eastwards (Box 4.2). However, in the current production areas the yield variability

Box 4.2 Wine quality

The quality of wine is determined by a wide range of conditions such as grape variety, rootstock, soil type, climate, and cultivation techniques. The combination of soils and climate has traditionally determined wine cultivation in Europe and together forms the basis for the concept of "terroir", which characterizes the ability of a given site to produce a given quality of food, in this case wine. However, whereas soils provide a constant background for the grapevine, the influence of climate varies from year to year, and this also leads to variation in the amount and quality of wines produced.

Wine production is limited to regions where the climate is conducive to growing grapes with a balanced composition of sugar and acids that both reflect the region and the particular grape variety. There are three primary climatic requirements for production of high quality wine: (1) low risk of frost damage; (2) accumulation of sufficient heat during summer; and (3) absence of extreme heat. A grapevine is resistant to summer droughts, provided the soil allows sufficiently deep rooting, and it is necessary to have dry conditions during harvest to increase sugar concentration and avoid disease development.

The climatic warming over the recent decades has in general led to improvements in wine quality throughout Europe.[16] This has mostly been the consequence of changing temperatures during the growing season, which affect grape quality in several ways. First, prolonged temperatures above 10°C initiates spring vegetative growth and thus determine the start of the growing season. Second, during the maturation stage, a high diurnal temperature range leads to the beneficial synthesis of grape tannins, sugars, and flavors. In the Alsace region the warmer temperatures during recent years has led to higher sugar accumulation in the grapes resulting in higher alcohol contents. In fact the potential alcohol content of harvested Riesling in Alsace has increased from about 9 vol-% in 1970 to 11 vol-% in 2005. However, there are also negative effects with higher temperatures. Thus, extreme heat during flowering and throughout the growth of the berries can cause grape mortality, failure of flavor ripening, and premature véraison (change of color and start of the accumulation of sugars).

Different grape varieties have different optimum temperatures. The cool season white grape varieties grown in the Alsace region generally give the highest quality for mean growing season temperatures of about 14°C, whereas the warm season red grapes grown in the Barolo region of Italy requires average temperatures of about 18.5°C. A climatic warming will therefore shift the region for wine growing and for specific grape varieties northwards in Europe. One of the effects would be that the traditional grapes and qualities of existing wine growing regions would be difficult to maintain, which has consequences for the current regional regulations in Europe for ensuring wine quality. Whereas wine cultivation would expand in northern Europe, parts of southern Europe could become too hot to produce wine of high quality.

(fruit production and quality) may be higher under global change than at present. Such an increase in yield variability would neither guarantee the quality of wine in good years nor meet the demand for wine in poor years, thus implying a higher economic risk for growers.[110] However, yields may be strongly stimulated by increased CO_2 concentration without causing negative repercussions on the quality of grapes and wine.

Several perennial crops are candidates for bioenergy crops. This includes willow for coppice, and reed canary grass and Miscanthus for solid biofuel crops to be used in providing biomass for fuel in combined heat and power plants, or for use in second generation bioethanol production. The climatic suitability for many of these perennial bioenergy crops is projected to increase over most of Europe in the 21st century.[21]

Grasslands

The response of grasslands to climate change will differ depending on their type (species, soil type, management). In general, intensively managed and nutrient-rich grasslands will respond positively to both the increase in CO_2 concentration and to a temperature increase, given that water supply is sufficient.[22] Nitrogen-poor and species-rich grasslands, which are often extensively managed, may respond differently to climate change and increase in CO_2 concentration, and the short-term and long-term responses may be completely different. Climate change is likely to alter the community structure of grasslands, in ways specific to their location and type, and these changes will often depend on complex interactions between soils, plants and animals. Species richness and management changes of grasslands may increase their resilience to change.[23]

Fertile, early succession grasslands have been found to be more responsive to climate change than more mature and/or less fertile grasslands. In general, intensively managed and nutrient-rich grasslands will respond positively to both increased CO_2 concentration and temperature, given that water and nutrient supply is sufficient. As a general rule, the productivity of European grassland is expected to increase, where water supply is sufficient. On the other hand an increased frequency of summer droughts will severely reduce grassland production in the affected areas.

How will livestock be affected?

Climate and CO_2 effects influence livestock systems through both availability and price of feed and through direct effects on animal health, growth, and reproduction. The impacts of changes in feed-grain prices or the production of forage crops are generally moderated by market forces because these commodities are traded on the world market. However, effects of climate change on grasslands will also have a direct effect on livestock living on these pastures. The effects may be both positive and negative depending on locality.

For animals, higher temperatures results in greater water consumption and more frequent heat stress,[24] which cause a decline in physical activities, including eating and grazing. Maintenance requirements are increased and voluntary feed intake is decreased at the expense of growth, milk production and reproduction. Livestock production may therefore be negatively affected in the warm months of the currently warm regions of Europe. Warming during the cold period for cooler regions may on the other hand be beneficial owing to reduced feed requirements, increased survival, and lower energy costs. Impacts will probably be minor for intensive livestock systems (e.g. confined dairy, poultry and pig systems) because climate is controlled to some degree.

An increase in the frequency of severe heat stress in Britain is expected to enhance the risk of mortality in pigs and broiler chickens which are grown in intensive livestock systems. Increased frequency of droughts along the Atlantic coast (e.g. Ireland) may reduce the productivity of grasslands such that they are no longer sufficient for livestock. Increasing temperatures may also increase the risk of livestock diseases by: (1) increasing the diffusion of insects (e.g. *Culicoides imicola*) that are the main vectors of several arboviruses (e.g. bluetongue and African horse sickness); (2) increasing the survival of viruses from one year to the next; or (3) improving conditions for new insect vectors that are now limited by colder temperatures.[25]

Other impacts on agroecosystems

Weeds, pests and diseases

The majority of pest and disease problems are closely linked with their host crops. This makes major changes in plant protection problems less likely. However, in current cool regions, higher temperatures favor the proliferation of insect pests, because many insects can then complete a greater number of reproductive cycles.[26] Warmer winter temperatures may also allow pests to overwinter in areas where they are now limited by cold, thus causing greater and earlier infestation during the following crop season. Climate warming will lead to earlier insect spring activity and proliferation of some pest species.[27] A similar situation may be seen for plant diseases leading to an increased demand for pesticide control.

Unlike pests and diseases, weeds are directly influenced by changes in atmospheric CO_2 concentration. Differential effects of CO_2 and climate changes on crops and weeds will alter the weed-crop competitive interactions, sometimes to the benefit of the crop and sometimes for the weeds. Interaction with other biotic factors and with changing temperature and rainfall may also influence weed seed survival and thus weed population development.

Changes in climatic suitability will lead to invasion of weeds, pests and diseases adapted to warmer climatic conditions.[28] The speed at which such species will invade depends on the rate of climatic change, the dispersal rate of the species and on measures taken to combat non-indigenous species. The dispersal rate of pests and diseases are often so high that their geographical extent is determined by the range of climatic suitability. The Colorado beetle, the European corn borer, the Mediterranean fruit fly and karnal bunt are examples of pests and diseases that are expected to show a considerable northward expansion in Europe under climatic warming, with some indication that this process has already begun.

Soils

Soils have many functions, of which water and nutrient supply to growing crops are essential for sustainable crop production systems. However, soils are also important in regulating the water and nutrient cycles, for carbon storage and for regulating greenhouse gas emissions. Soils are also habitats for many organisms that contribute to the functioning of soils. The perception of soil as an environmental medium providing substantial goods and services for all land and aquatic ecosystems has developed over the recent decades.[29]

Increasing temperatures will speed decomposition where soil moisture allows, so direct climate impacts on cropland and grassland soils will tend to decrease soil organic stocks for Europe as a whole.[31] This effect is greatly reduced by increasing carbon inputs into the soil because of enhanced plant productivity, resulting from a combination of climate change and increased atmospheric CO_2 concentration. However, decomposition becomes faster in regions where temperature increases greatly and soil moisture remains high enough to allow decomposition (e.g. north and east Europe), but does not become faster where the soil becomes too dry, despite higher temperatures (southern France, Spain, and Italy).

Any reduction in soil organic matter stocks can lead to a decrease in fertility and biodiversity (Box 4.3), a loss of soil structure, reduced water retention capacity and increased risk of erosion and compaction. All of this leads to lower productivity of crops growing on the soil. Changes in rainfall and wind patterns will lead to an increase in erosion in vulnerable soils, which often suffer from low organic matter content. Climate change will further increase the risk of desertification, which is already affecting southern Europe and is expected to move gradually northward.

Environmental impacts

Agricultural systems are important elements of the European landscape. With the increased intensification of the agricultural practices based on large inputs of fertilizers and pesticides, environmental concerns related to pollution from

---

Box 4.3 Soil degradation and desertification under climate change

Soil degradation involves a number of processes that reduce the ability of soils to support water and nutrient supply for plants. The most important processes include wind and water erosion, reduction of soil organic matter, soil compaction, pollution, salinization and desertification. Desertification is particularly severe land degradation in the arid, semi-arid, and dry sub-humid areas resulting from various factors, including climatic variations and human activities. Salinization is so severe because it is difficult to treat and can lead to desertification. Salinization from irrigation occurs where the irrigation water carries nutrients and solutes that accumulate in the soil.

Climatic conditions make the Mediterranean region one of the areas most severely affected by land degradation.[30] Much of the region is semi-arid and subject to seasonal droughts, high rainfall variability and sudden intense precipitation. Some areas, especially along the northwest coasts of the Black Sea, are classified as semi-arid. The level of soil degradation is severe in most of the region, and very severe in some parts, for example along the Adriatic, where soil cover has almost disappeared in some areas. In addition, other physical factors, such as steep slopes, the frequency of soil types susceptible to degradation and poor management practices, increase the vulnerability.

Soil loss, in turn, reduces the regeneration potential of ecosystems. The areas most sensitive to this are those with shallow soils, steep slopes and slow rates of recovery of the vegetative cover. The Mediterranean is particular sensitive to climatic changes that enhance the degradation processes. Summer warming and drying are expected to result in an increase in arid and semi-arid climates throughout the region. The projected decrease in summer precipitation in southern Europe, the increase in the frequency of summer droughts and the increased incidence of forest fires will probably induce greater risks of soil erosion throughout the region. This may very well lead to irreversible desertification in the most vulnerable zones in southern Europe.

---

agriculture has come onto the political agenda for protecting the natural environment of Europe. This issue is likely to become increasingly important under climate change. In particular, the negative impact of nitrate leaching on the quality of groundwater aquifers, rivers and estuaries is of concern. A warming is expected to increase the turnover rate of soil organic matter, which will also release more nitrogen into the soil. In large parts of Europe climate change will mean increased winter rainfall (Chapter 2), which is likely to lead to higher rates of nitrate leaching from agricultural systems.[32]

Increases in winter rainfall and in rainfall intensity will also lead to increased transport of phosphorus into streams, lakes and other aquatic ecosystems. Both nitrogen and phosphorous boost the frequency of algal blooms in lakes, fjords and in the sea. When these algae die and drop to the bottom of aquatic ecosystems they decompose and sometimes result in massive anoxia and dead fish ("dead seas"). Higher temperatures along with increased nutrient supply may also lead to increased growth of toxic cyanobacteria in lakes.[33]

Extreme events and climatic variability

Extreme weather events, such as spells of high temperature, heavy storms or droughts, can severely disrupt crop production. Individual extreme events will not usually have lasting effects on the agricultural system. However, when the frequency of such events increases agriculture needs to respond, either by adapting or by ceasing its activity.

Crops often respond in a nonlinear way to changes in their growing conditions, and have threshold responses, which greatly increase the importance of climatic variability and frequency of extreme events for absolute yield, yield stability and quality.[34] Thus an increase in temperature variability will increase yield variability and also result in a reduction in mean yield.[35] Therefore the projected increases in temperature variability over central and southern Europe (Chapter 2) may have severe impacts on the agricultural production in this region.

In addition to the linear and nonlinear responses of crop growth and development to variation in temperature and rainfall, short-term extreme temperatures can have significant yield-reducing effects.[36] This is particularly the case during flowering and fruiting periods, where short-term exposure to high temperatures (usually above 35°C) can greatly reduce fruit set and therefore yield. Exposure to drought during these periods may have similar effects.

The consequences of increased climatic variability are illustrated by the impacts of the 2003 heatwave (see Box 3.2). The combination of high temperatures and deficits in rainfall accompanying this unprecedented event led to reductions in crop and livestock production valued at 11 billion euros in central and southern Europe.[37] Although such a heatwave is statistically unlikely under current climate conditions, it is consistent with projected temperatures under climate change towards the end of this century (Chapter 2).

How can agriculture adapt to climatic change?

It is a fact that farmers are already adapting to climate change. This is an expected phenomenon since farmers constantly experiment with new cropping techniques, and the most successful ones quickly spread among the farming society where agricultural advisers and researchers are ready to take up and disseminate new results.

The results of the climatic warming over Europe can most easily be seen in the extent to which warm season crops are being grown. A typical example is the growing of maize in northern Europe. Maize is used to produce silage for feeding dairy cows in the intensive dairy farming systems in most of northwestern Europe. However, until the early 1990s very little silage maize was grown in Britain and Scandinavia. The climate was too cold, resulting in frequent poor harvests. Since then the climate has become warmer, and now almost all dairy farmers in southern Scandinavia feed their cattle on a diet of maize and grass.

To avoid, or at least reduce, negative effects and exploit possible positive effects, several agronomic adaptation strategies for agriculture have been suggested. Short-term adjustments include efforts to optimize production without major system changes (examples are given below). They are autonomous in the sense that no other sectors (e.g. policy, research, etc.) are needed for their development and implementation. On the other hand, planned adaptation most often involves a wider range of stakeholders and longer time horizons.

Adaptation only works to the extent that the basic resources for crop growth are still maintained and the climate allows proper soil and crop management to take place. In northern areas climate change may produce positive effects on agriculture through the introduction of new crop species and varieties, higher crop production and expansion of suitable areas for crop cultivation. Disadvantages may include an increase in the need for plant protection, the risk of nutrient leaching and the turnover of soil organic matter. In southern areas the disadvantages will predominate. The possible increase in water shortage and extreme weather events may cause lower harvestable yields, higher yield variability and a reduction in suitable areas for traditional crops. These effects may reinforce the current trends of intensification of agriculture in northern and western Europe and extensification in the Mediterranean and south-eastern parts of Europe.

Autonomous adaptation

The key autonomous adaptation options in crop production include changes in varieties, sowing dates and fertilizer and pesticide use.[38] In particular, in southern Europe, short-term adaptations may include changes in crop species (e.g. replacing winter with spring wheat), changes in cultivars and sowing dates (e.g. for winter crops, sowing the same cultivar earlier, or choosing cultivars with a longer crop cycle), and for summer irrigated crops, earlier sowing to prevent yield reductions or reduce water demand.

There are many plant traits that plant breeders can modify to better adapt varieties to increased temperature and reduced water supply.[39] The use of early ripening fruit tree varieties may reduce water consumption, as proper management practices may be applied to orchards to improve adaptation. However, the effectiveness of such traits in boosting yield depends on whether there is simultaneous change in climatic variability, and a combination of traits may be needed to stabilize yield in poor years, without sacrificing yield in good years.

Planned adaptation

The long-term or planned adaptations refer to major structural changes to over-come adversity caused by climate change. This involves changes in land allocation and farming systems, new land management techniques, and so on. Some land use changes will result from the farmer's response to the differentiated response of crops to climate change. The changes in land allocation may also be used to stabilize production. This means substituting crops with high inter-annual yield variability (e.g. wheat or maize) by crops with lower productivity, but more stable yields (e.g. pasture or sorghum). Crop substitution may be useful also for the conservation of soil moisture.

Other land use changes result from persistent changes in rainfall patterns. Where rainfall increases this may result in increasing groundwater levels, which would either require renewed drainage systems to maintain land in productive agriculture or the abandonment of the land for arable cropping. Where rainfall becomes increasingly scarce or infrequent, new irrigation systems will need to be put into place, or this land may also have to be abandoned.

Other examples of long-term adaptations include breeding of drought tolerant varieties, new land management techniques to conserve water or increase irrigation use efficiencies, and more drastic changes in farming systems (including land abandonment). However, increasing the supply of water for irrigation may not be a viable option in much of southern Europe, since a considerable reduction in total run-off is expected here (Chapter 5). The links with water availability may be among the most important ones, affecting the need for improving irrigation efficiencies or the need for terracing of sloping land surfaces.

Adaptation to climate change must be factored in, especially as part of the ongoing technological development in agriculture, including plant breeding, also using, for example, livestock feeding technologies, irrigation management, and application of information and communication technology. Such changes require that all available natural, human, financial and social resources are mobilized in a concerted effort to deal with the challenges (Table 4.2). In some cases such adaptation measures would make sense without considering climate change, because they help to address current climate variability. In other cases, the measures must be implemented in anticipation of climate change, because they would be ineffective if implemented as a reaction to climate change. One example is breeding for increased drought tolerance, because it takes considerable time before new varieties can be released. Another example is the need to maintain soil fertility, which will be crucial for maintaining high crop productivity in a future, more variable, climate. Building soil fertility is a long-term effort, and starting only when climate change has set in will not deliver the required benefits.

Table 4.2 Resource-based policies to support adaptation of European agriculture to climate change. Source: Olesen and Bindi.[40] Reprinted with permission from Elsevier Limited, Oxford.

Resource	Policy
Land	*Reforming agricultural policy to encourage flexible land use.* The great extent of Europe cropland across diverse climates will provide diversity for adaptation.
Water	*Reforming water markets and raising the value of crop per volume of water used to encourage more prudent use of water.* Deficient water management limits agriculture in some regions. Improvements in water management are crucial for adapting to a drier climate. This may be promoted by more efficient water markets, including water pricing.
Nutrients	*Improving nutrient-use efficiencies through changes in cropping systems and development and adoption of new nutrient management technologies.* Nutrient management needs to be tailored to the changes in crop production as affected by climate change, and utilization efficiencies must be increased, especially for nitrogen, in order to reduce nitrous oxide emissions.
Agrochemicals	*Support for integrated pest management systems (IPMS) should be increased through a combination of education, regulation and taxation.* There will be a need to adapt existing IPMSs to changing climatic regimes.
Energy	*Improving the efficiency of food production and exploring new biological fuels and ways to store more carbon in trees and soils.* Reliable and sustainable energy supply is essential for many adaptations to new climate and for mitigation policies. There are also many options for reducing energy use in agriculture.
Genetic diversity	*Assembling, preserving and characterising plant and animal genes and conducting research on alternative crops and animals.* Genetic diversity and new genetic material from nature will provide important basic material for adapting crop species to changing climatic conditions (e.g. by improving drought tolerance).
Research capacity	*Encouraging research on adaptation, developing new farming systems and developing alternative foods.* Greater investments in agricultural research may provide new sources of knowledge and technology for adaptation to climate change.
Information systems	*Enhancing national systems that disseminate information on agricultural research and technology, and encourage information exchange among farmers.* Fast and efficient information dissemination and exchange to and between farmers using the new technologies (e.g. internet), will speed up the rate of adaptation to climatic and market changes.
Culture	*Integrating environmental, agricultural and cultural policies to preserve the heritage of rural environments.* Integration of policies will be required to maintain and preserve the heritage of rural environments which are dominated by agricultural practices influenced by climate.

Increasing resilience to change

Some of the recent studies taking into account potential impacts, adaptive capacity, and the vulnerability of farmer livelihoods show that the agricultural sector in the Mediterranean region will be vulnerable to climate change under most scenarios.[41] A survey of vulnerability and adaptive capacity among farming systems in Europe shows that northern and northwestern parts of Europe are likely to gain most from short-term climate change, although they also are likely to be negatively affected by climate change over the longer term.[42] However, the most vulnerable agricultural areas are located in the Pannonian environmental zone (see Figure 2.1) rather than the Mediterranean, which is worrying considering the importance of agriculture in the economies of Romania, Bulgaria, Hungary, and Serbia and other countries in this region. Parts of Austria, Czech Republic and Slovakia are also expected to be negatively affected.

So far, research on climate change impacts in agriculture has placed little emphasis on changes in the frequency of extreme events. This is unfortunate because changes in climate variability may be particularly difficult for many farmers to adapt to, and strategies to cope with variability may be different from those dealing with changes in mean climate. Adaptation to extreme events includes measures to avoid periods of high stress. It also includes actions that increase the resilience of an agricultural system by adding diversity to the crop rotation and by improving soil and water resources.

Several adaptation measures may be used to increase resilience to climate change in cropping systems. However, when it relates to soil and water resources, building resilient systems may require not only long-term planning but also the need to act immediately. An example of this is the link between climate change and soil degradation, which is one of the greatest threats to global food production.[43] Most of the processes causing soil degradation are amplified by the higher temperatures, more intense rainfall and longer drought periods associated with climate change in Europe (Box 4.3). Soil degradation, in turn, leads to lower soil carbon stocks, increased soil erosion and desertification. Yet higher soil carbon contents and better soil structure will be critical in enabling cropping systems to cope with increased climate variability. In general, it is urgent to develop strategies to enhance the resilience of agricultural systems.

Irrigation of agricultural crops plays a important role in many European countries (Figure 4.4). In some Mediterranean countries more than 80% of the freshwater abstraction is used for irrigated agriculture. Increasing the supply of water for irrigation by increasing freshwater abstraction may not be a viable option in much of southern Europe, since the projections show a considerable reduction in total run-off.[44] Instead, focus needs to be shifted to using other sources of water, such as recycled or brackish water, and to improving irrigation use efficiencies, in particular, by reducing losses through transmission and application. To achieve such effects requires efficient institutions and efficient water markets in combination with public awareness of the need for long-term management of water resources.

How will fisheries be affected?

Marine ecosystems

The commercial fish stocks vary considerably across European seas, not only in response to climatic differences, but also because of variation in salinity, and depth

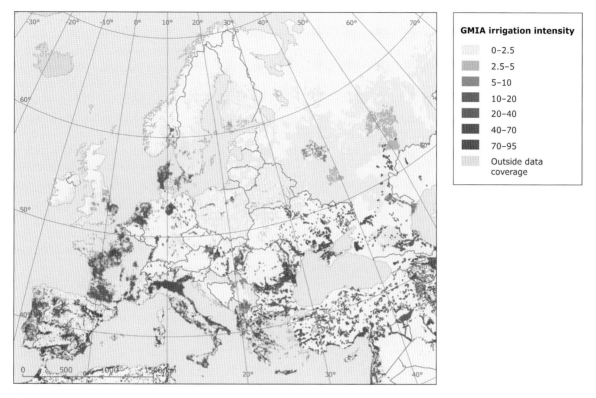

Figure 4.4 Irrigation intensity across Europe as illustrated by the percentage of area equipped with irrigation. Source: EEA.[45] Reprinted with permission from European Environmental Agency, Copenhagen. (See color plate.)

and structure of the sea floor. Climate changes will therefore vary across the European seas in response to local conditions for species migration and recruitment.

A changing ocean climate has led to a geographic shift in certain crustacean plankton assemblages of different species in European shelf seas over the past 40 years. This so-called "regime shift", first detected by British and French scientists in the late 1990s, was reflected by a northward extension of warm-water species and a decrease in the number of colder water species.[46]

Water temperatures in the seas adjacent to Europe are expected to rise throughout the 21st century; typical projections are for an increase of about 1.5–2.6°C up to the end of the 21st century relative to 1980–99 (Chapter 6). As noted, it is thought that warmer sea temperatures have already caused significant changes in zooplankton and phytoplankton populations, including changes in their abundance and distribution. Increasing regional sea temperatures has triggered a major northward movement of warmer water plankton in the northeast Atlantic and a similar retreat of colder water plankton to the north. This northward movement amounts to about 1100 km over the past 40 years, and appears to have accelerated in recent years.[47] Similar changes have been seen in shifts of fish species.

In some regions, changes in plankton biomass and seasonal timing of blooms have already been linked to the poor recruitment of young fish to several stocks of commercial interest and to the low breeding success of seabirds in recent years. This is because the life cycles of many fish and seabirds are timed to make optimum use of peaks of particular prey species; if the timings are "out of sync", because of changes in the plankton blooms, less food will be available for both fish and seabirds.[48] Such changes in plankton have been strongly implicated in accelerating the decline in North Sea cod stocks, caused initially by over-fishing.

An assessment of the vulnerability of the northeast Atlantic marine ecoregion concluded that climate change is very likely to produce significant impacts on selected marine fish and shellfish.[49] Temperature increase has a major effect on fisheries production in the North Atlantic, causing changes in species distribution, increased recruitment and production in northern waters and a marked decrease at the southern edge of current ranges.

Assessments of future movements of marine species under climate change have not yet been made. Projections about fish distribution changes over the next 20–50 years are subject both to the uncertainties in projections of ocean climate and uncertainties of fish community responses to those changes.

Aquaculture

Fish and shellfish aquaculture represented 33% of the value and 17% of the volume of the EU fishery in 2002. Warmer sea temperatures, increased growing seasons, growth rates, feed conversion, and primary productivity will benefit shellfish production.[50] Opportunities for new species will arise from expanded geographic distribution and range, but increasing temperatures will increase stress and susceptibility to pathogens. Ecosystem changes with new invasive or non-native species such as gelatinous zooplankton and medusa, toxic algal blooms, increased fouling and more frequent occurrences of low dissolved oxygen events will increase operating costs. Increased storm-induced damage to equipment and facilities will also affect capital costs. Moreover, aquaculture has a range of environmental impacts, particularly through organic wastes and the spread of pathogens to wild populations, which are likely to enhance climate-induced ecosystem stresses.[51]

Adaptation

There is evidence that the fish and shellfish farming industries are adapting their technology and operations to changing climatic conditions, for example, by expanding offshore and selecting optimal culture sites for shellfish cages.[52] However, autonomous adaptation is more difficult for smaller coastal-based fishery businesses, which do not have the option to sail long distances to new fisheries as compared to larger businesses with long-distance fleets.

Planned adaptation involves changes in fishing policy, primarily through institutions such as the European Common Fisheries Policy (CFP). Although CFP through its production quotas and technical measures provides an ideal platform for such adaptation actions, no consideration has yet been given to climate change. Another major adaptation option is to factor in the long-term potential impacts of climate change in planning for new Marine Protected Areas.[53] Adaptation strategies should eventually be integrated into comprehensive plans for managing coastal areas of Europe (Chapter 6).

How will food systems be affected?

The greater majority of studies in climate change impacts have focused on how the primary production in agriculture, horticulture, aquaculture and fisheries will be affected by climate change. However, the food supply in modern European societies is only to a small extent affected by local production. Still, the supply of fresh vegetables will to some extent be affected, since much of this supply is based on local products, at least during the growing season. With an expansion of the growing season in northern Europe, local supply of fresh vegetables can be extended to benefit the quality of supply to the consumers. In southern Europe temperatures during the summer period may become too high for cultivation of some vegetables (e.g. cauliflower),[54] resulting in a shift of production zones northward, and a increased reliance on imports from northern Europe.

The major beneficiary from agriculture, horticulture and fisheries is the food industry. This industry is becoming more oriented towards the global market, a trend that is being strengthened by the liberalization of world trade. Parts of the food industry will therefore be, in the future, less reliant on the local supply of produce and demand for products. However, a small part of the European food industry relies on local food brands (specialities), which have a long tradition rooted in local culture. Some of these are registered and protected by EU regulation. Such food specialities rely on favourable natural conditions to produce high quality products and may, therefore, be particularly susceptible to climate change.

Food systems will not only be affected by the impacts of climate change on primary production. Since food supply is a major contributor to greenhouse gas emissions (Chapter 3), strategies will also need to address how food supply and food consumption can contribute to reduced emissions. This will most likely revolve around a reduction in the consumption of meat and other livestock products, and a higher consumption of local produce and the seasonal vegetables. The entire food chain also needs to reduce its emissions, which could mean less packaging and a greater focus on local shopping as opposed to driving to large malls in the outskirts of cities.

Notes

1 Schmidhuber, J., Traill, W.B., 2006: The changing structure of diets in the European Union in relation to healthy eating guidelines. *Public Health Nutrition* 9, 584–595.

2 Gray, T., Hatchard, J., 2003: The 2002 reform of the Common Fisheries governance – rhetoric or reality? *Marine Policy* 27, 545–554.

3 Olesen, J.E., Bindi, M., 2002: Consequences of climate change for European agricultural productivity, land use and policy. *European Journal of Agronomy* 16, 239–262.

4 EEA, 2007: *Europe's environment. The fourth assessment.* European Environment Agency, Copenhagen.

5 Drake, B.G., Gonzalez-Meier, M.A., Long, S.P., 1997: More efficient plants: a consequence of rising CO_2? *Annual Review of Plant Physiology and Plant Molecular Biology* 48, 609–639.

6 Fuhrer, J., 2003: Agroecosystem responses to combinations of elevated CO_2, ozone, and global climate change. *Agriculture, Ecosystems & Environment* 97, 1–20.

7 Olesen, J.E., Bindi, M., 2002.

8 Ziska, L.H., 2001: The impact of elevated CO_2 on yield loss from a C_3 and C4 weed in field-grown soybean. *Global Change Biology* 6, 899–905.

9 Olesen, J.E., Jensen, T., Petersen, J., 2000: Sensitivity of field-scale winter wheat production in Denmark to climate variability and climate change. *Climate Research* 15, 221–238.

10 Olesen, J.E., Trnka, M., Kersebaum, K.C., Skjelvåg, A.O., Seguin, B, Peltonen-Saino, P., Rossi, F., Kozyra, J., Micale, F., 2011: Impacts and adaptation of European crop production systems to climate change. *European Journal of Agronomy* 34, 96–112.

11 Iglesias, A., Garrote, L., Flores, F., Moneo, M., 2007: Challenges to manage the risk of water scarcity and climate change in the Mediterranean. *Water Resources Management* 21, 775–788.

12 Quiroga, S., Iglesias, A., 2008: A comparison of the climate risks of cereal, citrus, grapevine and olive production in Spain. *Agricultural Systems* 101, 91–100.

13 Fronzek, S., Carter, T.R., 2007: Assessing uncertainties in climate change impacts on resource potential for Europe based on projections from RCMs and GCMs. *Climate Change* 81, 357–371.

14 Olesen, J.E., Carter, T.R., Diaz-Ambrona, C.H., Fronzek, S., Heidmann, T., Hickler, T., Holt, T., Minguez, M.I., Morales, P., Palutikov, J., Quemada, M., Ruiz-Ramos, M., Rubæk, G., Sau, F., Smith, B., Sykes, M., 2007: Uncertainties in projected impacts of climate change on European agriculture and ecosystems based on scenarios from regional climate models. *Climatic Change* 81 (suppl. 1), 123–143..

15 van Ittersum, M.K., Howden, S.M., Asseng, S., 2003: Sensitivity of productivity and deep drainage of wheat cropping systems in a Mediterranean environment to changes in CO_2, temperature and precipitation. *Agriculture, Ecosystems & Environment* 97, 255–273.

16 Jones, G.V., White, M.A., Cooper, O.R., Storchmann, K., 2005: Climate change and global wine quality. *Climatic Change* 73, 319–343.

17 Kimball, B.A., Kobayahsi, K., Bindi, M., 2002: Responses of agricultural crops to free-air CO_2 enrichment. *Advances in Agronomy* 77, 293–368.

18 Maracchi, G., Sirotenko, O., Bindi, M., 2005: Impacts of present and future climate variability on agriculture and forestry in the temperate regions: Europe. *Climatic Change* 70, 117–135.

19 Minguez, M.I., Ruiz-Ramos, M., Díaz-Ambrona, C.H., Quemada, M., Sau, F., 2007: First-order impacts on winter and summer crops assessed with various high-resolution climate models in the Iberian peninsula. *Climatic Change* 81 (suppl. 1), 343–355.

20 Bindi, M., Fibbi, L., Gozzini, B., Orlandini, S., Miglietta, F., 1996: Modeling the impact of future climate scenarios on yield and yield variability of grapevine. *Climatic Research* 7, 213–224.

21 Tuck, G., Glendining, M.J., Smith, P., House, J.I., Wattenbach, M., 2006: The potential distribution of bioenergy crops in Europe under present and future climate. *Biomass and Bioenergy* 30, 83–197.

22 Thornley, J.H.M., Cannell, M.G.R., 1997: Temperate grassland responses to climate change: an analysis using the Hurley pasture model. *Annals of Botany* 80, 205–221.

23 Duckworth, J.C., Bunce, R.G.H., Malloch, A.J.C., 2000: Modelling the potential effects of climate change on calcareous grasslands in Atlantic Europe. *Journal of Biogeography* 27, 347–358.

24 Turnpenny, J.R., Parsons, D.J., Armstrong, A.C., Clark, J.A., Cooper, K., Matthews, A.M., 2001: Integrated models of livestock systems for climate change studies. 2. Intensive systems. *Global Change Biology* 7, 163–170.

25 Colebrook, E., Wall, R., 2004: Ectoparasites of livestock in Europe and the Mediterranean region. *Veterinary Parasitology* 120, 251–274.

26 Bale, J.S., Masters, G.J., Hodkinson, I.D., Awmack, C., Bezemer, T.M., Brown, V.K., Butterfield, J., Buse, A., Coulson, J.C., Farrar, J., Good, J.E.G., Harrington, R., Harley, S., Jones, T.H., Lindroth, R.L., Press, M.C., Symrnioudis, I., Watt, A.D., Whittaker, J.B., 2002: Herbivory in global climate change research: direct effects of rising temperature on insect herbivores. *Global Change Biology* 8, 1–16.

27 Cocu, N., Harrington, R., Rounsevell, M.D.A., Worner, S.P., Hullé, M., 2005: Geographical location, climate and land use influences on the phenology and numbers of the aphid, Myzus persicae, in Europe. *Journal of Biogeography* 32, 615–632.

28 Baker, R.H.A., Sansford, C.E., Jarvis, C.H., Cannon, R.J.C., MacLeod, A., Walters, K.F.A., 2000: The role of climatic mapping in predicting the poten-

tial distribution of non-indigenous pests under current and future climates. *Agriculture, Ecosystems & Environment* 82, 57–71.

29 EEA, 2008: *Impacts of Europe's changing climate – 2008 indicator-based assessment*. EEA Report No. 4/2008.

30 EEA, 2008.

31 Smith, J., Smith, P., Wattenbach, M., Zaehle, S., Hiederer, R., Jones, R.J.A., Montanarella, L., Rounsevell, M.D.A., Reginster, I., Ewert, F., 2006: Projected changes in mineral soil carbon of European croplands and grasslands, 1990–2080. *Global Change Biology* 11, 2141–2152.

32 Olesen, J.E., et al., 2007

33 Moss, B., Mckee, D., Atkinson, D., Collings, S.E., Eaton, J.W., Gill, A.B., Harvey, I., Hatton, K., Heyes, T., Wilson, D., 2003: How important is climate? Effects of warming, nutrient addition and fish on phytoplankton in shallow lake microcosms. *Journal of Applied Ecology* 40, 782–792.

34 Porter, J.R., Semenov, M.A., 2005: Crop responses to climatic variation. *Philosophical Transactions of the Royal Society B* 360, 2021–2035.

35 Iglesias, A., Quiroga, S., 2007: Measuring the risk of climate variability to cereal production at five sites in Spain. *Climate Research* 34, 47–57.

36 Wheeler, T.R., Crauford, P.Q., Ellis, R.H., Porter, J.R., Vara Prasad, P.V., 2000: Temperature variability and the yield of annual crops. *Agriculture, Ecosystems & Environment* 82, 159–167.

37 Fink, A.H., Brücher, T., Krüger, A., Leckebusch, G.C., Pinto, J.G., Ulbrich, U., 2004: The 2003 European summer heat waves and drought – Synoptic diagnosis and impact. *Weather* 59, 209–216.

38 Howden, S.M., Soussana, J.F., Tubiello, F.N., Chhetri, N., Dunlop, M., Meinke, H., 2007: Adapting agriculture to climate change. *Proceedings of the National Academy of Science* 104, 19691–19696.

39 Sinclair, T.R., Muchow, R.C., 2001: Systems analysis of plant traits to increase grain yield on limited water supplies. *Agronomy Journal* 93, 263–270.

40 Olesen, J.E., Bindi, M., 2002.

41 Metzger, M.J., Rounsevell, M.D.A., Acosta-Michlik, L., Leemans, R., Schröter, D., 2006: The vulnerability of ecosystem services to land use change. *Agriculture Ecosystems & Environment* 114, 69–85.

42 Olesen, J.E., Trnka, M., Kersebaum, C., Peltonen-Saino, P., Skjelvåg, A.O., Rossi, F., Rossi, F., Kozyra, J., Seguin, B., Micale, F., 2008: Risk assessment and foreseen impacts on agriculture. In: Nejedlik, P. & Orlandini, S. (eds). Survey of agrometeorological practices and applications in Europe regarding climate change impacts. *COST Action* 734, 279–328.

43 Lal, R.. Follett, F., Stewart, B.A., Kimble, J.M., 2007: Soil carbon sequestration to mitigate climate change and advance food security. *Soil Science* 172, 943–956.

44 Lehner, B., Döll, P., Alcamo, J., Henrichs, H., Kaspar, F., 2006: Estimating the impact of global change on flood and drought risks in Europe: a continental, integrated analysis. *Climatic Change* 75, 273–299.

45   EEA, 2009: Water resources across Europe – confronting water scarcity and drought. EEA Report No 2/2009. European Environmental Agency, Copenhagen.

46   ICES, 2007: *Climate change: Changing oceans.* ICES, Copenhagen.

47   Brander, K., Blom, G., 2003: Changes in fish distribution in the Eastern North Atlantic: Are we seeing a coherent response to changing temperature? *ICES Marine Science Symposia* 219, 261–270: EEA, 2008.

48   Edwards, M., Richardson, A.J., 2004: Impact of climate change on marine pelagic phenology and trophic mismatch. *Nature* 430, 881–884.

49   Baker, T., 2005: *Vulnerability Assessment of the North-East Atlantic Shelf Marine Ecoregion to Climate Change.* Workshop Project Report, WWF.

50   Beaugrand, G., Reid, P.C., 2003: Long-term changes in phytoplankton, zoo-plankton and salmon related to climate. *Global Change Biology* 9, 801–817.

51   Alcamo, J., Moreno, J.M., Novaky, B., Bindi, M., Corobov, R., Devoy, R.J.N., Giannakopoulos, C., Martin, E., Olesen, J.E., Shvidenko, A., 2007: Europe. In: *Climate Change 2007: Impacts, Adaptation and Vulnerability. Contribution of Working Group II to the Fourth Assessment Report of the Intergovernmental Panel on Climate Change*, Parry, M.L., Canziani, O.F., Palutikof, J.P., van der Linden, P.J., Hanson, C.E. (eds), Cambridge University Press, Cambridge, UK, 541–580.

52   Pérez, O.M., Telfer, T.C., Ross, L.G., 2003: On the calculation of wave climate for offshore cage culture site selection: a case study in Tenerife (Canary Islands). *Aquacultural Engineering* 29, 1–21.

53   Soto, C.G., 2001: The potential impacts of global climate change on marine protected areas. *Reviews in Fish Biology and Fisheries* 11, 181–195.

54   Olesen, J.E., Grevsen, K., 1993: Simulated effects of climate change on summer cauliflower production in Europe. *European Journal of Agronomy* 2, 313–323.

5

Water security

Introduction

As the patterns of rain and snow take different forms throughout Europe, and the temperature rises on average, it is inevitable that the flows and stocks of water in Europe will also be transformed. Depending on the location, there will either be too much water (at least for people) or too little. Where precipitation increases, as expected in northern parts of Europe, flooding is likely to become more frequent, and the landscape will be transformed – more and higher dikes may appear alongside many rivers. Adjacent to other rivers the land may be cleared of development and conserved as a natural "floodway". While a river stays within its banks these floodways will be cultivated or used as pastureland or parkland. But when heavy rains cause a river to spill over its banks, these vast areas will be inundated and serve as temporary storage areas for floodwaters until a river settles down.

Life in Europe Under Climate Change, First Edition. Joseph Alcamo and Jørgen E. Olesen.
© 2012 Joseph Alcamo and Jørgen E. Olesen. Published 2012 by John Wiley & Sons, Ltd.

In the south, the problem is likely to be more frequent droughts with many years during which river run-off dwindles, the levels of lakes plummet, and remnants of old buildings reappear at the bottoms of dried-out reservoirs. Drought planning and awareness will be a major civic activity. To cope with water scarcity many cities may build desalination plants along the coastline.

Society already has to cope with the variability of climate since strong year-to-year variations in weather alternately cause surpluses or shortages in water supply. In fact, society in different parts of Europe has adapted over the centuries to significantly different patterns of both extreme climate events and average weather conditions. The temperatures occurring during summer heatwaves in the north are actually not that unusual to an inhabitant of a Mediterranean country, while average icy spells in the northern winter are difficult to cope with when they occur in the south.

But global climate change is expected to change not only long term climate patterns – making some regions drier, some wetter, and all warmer on the average – but also the variability of climate between years and between seasons.

Climate change and the great cycle of water

The water flowing out of household taps, accumulating in wells, or flowing in rivers, is all part of a great global freshwater system encompassing Europe and the rest of the earth. Climate change disrupts the cycle and affects all aspects of water resources. Of the many processes affecting river flows, two of the most important are evapotranspiration and precipitation, and both will be markedly altered by climate change.

Surface air temperatures will increase almost everywhere under climate change, and this will amplify the rate of evapotranspiration. (For an explanation of "evapotranspiration" see Box 5.1). Warmer air temperatures will also increase the water-holding capacity of the atmosphere. An increased rate of evapotranspiration sends more water moisture into the atmosphere and this often stimulates more precipitation. Hence it is sometimes said that climate change "accelerates the global water cycle".[1] Although the total amount of precipitation falling yearly on ocean and land surfaces is expected to increase, many parts of Europe and elsewhere are likely to experience strong decreases in their annual precipitation (see Chapter 2).

The most important variable affecting most water systems is precipitation, or more precisely, the amount, frequency, and type of rainfall, snow, hail, or other forms of precipitation. Of course, when the rate of rainfall or snowfall changes, either more or less water will recharge groundwater and flow into rivers and these changes will alter the hydrology of streams, rivers and lakes. This, in turn, will change the availability of freshwater. Where and how much the availability will change depends on a number of factors, including the way in which climate change plays out at different locations, for instance, whether precipitation increases

---

Box 5.1 Climate change and evapotranspiration

"Evapotranspiration" is the sum of water "evapored" from the ground surface and upper groundwater table (including the surface of vegetation) and "transpired" (i.e. conducted) out through the pores of plants. Evapotranspiration is important to the hydrology of a river basin because the higher the rate of evapotranspiration, the less water is left behind in liquid form in rivers, lakes and other water bodies for the use of society and ecosystems. Conversely, if evapotranspiration decreases, more flowing water could be present.

All things being equal, warmer temperatures at the surface of the earth might be expected to increase the rate of evapotranspiration. But there will be many exceptions in which local factors may actually hinder an increase in evapotranspiration. One exception is the situation in which the atmosphere is already saturated with moisture, in which case it would have little or no capacity to take up additional moisture produced by evapotranspiration. Another exception is the case in which soil is very dry from lack of precipitation which means there is no soil moisture to evaporate or transpire. Yet another example is the situation in which climate change increases the extent of daylight cloudiness over a particular location, which reduces the amount of solar radiation reaching the ground, which in turn, slows the rate of evapotranspiration. A reduction in the density of vegetation will also tend to reduce the rate of transpiration per unit area (assuming the vegetation type remains about the same). Finally the increased concentrations of CO_2 in the atmosphere will reduce the openness of plant pores and thereby reduce the amount of water transpired.

Despite these many exceptions, climate models tend to show that evapotranspiration may increase over large areas under climate change, apparently because the effect of warmer temperatures supersedes these other more local factors.

---

or decreases within a particular season. Another factor is the kind of water system responding to climate change: The response will be different for a lake, small river, large river, or other type of water body. The degree to which run-off is controlled, via dams and reservoirs, also has an important influence on how a river will respond to changes in precipitation. A highly regulated river with several well-managed reservoirs is likely to be less affected than an unregulated river.[2]

Will more or less water be available?

If we could account for all major processes affecting the water cycle, it would be possible to calculate the expected alteration in water availability in Europe because of climate change. In fact, hydrologic models are available for this task. Such models depict the major features of the water cycle in the form of mathematical equations. They require input data such as future climate and land cover at a particular location, and use these data to project future changes in groundwater recharge and river run-off.

Model results have suggested that the volume of groundwater recharge may decrease in Central and Eastern Europe,[3] with the larger reductions occurring in valleys[4] and lowlands (e.g. in the Hungarian steppes).[5]

Results from other modeling studies imply that water availability may increase over the Atlantic and northern parts of Europe[6] and decline in Central, Mediterranean and Eastern Europe.[7] These computations are driven by the future patterns of climate in Europe computed by climate models, as explained in Chapter 2.

How large a change in availability of water can we expect? According to one study shown in Figure 5.1, by the 2020s average water availability in parts of the northern part of Europe could increase by around 5 to 25%, relative to current climate conditions.[8] By the 2070s some areas could have an increase of more than 25%. Meanwhile in Southern Europe, the different models do not give a consistent picture for the 2020s. But by the 2070s the models agree that much of the southernmost part of Europe will experience a decrease in water availability of at least 5 to 25% relative to current climate conditions. The Iberian Peninsula may have a decrease of more than 25%, with one set of results showing a decline of water availability of more than 50% (Figure 5.1).

A proviso of these estimates is that it may be difficult to discern the "climate change signal" by the 2020s amid the strong year-to-year variation in climate and water availability (although this signal is expected to become stronger and stronger as the century progresses). Another qualification is that modeling calculations such as these are uncertain, so these numbers should be not be taken as predictions, but rather as plausible future scenarios.

With the above qualifications in mind, the trend still seems to be that more water is becoming available in the North to serve the needs of people, industry and nature. At the same time there may be less available in the South and East for public water supply, for manufacturing, for cooling electricity-generating turbines, and for aquatic ecosystems.

How will river flows change?

It is important to realize that climate change will not only alter the annual average availability of water, but also the seasonal patterns of stocks and flows of water. These patterns vary greatly in Europe depending on the type of climate holding sway in the region. Where a Mediterranean climate predominates, run-off is particularly low during the dry summer season. By comparison, rivers originating in moist continental or montane climates tend to have very high run-off in springtime because of snowmelt, and lower run-off in winter when precipitation is stored in snowpack. Under global warming, precipitation patterns may also

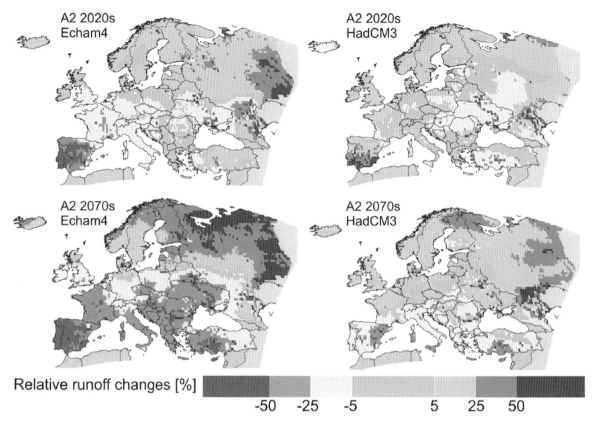

Figure 5.1 Change in annual river run-off in Europe between a 1961–90 baseline period and two future periods (2020s and 2070s). Calculations are produced by the WaterGAP model using climate data from two different climate models as input (left and right side of figure). Results from two different climate models are used in order to account for some of the uncertainty of future climate. (The two models are "Echam4" from the Max Planck Institute in Germany, and "HadCM3" from the Hadley Center in the UK.) Climate scenarios were computed by both models based on the "A2" emissions scenario of the Intergovernmental Panel on Climate Change (a higher emissions scenario). Source: Figure from Alcamo et al.,[9] which in turn was based on underlying data sets from Alcamo et al.[10] Reprinted with permission from Intergovernmental Panel on Climate Change, Geneva. (See color plate.)

change seasonally, with some regions experiencing a net increase in annual precipitation and, at the same time, a decrease at critical junctures in the year such as during the crop growing season. Studies have shown that under some scenarios winter flows may increase and summer flows decrease in the Rhine,[11] Slovakian rivers,[12] the Volga and in various rivers in central and eastern Europe.[13] A decline in summer precipitation may lower average summer flows by up to 50% in central Europe,[14] and up to 80% in some rivers in southern Europe.[15]

A special situation exists for rivers fed by Alpine glaciers. The faster melting of glaciers at first will increase the late spring and early summer run-off of these rivers. This should be an impressive sight since these rivers already rush down the mountains in springtime as a torrent of milky-colored water (tinted by its fine-grained suspended sediment). But gradually, after several decades the glaciers will shrink so much that they will no longer feed downstream rivers. As a consequence, their average summer run-off is likely to drop considerably,[16] perhaps by as much as 50%.[17]

Will the frequency of floods change?

As we try and imagine life in Europe under climate change a key consideration is certainly whether floods and droughts will become more or less frequent.

Between 1998 and 2009 Europe suffered over 175 damaging floods with some river basins experiencing six or more high-water episodes (see Figure 5.2).[18] Among these was the devastating Elbe flood of 2002 (see Box 5.2). Floods during this period led to at least 700 deaths, the dislocation of around half a million people and a minimum of 25 billion euro in insured economic losses.[19]

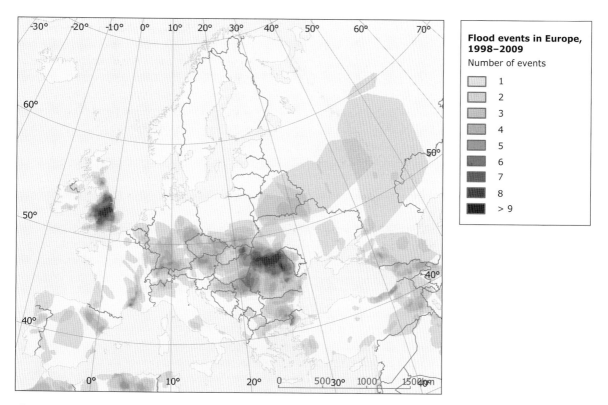

Figure 5.2 Recurrence of flood events in Europe, 1998–2009. Source: EEA.[20] Reprinted with permission from European Environmental Agency, Copenhagen. (See color plate.)

Should we expect more or fewer destructive floods under climate change? An important trend to consider is that mean winter precipitation increased during most of the 20th century in northern Europe, and climate projections suggest that this trend may continue. But flooding is not only related to heavy precipitation. Changing land cover could also be a major factor. Many cities in Europe are extending their boundaries into existing vegetation, and enlarging the area covered by impermeable surfaces.[21] Replacing vegetation with pavement prevents heavy precipitation from seeping into the ground and causes water to run off even faster into rivers, bringing about flooding downstream.

Another reason for more frequent high river flows could ironically be the actions taken to protect city populations from flooding. Earlier, many of Europe's rivers had wide and relatively unpopulated floodplains where high water could overflow and be retained until it evaporated or percolated into the soil. As settlements along rivers grew it became customary to channel and straighten waterways to protect their populations against high river flows. Unfortunately, protection for one city means greater danger for the next town downstream. As compared to natural, meandering rivers, a straight and smooth concrete channel sends floodwaters rapidly downstream rather than dampening and slowing the flood wave, and also reduces the overall volume available for storing floodwaters.

Despite these complicating factors, climate change is expected to increase the frequency of threatening floods in some areas and decrease it in others. Extremely high river discharges, which are now very seldom, could occur much more often in parts of northern and western Europe. One study, depicted in Figure 5.3, shows that flood flows now occurring on average once per century in some parts of Europe could occur every 10 to 50 years.[22] Another modeling study confirms the changing frequency of flood flows, although with a different spatial pattern of changes over the continent.[23]

Many parts of Europe are likely to experience an increase in annual precipitation. Equally important, a large European area is also likely to see an increase in the intensity of rainfall (the volume of rain per event). We know from everyday experience that short but intense rain showers can cause temporary flooding because soils are unable to soak up rainfall quickly enough. Sometimes the rainfall is so intense that it floods city streets rapidly, and eventually runs into nearby waterways, which then overflow their banks, causing "flash floods". As the intensity of rainfall increases it is likely that flash floods will become more common throughout much of Europe.[25]

Apart from the danger that more frequent floods pose to human life, livelihoods and property, they could also accelerate the erosion of dikes along rivers and tax the capabilities of existing upstream reservoirs to store potential flood waters. Another side effect could be more intense water pollution, as in the case when severe flooding causes sewers and sewage treatment plants to exceed their capacity and overflow into waterways. A well-publicized example is the storm event in August 2004, which caused London's sewers to overflow, dumping raw sewage into the Thames River and causing a major fish kill.[26]

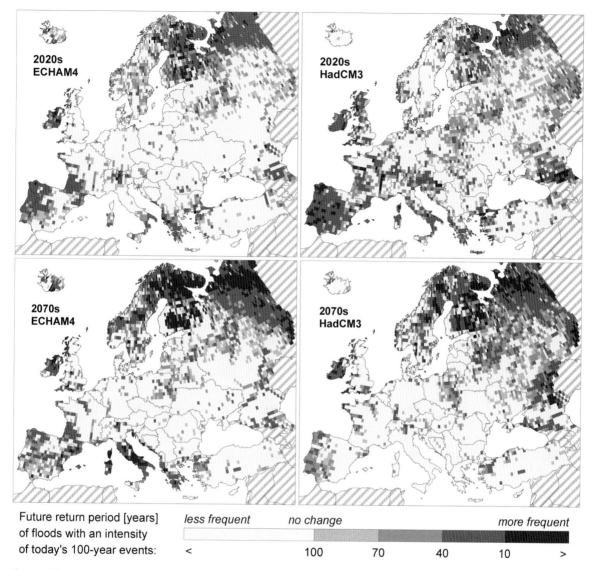

Future return period [years]
of floods with an intensity
of today's 100-year events:

less frequent   no change   *more frequent*

<   100   70   40   10   >

Figure 5.3 Change in the recurrence interval of 100-year floods in Europe between a 1961–1990 baseline period and two future periods (2020s and 2070s). Calculations are from the WaterGAP model using climate data from two different climate models (left and right side of figure). See Figure 5.1 caption for explanation of climate models and climate scenarios. Source: Lehner et al.[24] Reprinted with permission from Springer Science + Business Media.

It is worth noting that estimates of changes in the future frequency of floods are particularly uncertain. This is because the type of weather event directly causing a flood, whether it be a sudden warm period triggering rapid snowmelt or an episode of intense rainfall, usually last only a few hours or days. Unfortunately, despite the fact that regional climate models have recently improved their fine-scale predictions, these are the kind of events that climate models cannot yet simulate very well (see Chapter 2).

Box 5.2 The Great Elbe Flood of 2002

On 10 August 2002, a large low pressure system moved slowly over Central and Eastern Europe. The storm had begun a few days earlier near the British Isles and then swung south where it had absorbed plentiful amounts of moisture from the warm Mediterranean. While making slow progress across the middle of the continent it released its moisture, causing a sustained rainfall at some locations of nearly 25 cm over four days.[27] Flooding first began in the Czech Republic where the Vlatava River, a major tributary of the Elbe, washed over into the medieval villages of Cesky Krumlov and Ceske Budejovice. The next casualty was the low-lying area of the city of Prague. The flood wave continued into the Elbe and on 15 August caused major flooding in the city of Dresden. As the flood wave traveled further downstream the Elbe overflowed its banks at Wittenberg, Dessau and other Saxon cities. High water reached Hamburg near the mouth of the Elbe on 24 August. While the heavy rains were causing major flooding along the Elbe, the level of the Danube also rose precipitously causing flooding in many towns in Austria and Slovakia.

As the Elbe and Danube overflowed they inundated the rich croplands along their length and destroyed crops nearing harvest. Enormous river flows damaged or destroyed bridges, roads and other infrastructure. Muddy waters flooded low-lying areas and forced many businesses and factories along the river to halt operations. Homes and other private property were lost under the floodwaters. The total cost of flood damage in Dresden and other Saxon towns was estimated to be around 6 billion euros. When other areas affected in Germany were included, cost estimates rose to about 9 billion euros. Summing up all flooded areas of Central and Eastern Europe, the economic damage of the great August 2002 flood was estimated to be about 15 billion euros.[28]

Will the frequency of droughts change?

While it is likely that some parts of Europe will have to contend with more frequent flooding others will have to confront dry periods more often. As pointed out in Chapter 2, climate change is likely to increase the frequency of so-called "meteorological droughts" throughout much of Europe. These are periods of little or no precipitation lasting from months to years. The lack of rainfall and snowfall during this time causes "hydrological droughts" which are periods of very low flow in rivers and diminished recharge of the aquifers that serve as underground reservoirs of precipitation.

According to model calculations, the Mediterranean area is likely to have the highest risk in Europe of more frequent meteorological and hydrological droughts. While droughts are not uncommon in the region, very severe hydrological droughts are now relatively seldom. Under climate change, however, these severe events could become more frequent. By the 2070s the intense droughts that now occur on average once per century may occur every 40 years, or even every 10 years, in some regions (Portugal, Mediterranean countries, Hungary, Bulgaria, Romania, Moldova, Ukraine, southern Russia) (Figure 5.4).[29] Drought risk may also increase in parts of western Europe. Meanwhile, because of increasing precipitation, northern Europe may experience a decline in the frequency of severe droughts.

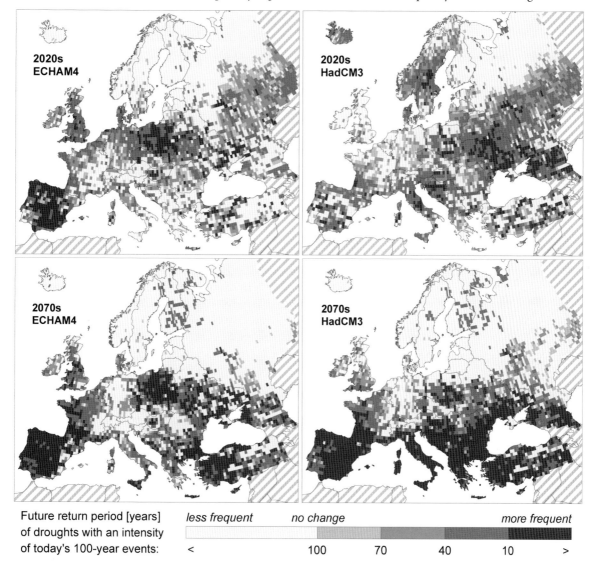

Figure 5.4 Change in the recurrence interval of 100-year droughts in Europe between a 1961–90 baseline period and two future periods (2020s and 2070s). Calculations are from the WaterGAP model using climate data from two different climate models (left and right side of figure). Source: Lehner et al.[30] Reprinted with permission from Springer Science + Business Media. (See color plate.)

Changing lakes and groundwater

Not only rivers, but other bodies of freshwater will also be affected by changing climate. The volume of a lake is determined by the flow rate of the various streams feeding into it, by evaporation from its surface, and by precipitation falling directly into the lake or onto its watershed. As climate changes, so too will these factors. Where climate becomes drier, lakes will diminish or dry up. Where it becomes wetter, the lakes volume may increase or even overflow. Small closed lakes are particularly sensitive to climate change because they rely on smaller flows of water for their volumes, and even small changes in these flows are magnified over long periods of time.

Changes in climate will almost certainly alter the volume of precipitation that seeps into soil. This will change the amount of water renewed and stored in aquifers and later pumped to the surface for household water supply, for irrigation, and sometimes for industry. The impact of climate change on aquifers should not be underestimated since they are a critical source of water supply in Europe. The withdrawal of water from groundwater aquifers accounts for a full 29% of all water extracted for all purposes in Europe.[31] Likewise, global warming will cut into water supply by increasing the evaporation rate of reservoirs and thus make less water available for potential users.

Climate change will also affect the water quality of lakes and rivers. Warmer air temperatures will translate into warmer water temperatures and this will have a variety of side impacts on water quality.[32] For example, an increase in the temperature of water leads to a decrease in its oxygen-holding capacity, and it absorbs less oxygen from the atmosphere. This will lower the amount of oxygen available for aquatic fauna. Warmer water temperatures may also stimulate algal blooms. Furthermore, changing water temperature may alter the habitats of some aquatic fauna because many organisms seek out particular temperature zones in lakes and streams.

In southern Europe, where precipitation is expected to decrease, a lower volume of water may be available in lakes and streams to dilute the pollutants washed off the surfaces of agricultural fields and urban areas and discharged by manufacturers and municipalities. In other parts of Europe, an increase in the intensity and volume of precipitation may amplify the amount of contaminants washed into receiving waters from croplands (pesticides and fertilizer) and from overflowing sewers (organic wastes and bacteria).

Impact on household and irrigation water demand

Warmer air temperatures may increase the demand for water in the household, especially in summer. People will tend to drink more water, take more showers and water the garden more often.[33] As an example, during the August 2003 heatwave in Europe (see Chapter 2) municipal water use in the Netherlands was 15% higher than the multi-year average for this month.[34]

Agricultural water use will also be affected. Warmer temperatures and lower summer precipitation in southern and central Europe are likely to boost water requirements for irrigation (the volume of water needed for irrigation per hectare), all other factors being equal.[35] Although the sight of irrigation spigots spraying water over a crop field is now relatively rare in northern and western parts of the continent, it could become much more common over the coming decades.[36]

Changing potential for hydroelectricity

Hydroelectricity is an important source of energy in Europe, accounting for about 15% of its total electricity production and for more than 50% of the electricity produced in Norway, Iceland, Austria and Switzerland.[37] Moreover, it is gaining in popularity because it has the image of being a "green energy source" on account of its relatively small indirect emissions of greenhouse gases. Unfortunately, this image does not factor in other social and environmental impacts of hydropower such as the inundation of terrestrial ecosystems in the valley behind hydroelectric dams, or the expelling of residents from these valleys, or the disruption or elimination of aquatic ecosystems caused by the fragmentation of a stream into a series of dams and reservoirs. Other drawbacks of large hydropower facilities are described in Box 10.2 of Chapter 10.

Climate change is likely to influence the potential for producing electricity at hydroelectric power facilities because this potential is strongly related to river discharge (roughly speaking, the greater the discharge the more electricity produced by the turbines at a hydroelectric facility). One study estimated that increasing precipitation and higher river run-off might increase hydropower potential by 15–30% by the 2070s over large areas of northern and eastern Europe.[38] However, during the same period the hydropower potential of existing facilities throughout much of southern and southeastern Europe could decrease by 25% or more, with a 20–50% decrease around the Mediterranean.[39] Of the 40 European countries examined, 22 were found to have a net decrease in hydropower potential at current hydroelectric facilities and 14 of these countries could experience a decline of more than 25% (according to at least one climate scenario).[40] European countries may be faced with an ironic situation: on one hand they may try to expand their production of hydroelectricity as a strategy for reducing their greenhouse gases; on the other, they may be hindered in doing so by the impacts of greenhouse gases on global climate.

To worsen the situation, most scenarios show a sharp increase in the demand for electricity in the coming decades in eastern and southeastern Europe, and it is logical to assume that part of this electricity production would be needed from the expansion of hydroelectric facilities. In other words, while hydroelectric facilities in these regions are under pressure to expand their output, climate change may at the same time hamper their ability to produce electricity.

Impacts on thermal power plants

The impacts of climate change on water resources may also indirectly affect electricity production at thermal power plants (facilities that burn fossil, nuclear, or other fuels to generate steam and run turbines). These facilities are often located on rivers or other large water bodies because they require substantial volumes of intake water to produce steam and cool turbines. Hence, sinking river levels in southern Europe could endanger the source of cooling water for these facilities.

Furthermore, warmer air temperatures will lead to warmer river temperatures, which could affect the use of this water for cooling turbines. The warmer river temperatures will also make it more difficult for cooling towers to cool down the water before it is discharged back into a river. Cooling of hot water after its use in power plants is a legal requirement over most of Europe in order to avoid "thermal pollution" of rivers – the condition by which hot water discharges artificially elevate stream temperatures and disrupt aquatic ecosystems.

Effects of sea-level rise on water supply

Apart from its direct impacts on hydrology, climate change is also likely to have an important indirect impact on coastal water supply because of rising sea level. As the atmosphere heats up, the upper layers of the ocean will also become warmer. Since water expands when heated, the volume of ocean water will expand, raising global average sea levels.[41] Adding to this effect will be the melting of mountain glaciers, and perhaps parts of the Greenland, and other ice caps. During the 21st century, the sum of these effects could lead to an increase in global average sea level of 19–58 cm (see Chapter 6).

Sea-level rise will have at least two notable impacts on coastal water resources. First, it will increase the risk of saltwater intrusion into aquifers lying along the coast, and so increase the risk of contaminating community water supplies that rely on this groundwater. At greatest risk will be low-lying coastal areas with shallow aquifers. Second, sea-level rise will thrust the ocean's saltwater further upstream in river estuaries, and in this way threaten municipal and other withdrawals of freshwater from estuaries. It is possible that cities located on estuaries will have to implement early warning systems for anticipating an unsafe salt contamination, or move their water intakes further upstream or out of the estuary entirely.

Adapting to climate impacts on water resources

What are the options for coping with changing river flow, more frequent floods, and more frequent droughts? Since floods and droughts are already part of life in

Europe, managers of the continent's water resources have long adapted to their happening. To a large degree the practice and discipline of water resources engineering and planning was developed to deal with the variability of climate, and so it can be argued that water managers and engineers might be among the best prepared for the impacts of climate change.

But it is not only water managers who are involved in adapting to climate change. Over the past few years a worldwide "Dialogue on Climate and Water" has been held to exchange information between water managers, climate scientists, public interest groups, community groups, farmers, and a host of other stakeholders concerned about coping with climate variability in river basins. The participants of this dialogue identified what they saw as options for action:[42]

- *policy instruments* such as "regional strategic action plans" that consolidate and document the actions that government and other actors can take to prepare for climate change;
- *technological and structural instruments* such as building new reservoirs or beginning a groundwater recharge program to increase long-term supplies as a hedge against increasing frequency of drought;
- *risk sharing and spreading of risk* through new forms of insurance dealing with extreme climatic events, which are also affordable to vulnerable poor communities;
- *change of use, activity or location* including special land use zoning to protect the population from changes in flooding.

Table 5.1, from the European Environment Agency, summarizes the numerous actions currently being taken or planned in the European water sector for coping with climate change.[43] As we see in the following paragraphs, these options are many.

Protecting against floods

Technical flood protection – For centuries, dikes, dams and other structures have protected inhabitants of lowlands from flooding and it is likely that they will continue to play an important role in flood protection. Throughout Europe, river and coastal dikes are already being continuously repaired and drainage systems improved to protect against recurring river and coastal floods. Going beyond this traditional maintenance, some federal states in Germany as well as other regional governments in Europe are building their dikes wider and higher as a hedge against the additional threat from climate change. As a possible model for Europe, Japan has even begun building "super-levees" to protect its population from levee failure (see Chapter 6).

Natural retention of flood water – An important factor to consider is that more frequent flooding events in some parts of Europe may increase the potential costs of dams, dikes and other technical measures to protect society. Because of rising costs, another approach has been gaining in popularity: rather than trying to hold back floodwaters with dikes, some countries (including Denmark, Germany, the

Table 5.1 Catalog of adaptation measures for water resources to climate change. For explanations see text. Source: Responses to questionnaires sent to water management officials in EU countries by the European Environment Agency, as reported in Footitt, A., Hedger, M., 2007.[44] Reprinted with permission from European Environmental Agency, Copenhagen.

Protection category	Adaptation measure
Flood protection	Technical flood protection
	Natural retention of flood water
	Restriction of settlement/building development in risk areas
	Standards for building development
	Improving forecasting and information
	Improving insurance schemes and information
Drought/low flow protection	Technical measures to increase supply
	Increasing efficiency of water use
	Economic Instruments
	Restriction of water uses
	Landscape planning measures to improve water balance
	Improving forecasting, monitoring, information
	Improving insurance schemes against drought damage
Coastal zones	Reinforce or heighten existing coastal protection infrastructure
	Retreat strategies e.g. managed realignment of dams
	See also Chapter 6
General adaptation measures	Awareness raising or information campaigns

Netherlands, and Sweden) are beginning to make room for the river by setting aside "floodways", which are intentionally inundated during high water periods. The idea is to route high river flows to designated unpopulated areas in the flood plain set aside to temporarily store floodwaters. These are mainly recreational, farming, or other undeveloped areas that could be inundated without risk to people or their structures. Storing floodwaters in floodways reduces the risk that the river will overtop its banks or dikes at populated areas downstream.

Restrictions of settlement/building development in risk areas – Another way to protect society from the risk of flooding is to adopt long-term land use policies that prohibit the expansion of settlements at risk from floodwaters. Austria has developed "hazard zone plans", France "flood risk maps", and Switzerland "hazard maps" for most of their territories that depict the location of particular flood risk to inhabitants. These maps have an indirect effect of restricting settlements in high risk areas, because insurance companies raise the price of insurance premiums in these locations. In some cases, flood insurance becomes so expensive that further development in high risk areas is minimal.

Standards for building development – Some countries are giving special attention to strengthening building codes in high-risk flood plains so that structures are as flood-resistant as possible. A few cities in Sweden and Denmark are accounting for a future increase in flooding by modifying building codes and requiring higher minimum floor levels and, in some cases, restricting the location of buildings. Some cities are also mandating larger capacities of sewer systems to handle the expected increase in precipitation.

Improving forecasting and information – Even where flood protection structures are strong, the threat of damage from flooding remains so that an important measure for coping with climate variability and change is to establish early-warning systems that communicate flood forecasts and emergency information to the public. These are large-scale complex alert systems combining scientific information and state-of-the art communications (see Box 5.3).

Improving insurance schemes and information – To compensate for the inevitable damage caused by floods it is possible to improve the type and coverage of insurance. Recent legislation passed in Belgium requires flood damages to be included in household fire insurance policies. Insurance for flood damage is also available in Germany. On average, France spends around 250 million euros each year to reimburse damage caused by floods.[45]

Coping with droughts and low flow periods

Implementing technical measures to increase supply – An age-old hedge against droughts is storing water in reservoirs, tanks, and sometimes in aquifers, for use during dry periods. As changing climate reduces precipitation in some regions, this approach will again be called upon. For example, officials in the Netherlands are investigating the possibility of storing freshwater in Lake Ussemeer, and water managers in Greece are looking into recharging aquifers as a hedge against drought.

Increasing efficiency of water use – A comprehensive way of dealing with drought is to reduce the exposure of society to these events by decreasing its reliance on large water withdrawals. This can be done by increasing the efficiency of water use, or put another way, gaining the same water service (drinking water, irrigation or other) by withdrawing a smaller volume of water (see Box 5.4).

Restricting water uses – Another approach taken by governments under severe drought conditions is mandating restrictions on water use or shifts in allocations. For example, local authorities are allowed by the Drought Management Plan of the Guadalquivir River to reallocate water from irrigation water users to urban and industrial users.[48]

Adopting landscape planning measures to improve water balances – Land cover has a significant impact on the water balance of a watershed. Rainfall tends to pool on surfaces with bare, compacted soils and then evaporate quickly. Vegetated surfaces, by comparison, tend to have an upper layer of decomposing organic matter (humus) which better absorbs precipitation and allows water to percolate

Box 5.3 Early warning systems: reducing the immediate risk of floods and droughts

One of the most effective ways to allay the immediate threats to society posed by floods and droughts are "early warning systems". First and foremost, these systems warn the public and private sectors about an impending flood or drought and inform them about procedures for coping with the imminent threat. Early warning systems consist of three main elements: (i) a modeling system by which local scientists/experts make forecasts of flood or drought events; (ii) a procedure by which authorities decide on issuing an alert to the public; (iii) and a set of actions that can be taken in an emergency (for example, curtailment of water use or evacuation of the population). Early warning systems require close cooperation between scientists, decision makers and those responsible for communication networks.

Several river basins in Europe have flood early warning systems that combine meteorological forecasts with knowledge about the hydrology of rivers and topography. This combination of information makes it feasible to forecast a flood event some days in advance. An example is the FEWS-Rhine (**F**lood **E**arly **W**arning **S**ystem for the River Rhine) operated by Germany, the Netherlands, and Switzerland.

While these systems are now being implemented on the municipal and river basin level, they are also an effective way to cope with natural emergencies on the national and continental scale. On the national level, France has inaugurated a national early warning center for flooding covering 22 river basins. Meanwhile, on the continental scale, a new system for flood warning is currently being tried out in Europe. The so-called European Flood Alert System (EFAS) is intended to provide forecasts 3–10 days in advance of high water events throughout Europe. EFAS will be run by the European Commission's Joint Research Center at Ispra in direct cooperation with river basin authorities.[46]

For early warning of droughts, Spain has organized a national "hydrologic indicators system" (river inflows, reservoir levels) to anticipate and prepare for significant droughts. On the pan-European level, the European Commission has proposed setting up a "European Drought Observatory" and early warning system for droughts.[47] This would be an apparatus for combining meteorological forecasts, hydrologic data, and other information to make medium-term forecasts of droughts in Europe and to give authorities enough time for planning emergency measures.

Box 5.4 Water conservation as a hedge against drought

Roughly speaking, the higher the dependence of a population on its regional water resources, the greater its exposure to the impacts of drought. Or to put it positively, the lower the water use relative to water availability, the more resilient the population against drought. This follows from the simple logic that a community using, say, three-quarters of its available water resources on a continuous basis will have a more difficult time coping with an occasional drought than a community dependent on only one-quarter of its long-term average water resources.

It can be argued that Europeans have done fairly well in slowing down or even stopping their growth in water consumption. National water use has more or less leveled off or even declined since the 1990s in many northern and western European countries.[49] But the potential for saving water and thereby lowering exposure to drought is still enormous.[50] There are four basic approaches to water conservation:

1 *Reducing water demands* – The most direct way of conserving water is simply to reduce the amount currently used for the same purpose. In the household this means doing things differently, for example by installing a low-flow showerhead that uses less water per minute than a standard faucet. Alternatively, water-wasting appliances and devices can be replaced with ones using less water. For example, in Europe washing machines have become steadily more water efficient, decreasing from 175 to 50 liters per cycle between 1970 and 1998.[51]

2 *Reducing water losses* – The overall use of water by industry, agriculture and municipalities is not as efficient as it could be. For example, there is great potential to save water in irrigated agriculture. This is because many European farmers continue to use traditional "gravity" irrigation which has much greater water losses than more modern set-ups such as sprinkler, drip or other types of irrigation that deliver water under pressure to the crops. In a survey of 39 Spanish irrigation schemes, it was found that projects with traditional gravity irrigation had an efficiency of only 60% (that is, 60% of the total volume of water abstracted from its source was finally used by crops) as compared to the 80% efficiency achieved by modern pressure-type irrigation projects.[52] In the municipal sector, water is often lost unnecessarily between the place where it is abstracted and the place to which it is delivered. Although the losses in Zurich's municipal water system were not large compared to other cities (about 10%), municipal authorities managed to reduce them down to 5% over a ten year period through a program of leakage control.[53]

3 *Recycling water* – Water recycling is the use of water repeatedly for the same purpose. In effect, wastewater coming from an industrial process is used again as intake water for the same process. Many industries throughout Europe, from auto manufacturers to dairies, have achieved substantial water savings by modifying their processes in order to recycle water.[54]

Continued

4 *Reusing water* – Water reuse is slightly different than recycling in that water is first used for one purpose and then for another. For example, in one Polish community water is first used for household purposes and then the wastewater is collected and delivered to a coke-manufacturing plant for quenching coal.[55] Another example is the use of treated municipal wastewater for farm irrigation in France and other countries.

Now that we have reviewed the four main technical approaches to water conservation, the question arises, how can these approaches be implemented in practice? One option is to enforce these measures through government mandate. Another, less forcible approach, would be to encourage them through public education. A third option is to stimulate water conservation through economic signals – by levies on water services or by increasing the sale price of water. Water pricing is already wide spread in Europe. In France, for example, 85% of the costs of water are recovered from households and industry through metering and various tariffs.[56] On the other hand, water is treated differently in the irrigation sector. European governments have been hesitant to monitor and realistically price irrigation water use because low water prices are a kind of subsidy for domestic agriculture. France is trying out a compromise solution by requiring farmers to meter their water withdrawals when their abstraction rate is above a particular threshold.[57] In principle, tariffs for all sectors can be made higher during drought periods to discourage water usage, but this practice is still rare in Europe.

down to aquifers where it can eventually be retrieved. The need for the land surface to absorb precipitation is taken very seriously in the Netherlands. Builders here are required to estimate any losses to infiltration capacity that may be caused by their projects and must compensate for these losses.

Improving the forecasting of droughts – Early warning systems are being built not only for floods, but also droughts (see Box 5.3). But there is a basic difference in how these events are forecast. While it is relatively easy to make the short-term weather forecasts needed for anticipating floods, droughts are a longer-lasting phenomenon and therefore require weather forecasts many months in advance. But new scientific understanding about the climate system and its long-term cycles, such as the "North Atlantic Oscillation", is enabling more accurate medium-term weather forecasts and this will eventually improve the performance of drought early warning systems.

Improving insurance schemes against drought damage – Insurance can also be used to help victims of drought in the same way as it is used to compensate flood victims. Nevertheless, as the scale of climate impacts increases, the insurance industry may hesitate to take on large new risks related to nature. Chapter 9 examines the reaction of the insurance industry to climate change.

Adaptation: limits and integration

While there are many different options for responding to changing climate, the question is, will every option be equally affordable and feasible? A key factor here is the rate of climate change, because the faster the tempo, the shorter the time for devising a sufficient adaptation plan. Likewise, the faster the climate change, the shorter the period for spreading out the costs of adaptation. For example, if long-term average annual precipitation decreases slowly, then it is possible that slow adjustments in the water sector (such as gradually reducing per capita water use) could be sufficient to cope with a reduction in water availability. But if the changes are rapid, then reducing water demand may not be sufficient, and new infrastructure might be needed for storing water.

There are also many other possible barriers to adaptation:[58]

1 An adaptation measure could have physical limits. For instance, it may not be possible to build additional reservoirs for storing water because suitable sites may already be occupied.
2 It may not be politically or socially responsible to carry out an adaptation measure. It may not be politically acceptable to reduce the standards of service or build new reservoirs to store water.
3 Adaptation measures may not be economically affordable. In some cases, the costs of building new infrastructure will simply not fit into municipal or state budgets.
4 Water management agencies may be unable or unwilling to take on new adaptation measures. Reasons for this might be the low priority given to water management or perhaps competition between municipal agencies.

Taken together, these factors emphasize the fact that adaptation measures cannot be taken in isolation of the greater challenge of dealing with water scarcity and flood hazards in Europe. Actions for adaptation have to compete with the agendas of the many institutions concerned with water locally, or globally. For this reason it is advisable to try and "mainstream" measures for coping with climate change into existing water-related institutions, policies and planning.[59] (This mainstreaming policy applies equally well to adaptation measures discussed in other chapters of this book.) The objective would be to fine-tune existing planning procedures to make them more responsive to climate change. An example of mainstreaming is for state agencies to factor in climate change impacts when designing new infrastructure such as roads, reservoirs and coastal protection structures. There are already many examples of such mainstreaming in Europe, such as regional governments that take into account rising sea level in their designs for new river dikes or highways. Another example is adapting the operation of existing reservoirs to account for more frequent floods or droughts.

Although mainstreaming helps to find an appropriate place for adaptation strategies, it has the disadvantage of adding yet another dimension to the already complex practice of water management. Adaptation measures might have an even higher chance of success if they could fit into a still bigger concept for rationally managing water in a particular river basin. Such an approach exists, called "integrated water resources management" (IWRM). This "integrates" many different aspects of river basin development by promoting a long-term perspective to planning, by encouraging the participation of diverse interest groups in the planning process, by reconciling the water needs of human users together with needs of aquatic ecosystems, and by advocating improving water use efficiency as an alternative to expanding water supply.

A commonly cited definition of IWRM comes from the Global Water Partnership[60] which says:

> IWRM is a process which promotes the coordinated development and management of water, land and related resources, in order to maximize the resultant economic and social welfare in an equitable manner without compromising the sustainability of vital ecosystems.

If this definition seems vague, it is because IWRM is more of a strategy or viewpoint than a procedure for managing water. Nevertheless, it has stimulated discussion worldwide on how to rationally and comprehensively manage water. Indeed, much has been written about IWRM, and it has also been criticized from different directions. One critique of particular relevance to climate change is that it does not deal enough with the uncertainties of water management. With this in mind, researchers from the EU Project "NeWater" have suggested that IWRM should take into account factors such as extreme climate events and the special responses these events require.[61] They recommend that in general IWRM should move in the direction of "adaptive management", which would permit water managers to respond more flexibly to unknown future demands on the water system.

But regardless of how IWRM will evolve over the coming years, this kind of global systems thinking provides a sensible way to combine the need for climate adaptation with the greater need to provide sufficient water for nature and society in Europe.

Notes

1   Framing Committee of the Global Water System Project, 2005: *The Global Water System Project: Science Framework and Implementation Activities. Earth*

System Science Partnership. Global Water System Project Office. Bonn, Germany. www.gwsp.org.

2 Arnell, N.W., Liu, C., 2001: Hydrology and water resources. In: *Climate change 2001: Impacts: adaptation and vulnerability*, McCarthy, J.J., White, K.S., Canziani, O., Leary, N., Dokken, D.J. (eds): Cambridge: Cambridge University Press. UK. 191–234

3 Eitzinger, J., Stastna, M., Zalud, Z., Dubrovsky, M., 2003: A simulation study of the effect of soil water balance and water stress in winter wheat production under different climate change scenarios. *Agricultural Water Management* 61, 195–217.

4 Krüger, A., Ulbrich, U., Speth, P., 2002: Groundwater recharge in Northrhine-Westfalia by a statistical model for greenhouse gas scenarios. *Physics and Chemistry of the Earth, Part B: Hydrology, Oceans and Atmosphere* 26, 853–861.

5 Somlyódy, L., 2002: *Strategic issues of the Hungarian water resources management*. Academy of Science of Hungary, Budapest, 402 pp. (in Hungarian).

6 Werritty, A., 2001: Living with uncertainty: climate change, river flow and water resources management in Scotland. *The Science of the Total Environment* 294, 29–40. Andréasson, J., Bergström, S., Carlsson, B., Graham, L.P., Lindström, G., 2004: Hydrological change – climate impact simulations for Sweden. *Ambio* 33, 228–234.

7 Menzel, L., Bürger, G., 2002: Climate change scenarios and run-off response in the Mulde catchment (Southern Elbe, Germany). *Journal of Hydrology* 267, 53–64. Etchevers, P., Golaz, C., Habets, F., Noilhan, J., 2002: Impact of a climate change on the Rhone river catchment hydrology. *Journal of Geophysical Research* 107, 4293, 18 pp. Chang, H., Knight, C.G., Staneva, M.P., Kostov, D., 2002: Water resource impacts of climate change in southwestern Bulgaria. *GeoJournal* 57, 159–168. Iglesias, A., Estrela, T., Gallart, F., 2005: *Impactos sobre los recursos hídricos. Evaluación Preliminar de los Impactos en España for Efecto del Cambio Climático*, Moreno, J.M. (ed.), Ministerio de Medio Ambiente, Madrid, Spain, 303–353.

8 These and other results in this paragraph are based on a range of emission scenarios and results from two different climate models. *See caption of Figure 5.1.*

9 Alcamo, J., Moreno, J.M., Novaky, B. (convening lead authors), Bindi, M., Corobov, R., Devoy, R.J.N., Giannakopoulos, C., Martin, E., Olesen, J.E., Shvidenko, A. (lead authors), 2007: *Europe. Chapter 12 in: Climate Change 2007: Impacts, Adaptation and Vulnerability. Contribution of Working Group II to the Fourth Assessment Report of the Intergovernmental Panel on Climate Change (IPCC)*, Parry, M.L., Canziani, O.F., Palutikof, J.P., van der Linden, P.J., Hanson, C.E. (eds), Cambridge University Press, Cambridge, UK. www.gtp89.dial.pipex.com/12.pdf

10 Alcamo, J., Floerke, M., Maerker, M., 2007: Future long-term changes in global water resources driven by socio-economic and climatic changes. *Hydrological Sciences* 52 (2), 247–275.

11 Middelkoop, H., Kwadijk, J.C.J., 2001: Towards an integrated assessment of the implications of global change for water management – the Rhine experience. *Physics and Chemistry of the Earth, Part B Hydrology, Oceans and Atmosphere* 26, 553–560.

12 Szolgay, J., Hlavcova, K., Kohnová, S., Danihlik, R., 2004: Assessing climate change impact on river run-off in Slovakia. Characterisation of the run-off regime and its stability in the Tisza Catchment. Proceedings of the XXIInd Conference of the Danubian Countries on the Hydrological Forecasting and Hydrological Bases of Water Management. Brno (30 August–2 September). CD-edition.

13 Oltchev, A., Cermak, J., Gurtz, J., Tishenko, A., Kiely, G., Nadezhdina, N., Zappa, M., Lebedeva, N., Vitvar, T., Albertson, J.D., Tatarinov, F., Tishenko, D., Nadezhdin, V., Kozlov, B., Ibrom, A., Vygodskaya, N., Gravenhorst, G., 2002: The response of the water fluxes of the boreal forest region at the Volga source area to climatic and land-use changes. Physics and Chemistry of the Earth, Parts A/B/C 27, *Hydrology, Oceans and Atmosphere*, 675–690.

14 Eckhardt, K., Ulbrich, U., 2003: Potential impacts of climate change on groundwater recharge and streamflow in a central European low mountain range. *Journal of Hydrology*, 284, 244–252.

15 Santos, F.D., Forbes, K., Moita, R. (eds), 2002: *Climate Change in Portugal: Scenarios, Impacts and Adaptation Measures.* SIAM project report, Gradiva, Lisbon, 456 pp.

16 Hock, R., Jansson, P., Braun, L., 2005: *Modelling the response of mountain glacier discharge to climate warming. Global Change Series,* Huber, U.M., Reasoner, M.A., Bugmann, H. (eds), Springer, Dordrecht, 243–252.

17 Zierl, B., Bugmann, H., 2005: Global change impacts on hydrological processes in Alpine catchments. *Water Resources Research* 41, 13 pp.

18 European Environment Agency (EEA), 2010. The European environment state and outlook 2010. Water resources: quantity and flows. www.eea.europa.eu/soer/europe/water-resources-quantity-and-flows/. Footitt, A., Hedger, M., 2007: *Climate change and water adaptation.* European Environment Agency. EEA Technical Report No 2/2007. eea.europa.eu

19 EEA, 2011: www.eea.europa.eu/themes/water/water-resources/floods (retrieved April 2011).

20 EEA, 2011. www.eea.europa.eu/data-and-maps/figures/occurrence-of-flood-events-in-europe-1998/map-5-24-climate-change-2008-occurence-of-floods.eps/Map%201.1%20Water%20resources_101.eps.75dpi.png/at_download/image (retrieved April 2011).

21 de Roo, A., Schmuck, G., Perdigao, V., Thielen, J., 2003: The influence of historic land use change and future planned land use scenarios on floods in the Oder catchment. *Physics and Chemistry of the Earth*, Parts A/B/C 28, 1291–1300.

22 Lehner, B., Döll, P., Alcamo, J., Henrichs, H., Kaspar, F., 2006: Estimating the impact of global change on flood and drought risks in Europe: a continental, integrated analysis. *Climatic Change* 75, 273–299.

23 Dankers, R., Feyen. 2008: Climate change impact on flood hazard in Europe: An assessment based on high-resolution climate simulations. *Journal of Geophysical Research* 113, 17 pp.

24 Lehner, B., Döll, P., Alcamo, J., Henrichs, H., Kaspar, F., 2006: Estimating the impact of global change on flood and drought risks in Europe: a continental, integrated analysis. *Climatic Change* 75, 273–299.

25 EEA, 2004: *Impacts of Europe's changing climate: An indicator-based assessment. EEA Report No 2/2004*, European Environment Agency, Copenhagen (or: Luxembourg, Office for Official Publications of the EC), 107 pp.

26 Thames Water, 2005. *Thames tideway strategic study. Steering group report.* Thames Water, Reading, UK.

27 Petrie, G., 2002: Flooding in Middle Europe observed from space. Geoinformatics. October/November 2002

28 Mueller, M., 2003: Damages of the Elbe Flood 2002 in Germany – a review. Geophysical Research Abstracts 5, 12992.

29 Lehner et al., 2006.

30 Lehner et al., 2006.

31 EEA, 1999: *Sustainable water use in Europe. Environmental Assessment Report No. 1. Part 1: Sectoral use of water.* EEA. Kongens Nytorv 6. Copenhagen. DK-1050. 91 pp.

32 Kundzewicz, Z.W., Mata, L.J., Arnell, N.W., Döll, P., Kabat, P., Jiménez, B., Miller, K.A., Oki, T., Sen, Z., Shiklomanov, I.A., 2007: *Freshwater resources and their management. Climate Change 2007: Impacts, Adaptation and Vulnerability. Contribution of Working Group II to the Fourth Assessment Report of the Intergovernmental Panel on Climate Change*, Parry, M.L., Canziani, O.F., Palutikof, J.P., van der Linden, P.J., Hanson, C.E. (eds), Cambridge University Press, Cambridge, UK, 173–210.

33 Arnell, N.W., Liu, C., 2001.

34 Footitt, A., Hedger, M., 2007.

35 Döll, P., 2002: Impact of climate change and variability on irrigation requirements: a global perspective. *Climatic Change* 54, 269–293. Donevska, K., Dodeva, S., 2004: Adaptation measures for water resources management in case of drought periods. In: *Proceedings, XXIInd Conference of the Danubian Countries on the Hydrological Forecasting and Hydrological Bases of Water Management.* Brno (30 August–2 September 2004), CD-edition.

36 Holden, N.M., Brereton, A.J., 2003: Potential impacts of climate change on maize production and the introduction of soybean in Ireland. *Irish Journal of Agricultural and Food Research* 42, 1–15.

37 Energy portal of the European Union (retrieved 17 September 2008), www.energyportal.eu/reviews/hydro-energy/energy-from-water-the-different-forms-of-hydro.html

38 Lehner, B., Czisch, G., Vassolo, S., 2005: The impact of global change on the hydropower potential of Europe: a model-based analysis. *Energy Policy* 33, 839–855.

39 Lehner et al., 2005.

40 Lehner et al., 2005.

41 In Scandinavia and some other locations, continents are slowly rising which compensates somewhat for the increase in the volume of seawater.

42 Kabat, P., van Schaik, H., 2003: *Climate changes the water rules: How water managers can cope with today's climate variability and tomorrow's climate change.* Report from the Dialogue on Water and Climate, 120 pp.

43 Footitt, A., Hedger, M., 2007.

44 Footitt, A., Hedger, M., 2007.

45 Footitt, A., Hedger, M., 2007.

46 European Commission. Joint Research Center, ISPRA. http://efas-is.jrc.ec.europa.eu/ (retrieved April, 2011).

47 EC (European Commission). 2007: *Addressing the challenge of water scarcity and droughts in the European Union.* Communication from the Commission to the European Parliament and Council.

48 Ecologic, 2007. EU Water saving potential (Part 2 – Case Studies) ENV.D.2/ETU/2007/0001r. Ecologic, Institute for International and European Environmental Policy. Pfalzburger Str. 43–44, D – 10717 Berlin, www.ecologic.de

49 EEA, 1999.

50 See, e.g. EEA, 1999. EEA, 2001: *Sustainable water use in Europe. Part 2: Demand management.* Environmental Issue Report No. 19. EEA. Kongens Nytorv 6. Copenhagen. DK-1050. 94 pp. Flörke, M., Alcamo, J., 2004. *Background document: water use and water stress outlook European outlook on Water Use,* Center for Environmental Systems Research – University of Kassel, Final Report, EEA/RNC/03/007. http://scenarios.ew.eea.europa.eu/reports/fol949029/fol040583

51 EEA, 2001.

52 EEA, 2001.

53 EEA, 2003: Indicator Fact Sheet (WQ06) Water use efficiency (in cities): leakage, version 01.10.2003 as quoted in Ecologic, 2007: *EU Water saving potential* (Part 1 –Report) ENV.D.2/ETU/2007/0001r. Ecologic, Institute for International and European Environmental Policy. Pfalzburger Str. 43–44, D – 10717 Berlin, www.ecologic.de

54 EEA, 2001.

55 EEA, 2001.

56 EC, 2007.

57 EC, 2007.

58 Arnell, N., Delaney, E. 2006: Adapting to climate change: Public water supply in England and Wales. *Climatic Change* 78, 227–255.

59   Kabat, P., van Schaik, H., 2003.

60   Global Water Partnership, 2000: SE -105 25 Stockholm, Sweden Technical Advisory Committee Background Paper No. 4. *Integrated Water Resources Management*, 71 pp. www.gwpforum.org/gwp/library/ Tacno4.pdf

61   Pahl-Wostl, C., Möltgen, J., Sendzimir, J., Kabat, P., 2005: *New methods for adaptive water management under uncertainty – The NeWater project*. EWRA Conference Paper. www.newater.info/downloadattachment/1133/74/ewra_ newater.pdf

6 Sea and coastline

Introduction

Chapter 5 examined how climate change is likely to alter the cycling of freshwater in Europe and how these changes will affect the availability of water and frequency of floods and droughts. This chapter moves from freshwater to seawater and explores the future state of Europe's lengthy coastline and adjacent seas. Climate change and its threat to the coast is a far from trivial matter since roughly one-third of Europe's population lives within 50 km of a major sea.[1] Nearly 100% of Denmark's population lives within this coastal strip and 75% of the populations of the United Kingdom and the Netherlands.[2]

Another way of looking at the vulnerability of Europe's coastline is to consider how much of it is below or close to sea level. The European Environment Agency

Life in Europe Under Climate Change, First Edition. Joseph Alcamo and Jørgen E. Olesen.
© 2012 Joseph Alcamo and Jørgen E. Olesen. Published 2012 by John Wiley & Sons, Ltd.

estimates that 9% of Europe's coastal zones (defined as a 10-km wide strip reaching inland from the sea) are 5 meters or more below sea level and therefore "potentially vulnerable" to sea-level rise and "related inundations".[3] In the most vulnerable category are the Netherlands, Belgium, Germany, Romania, Poland and Denmark. The most vulnerable stretches are the low river deltas and estuaries, coastal wetlands, low-lying coastal plains and the various islands and barrier formations scattered along the coastline.

Life along Europe's shoreline will certainly be different under climate change. The steadiness of the winds will change, becoming either more or less consistent. The level of the ocean will be noticeably higher, and walking along the shore one will see remnants of old buildings, wharfs and jetties surrounded by the rising sea. Some stretches of shoreline will be less frequently buffeted by coastal storms, and stories of devastating tempests will recede into the local folklore. But along other stretches of coast, strong storms will occur even more frequently than in the 20th century and will send waves higher than ever seen before crashing into the shoreline. Here the impact of storms will erode beaches and shoreline so quickly that the coastline will be several meters away from its 20th century location. Where the population has prepared itself for climate change, huge thick levees will defend the shoreline, at least for a time, and protect people and their buildings against the rising sea and stronger storms. The temperature of the ocean will be markedly warmer than in the 20th century and sea ice will form much less frequently. Some northern coastlines normally locked in by sea ice through much of the winter will be ice-free; ships will come and go throughout the entire year. Not only the temperature but even the chemical characteristics of the sea will change, as it becomes steadily more acidic, absorbing more and more carbon dioxide from the atmosphere. Many changes will occur, and these changes will force Europeans to forge a new relationship with their surrounding seas.

How much ocean warming?

As Chapter 2 shows, the oceans play an important role in the heat balance and climate system of the earth. By absorbing large amounts of heat and carbon dioxide they tend to slow down the impact of greenhouse gases on the atmosphere. But by absorbing heat the oceans have also become warmer. Over the 20th century the pace of warming in the world's oceans has generally followed the increasing trend in global air temperature with some regional variations, but generally at a slower pace than the increases in air temperature over land. Since 1950, average global sea surface temperatures have increased about 0.6 degrees Celsius.[4] While the North Atlantic and Mediterranean Sea have warmed by this amount, not all of Europe's seas have followed this pattern.[5] A warming tendency has not been apparent in either the Arctic or Norwegian Seas, probably because sea temperatures in this region are closely related to "local" climate variations such as the "North Atlantic Oscillation" and the "Arctic Oscillation" (see Chapter 2).[6]

Meanwhile, only a slight warming trend has been observed in the Baltic and North Seas since the 1980s (especially in winter).[7]

While these have been the trends up to now, computer models project significantly warmer ocean temperatures up to end of the 21st century, averaging about 1.5–2.6°C warmer than today (relative to 1980–99).[8] Again, this could vary substantially from sea to sea because of local influences on coastal temperature and other factors such as changes in ocean circulation patterns or run-off from cooler or warmer inland sources.

The consequences of ocean-warming

Rising temperatures will have various consequences on marine ecosystems along Europe's coastline. First of all, warmer temperatures will extend the growing periods and stimulate the growth rate of plankton (single-celled floating plants and animals) which make up most of the biomass of the ocean. A lengthening of the growing season of plankton and an increase in their annual production (the amount of biomass growing over a period of time) has already been observed over the last few decades in the North Atlantic and North Sea. Since the 1940s algal blooms (short-term huge increases in population) have been occurring more frequently during spring and autumn.[9]

While global warming undoubtedly played an important role in the observed increase of plankton, other factors were also important. For example, increasing quantities of nitrogen have been dumped into Europe's seas and coastal zones from its wastewater treatment plants, leached from agricultural fields, and deposited from the atmosphere from distant air pollution sources. This nitrogen has acted as a fertilizer to stimulate the growth of single-celled plants, "phytoplankton", which when combined with warmer temperatures and other factors, has led to the noticeable increase in plankton production.

Rising sea surface temperatures during the 21st century will further stimulate the growth rate of plankton and increase the length of their growing season. But the potential impacts of these changes on Europe's marine fisheries are not clear. On the one hand, more plankton means more food for fish and other marine organisms. On the other hand, higher plankton production could also come in the form of more frequent algal blooms. These algal blooms would have a devastating effect on fish and other organisms because algae die rapidly after they bloom and when they die their decomposition can deplete the dissolved oxygen over a large zone of the sea. Fish and other higher marine organisms depend on dissolved oxygen and cannot survive without it. Another possibility is that warmer temperatures will promote the growth of phytoplankton species that are toxic to fish and other marine organisms.[10]

The feeding of fish will also be affected by the migration of warm water species of aquatic organisms to higher and colder latitudes. A northward shift of some zooplankton species (free-floating microscopic animals which fish feed on) has been observed since the 1980s in the North Sea and southwest of the British Isles.[11] Warm-water fish species have also extended their range into this region[12] (see Chapter 7).

The warming of the seas and oceans will also have a more subtle but important effect on the global climate system. This effect has to do with the continuous exchange of CO_2 between the ocean and atmosphere. Over centuries this exchange was held roughly in balance until society began burning large quantities of fossil fuels which sent significant new amounts of CO_2 into the atmosphere. The oceans compensated by increasing their absorption of CO_2 from the atmosphere. Currently, the world's oceans take up more CO_2 than they exhale back to the atmosphere. By storing CO_2 (serving as a "sink" of atmospheric CO_2) the ocean slows the atmospheric build-up of this greenhouse gas and likewise dampens the tempo of climate change. But the warming of the ocean will hamper its ability to absorb carbon dioxide so that during the course of the 21st century it will become a much less effective sink for CO_2 from the atmosphere.[13]

Acidification of the ocean and seas

When it is said that the ocean "absorbs" CO_2 what this means is that the gas dissolves in seawater and combines chemically with other substances. The main form it takes is a weak solution of carbonic acid. Since the ocean has absorbed CO_2 at an increasing rate over the 20th century the concentration of carbonic acid has also become steadily stronger and the oceans more acidic. An indication of the increasing acidity is that average pH has dropped by about 0.1 units over this period (from its level of pH = 8.2 which it has maintained since the last ice age). This may not seem like much, but it represents a 26% increase in hydrogen ion concentration, or a noteworthy elevation in acidity. As the concentration of CO_2 continues to rise during the 21st century, the acidification of the ocean is expected to accelerate and pH is likely to drop by an additional 0.3 to 0.4 units.[14] The consequences of this higher acidity could be the disruption of the various life processes of many different marine organisms, but these are very uncertain conjectures.[15]

The ocean overflowing: sea-level rise

There is an important connection between global warming and the seas bordering Europe which has critical implications for the future of Europe's coastlines. As the atmosphere heats up, the upper layers of the ocean also become warmer and eventually this heat will be transmitted down to deeper layers of the ocean. Since water expands when heated, the volume of ocean water also increases, raising global average sea levels.[16] Adding to this effect will be the future melting of

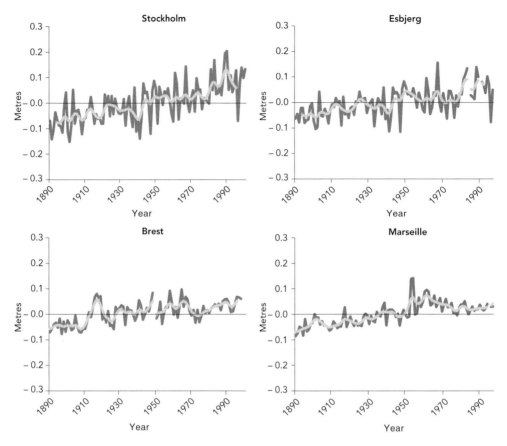

Figure 6.1 Sea-level rise at four gauging stations. Source: Liebsch et al.[20] Reprinted with permission from European Environmental Agency and Technische Universität Dresden.

mountain glaciers and parts of the Greenland, Antarctic and other ice caps caused by warmer air temperatures.

Looking at long-term coastal records, it becomes clear that sea-level rise is already a reality. During the 20th century the level of the sea rose by a global average of about 17 mm per decade.[17] Based on measurements at various gauging stations in Europe (see Figure 6.1), sea levels rose during this period by about 8 to 30 mm per decade.[18] Part of this rise can be attributed to global warming because ocean temperatures rose over the 20th century and hence some thermal expansion of the oceans has already taken place. But glaciers have also begun melting, as shown in Chapter 2, and much of the melt water has made its way into the oceans. It is important to know that part of the rise observed up to now in Europe has to do with factors other than climate change. Some stretches of coastline have experienced changes in prevailing winds and ocean currents which have brought masses of water onshore and, over time, raised the average sea level. At other locations the continental landmass is subsiding (sinking relative to the mean sea level) because of changes in the vertical position of the continents. As the landmass subsides it appears that sea level is rising. On the other hand, in the north and other parts of Europe, the landmass has been doing the opposite of subsiding: it has been springing upwards ever since being released from the tremendous weight of ice sheets at the end of the last ice age. This uplift actually compensates for sea-level rise. But it is worth noting that this uplift is slowing down in places, and in coming decades will compensate less and less for the rise of the ocean.[19]

While past sea-level rise has been slow and can be attributed to a variety of factors, it seems likely that it will be much more rapid over the coming decades and centuries, and will be strongly driven by climate change. The IPCC in 2007 projected an increase of around 19 to 58 cm in global average sea level during the 21st century.[21] These projections do not include rising sea levels from melting and mass loss from the ice caps on Greenland and Antarctica. Although the IPCC authors were aware of publications relating to recent changes in Greenland and Antarctica ice balance they could not extrapolate meaningfully from these data to future sea-level rise. Such projections are now available and they show sea-level rises by 2100 of 50–140 cm.[22] When the very recent increases in glacier discharge rates in Greenland are included, global sea level rise is projected as 80–200 cm.[23] Recent data thus consistently indicate that the ice caps on Greenland and Antarctica even in this century could become a major cause of sea level rise, in addition to the rising sea level from expansion of water volume at higher temperatures.

Because, as pointed out above, sea level rise is also affected by other more local factors, such as the persistence of onshore winds and subsidence of the coastline, the average rise in sea level in some parts of Europe could conceivably be 50% greater than the global average.[24] However, if a major part of the sea level rise comes from mass loss from the Greenland ice cap, sea level rises in Europe could be smaller than global values, owing to mass attraction effects. Since there is a long lag time in the response of the oceans to global warming, sea level will continue to rise well beyond the time when air temperature stops rising. A best estimate is that mean global sea level will continue to rise during the 21st century by up to 2.4 times faster than it did during the 20th century.[25] Taking all of these factors into account, we might expect the level of the sea in Europe this century to rise by a minimum of 20 cm (about half way up to the knees of an average adult) or at the most, more than one meter. Regardless of how much exactly, a continually rising sea level will be a fact of life for the remainder of this century and beyond for those living along European shorelines.

Coastal flooding and other impacts of sea-level rise

The steady rise in sea level will have many spin-off effects on Europe's coastlines. The risk of storm surges will increase, especially in areas where the tidal range is low, where the coast is subsiding, or where tectonic activity is going on, as along the Mediterranean and Black Seas.[26] As the sea level gets steadily higher, continuous wave and wind action will push beaches and low-lying sedimentary coasts gradually inland,[27] that is, unless they are continuously replenished and raised artificially. The retreat rate of beaches, which is already about one-half to one meter per year for particularly stormy stretches of the Atlantic coast, is likely to accelerate because of sea level rise.[28]

Over the past century, 67% of the low-water mark of the eastern coastline of the United Kingdom has retreated landward.[29] In fact, most of the sandy coastlines of the world have moved inland since the beginning of the 20th century.[30] Yet it is not clear how much of past beach retreat was caused by sea level rise rather than other factors. One other factor is upstream channelization and damming of rivers which has greatly reduced the supply of sediments delivered to coastal waterways. Other factors include commercial sand mining, which has reduced the supply of sand, and drainage of coastal wetlands which has disrupted coastal circulation patterns.

Under future climate change, it is not only sea level rise that will be responsible for carving away Europe's coasts. In a typical winter the cover of sea ice along the Baltic and Arctic seas prevents waves from reaching the coastline and in this way protects it from erosion in winter. Under climate change, sea temperatures will rise and sea ice will become rare or disappear entirely, changing the character of the winter shoreline. Waves will reach the shoreline during all seasons and coastal erosion will quicken its pace.[31] The ice cover of the Baltic has already noticeably decreased over the past few decades.[32]

Sea level rise will also have two particularly important impacts on water supply. First, as the level of the sea rises seawater will seep into freshwater aquifers lying along the coast ("saltwater intrusion") and contaminate the water supply for communities relying on this groundwater. At greatest risk will be low-lying coastal areas with shallow aquifers.[33] Second, sea level rise will thrust the saltwater from the ocean further upstream into river estuaries, and this could bring saltwater to the vicinity of the intake pipes for the water supply to municipalities and industries. The consequence could be undrinkable water and unsuitable industrial water supply.

The stretches of coastline most vulnerable to sea level rise and coastal flooding will be those with low-lying, small tidal ranges and high population densities (Figure 6.2). In one estimate, a 34 cm average sea level rise in Europe by the 2080s would put a further 2 million people at risk from coastal flooding.[34] This estimate assumes improved coastal protection against current climatic threats but not against sea level rise. Assuming a higher but still highly feasible sea level rise of 1 m, the threat to people and their property could be catastrophic in Europe, unless actions are taken to adapt to this threat. Though not life-threatening, the retreat of beaches will obviously affect the tourism industries and force people to find other sources of recreation.

Flooding and inland migration of the coastline will also leave its mark on natural areas. One estimate is that 20% of Europe's wetland areas could disappear

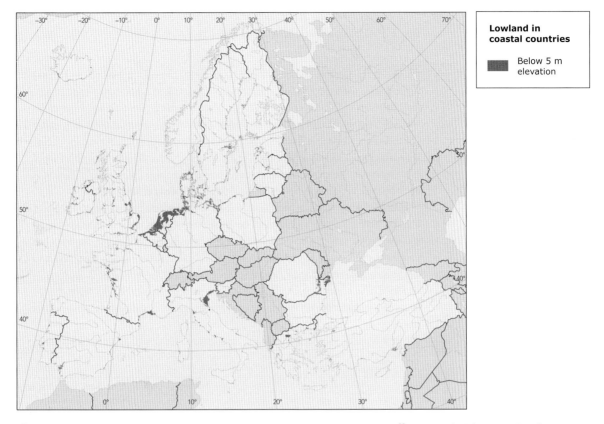

Figure 6.2 Coastal lowlands (elevation below 5 meters) in Europe. Source: EEA.[35] Reprinted with permission from European Environmental Agency, Copenhagen. (See color plate.)

because of sea level rise by the 2080s.[36] Also vanishing will be the unique plant and animal communities that depend on these wetlands (See Box 6.1).

As the threat to people, buildings and livelihoods gets greater, the costs of increased coastal protection will become an ever greater economic burden.[40]

Storms and waves

While society has certainly made an imprint on Europe's coastlines with its buildings, levees and other edifices, the forces of nature continue to be a more important factor in shaping the coastline. Hence, to anticipate the future condition of the coastlines we need to foresee changes in nature's forces such as the persistent waves that lap onto the shore, the patterns of the winds that form its dunes, and the occurrence of occasional but severe storms that inundate and reshape stretches of the coastline.

Box 6.1 Ecological impacts of sea level rise

For centuries steady development has destroyed the habitat of flora and fauna along Europe's coastline. Now climate change is poised to further undermine these ecosystems. As the sea rises, more saltwater will be forced into estuaries and this will tend to displace existing plant and animal communities further inland. These communities might persist if their migration route upstream is not blocked by urban areas or infrastructure, and if the rate of change is not too rapid.[37] Alternatively, freshwater flows into the estuary might increase, in which case average salinity levels would decrease, and aquatic ecosystems would this time be displaced seaward.

Sea level rise will not necessarily lead to a decline in salt marshes, especially where tides are significant, because they can maintain their elevation relative to the sea by depositing sediments on their bed.[38] However, in other cases coastal wetlands will be "squeezed" between the encroaching sea and human development along the coastline and are likely to shrink in size.

The disruption of plant and animal communities caused by either changing salinity in estuaries or physical elimination by coastal squeeze can have unforeseen consequences on coastal ecosystems, especially if they involve the removal of "keystone species". These are the plant or animal varieties that play a key role in the food web or nutrient cycles of an ecosystem.[39]

It may turn out that slowly rising sea level will not pose the main danger to the coast, but that it will come from the passage of occasional but severe storms along the coast. Simulations with climate models for various greenhouse gas emission scenarios show an increase in storm intensity and wind speeds in the northeastern Atlantic in the coming decades of this century (2010–30) with the centers of storms moving closer to the coasts of Europe.[41] More severe storms usually bring higher waves, and by the 2080s average wave heights in the northeastern Atlantic could be, on average, 40 cm higher than now.[42]

Although more frequent storms may occur locally over parts of the Black Sea, Aegean Sea and Adriatic Seas, other shifts in climate patterns over Europe can also lead to a softer impact on shorelines. For example, a change in large-scale weather patterns may lead to less storminess and lower average wind intensity over much of the eastern Mediterranean.[43]

The risk of flooding is particularly high when storm surges coincide with high tides, and climate change will contribute to this by affecting both the overall sea level and the frequency and severity of storm surges (Box 6.2). Climate change impact analyses show that under future climate conditions storm surge extremes may increase along the North Sea coast during the 21st century.[44] This is an area with a large population living within the coastal zone and much of the coastal defenses are designed to protect against such storm surges. The combination of

---

Box 6.2 The Thames barrier

The Thames barrier was built in response to the catastrophic floods of 1953 (which killed 1835 people) and was finally ready for use in 1982. It is the world's second largest movable flood barrier. Its purpose is to prevent London from being flooded by an exceptionally high tide moving up from the sea, often exacerbated by a storm surge. A storm surge generated by low pressure in the Atlantic Ocean may sometimes track eastwards north of Scotland and down the North Sea which narrows towards the English Channel and the Thames Estuary. Dangerously high water levels can occur in the Thames Estuary when such a storm surge coincides with high tide.

The Thames barrier protects $150 \, km^2$ of London and property worth at least £80 billion.[45] It was built to withstand a 1 in 2,000-year flood. With increased sea level, due to global warming, this risk by 2030 has been projected to increase to a 1 in 1,000-year event. However, this does not account for the recently observed increase in rate of sea level rise. For example, between 1982 and 2001 it was closed 63 times. In the winter 2000/1, it was closed 24 times. In 2003, the barrier was closed for 14 consecutive tides, and in November 2007 it was closed twice for a storm surge the same size as the one that occurred in 1953. The UK Environment Agency now has plans for coping with significant sea level rise, including a prospective plan for a 4.5 m rise, which would involve a 16 km barrier between Essex and Kent.

---

rising sea levels and more severe storm surges puts considerable stress on the current coastal defenses, and most need to be considerably improved to deal with the higher risks of flooding.

Coastal erosion is the wearing away of land, beaches or dunes from the action of waves, tidal and wave currents or drainage. Waves cause particularly severe erosion because of the large amount of energy contained in high waves. The erosion is often just a redistribution of sediment along a coastline. However, this may change with sea level rise, where coastal erosion will affect different or larger parts of the coastline owing to new exposure to the erosion processes. With more intense storms waves will become higher and this will then exacerbate the erosion processes along all shorelines. The combination of higher sea levels and more intense storms under climate change will therefore accelerate coastal erosion along European coasts.

Natural adaptation and natural thresholds

While coastlines seem to be vulnerable to the coastal storms, they are – within limits –relatively robust to the high winds and waves accompanying these storms.

This is confirmed by the quick recovery of a beach or other stretch of shoreline shortly after a storm passes. For example, if part of a shoreline is scoured out by heavy waves during a storm, it is common that currents will quickly redeposit sediment to fill in the gap after the storm passes. These are called "morphological adjustments", and they occur naturally throughout the world. If sea level and other climate changes happen slowly, then some salt marshes and lagoons will be able to adjust naturally by depositing new sediments. But only to a point. Estuaries and coastlines have thresholds above which they can no longer adjust to the disturbances caused by storms and other external forces. Exceeding these thresholds leads to an irreversible process of submergence.[46] It has been estimated that such a threshold exists for the tidal salt flats known as the Wadden Sea, along the Netherlands, Germany and Denmark. Experts on the physical processes of coasts estimate that a sea level rise greater than 10 mm per year would lead to permanent submergence of this ecologically important saltwater tidal area.[47]

Options for adapting to coastal impacts of climate change

While the coastline might have some capacity to self-adapt to the initial effects of climate change, it is likely that this capacity will be quickly overcome by the pace of climate change. For this reason it is imperative for society to actively begin adapting to changing conditions.

What are the options for dealing with changing sea level, higher risk of coastal flooding and other changes along the coastline? Since the beginning of the 1990s[48] public discussion has centered around three main options:

1 *Protect* – Hold the line by strengthening flood protection structures or building new ones.
2 *Accommodate* – For example, improve the resilience of buildings and other infrastructure along the coast.
3 *Retreat* – Adapt to coastal inundation by moving people away from the coastline and re-establishing wetlands and other ecosystems at other locations.

These options have been elaborated over the past years as shown in Figure 6.3.

Protecting the coast

Many European countries have begun to factor in climate change in their coastal protection plans. As the levees protecting 65 km of Belgium's coastline are gradually renovated, they are built to withstand a 60 cm rise in sea level.[50] In 1999 the United Kingdom's Department for Environment, Food and Rural Affairs specified

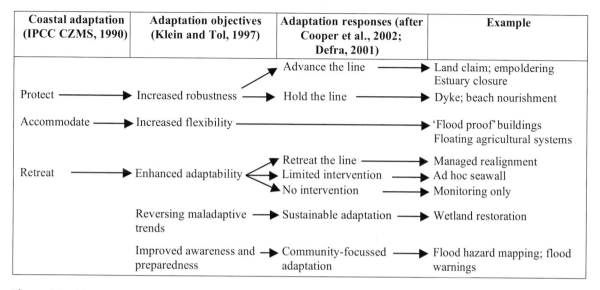

Coastal adaptation (IPCC CZMS, 1990)	Adaptation objectives (Klein and Tol, 1997)	Adaptation responses (after Cooper et al., 2002; Defra, 2001)	Example
Protect	Increased robustness	Advance the line	Land claim; empoldering Estuary closure
		Hold the line	Dyke; beach nourishment
Accommodate	Increased flexibility		'Flood proof' buildings Floating agricultural systems
Retreat	Enhanced adaptability	Retreat the line	Managed realignment
		Limited intervention	Ad hoc seawall
		No intervention	Monitoring only
	Reversing maladaptive trends	Sustainable adaptation	Wetland restoration
	Improved awareness and preparedness	Community-focussed adaptation	Flood hazard mapping; flood warnings

Figure 6.3 Elaboration of options for adapting to climate change in the coastal zone. From Nicholls et al.[49] Reprinted with permission from Intergovernmental Panel on Climate Change, Geneva.

that new coastal flood protection structures should allow for a rise of 4.5 mm per year in mean sea level over the next 40 to 50 years.[51] In Germany, coastal protection structures in Mecklenburg-West Pomerania take into account a 25–30 cm rise in mean sea level over 100 years and in Lower Saxony an extra 60 cm.[52]

Natural features such as dunes and beaches also play an important role as a first line of defense against coastal flooding. Hence their replenishment qualifies as an adaptation measure against climate change. Belgium has had a policy of replenishing sand on its beaches since 1960 and this approach is also becoming more popular in Denmark. The level of the beach at the major tourist city of Ostend was recently raised to provide at least temporary protection for the city's inhabitants.

As the threat of coastal and river flooding increases, engineers are beginning to question the resilience of conventional dikes. This is because the pressure and scavenging action of a major coastal or river flood can quickly erode and undermine a dike. A single breach in the line of defense, as in New Orleans during Hurricane Katrina in 2006, is enough to submerge an enormous inland area. A safer, but certainly more expensive strategy, is to build "super-dikes", appropriately named because of their enormous size compared to conventional barriers (Box 6.3).

While the strengthening of barriers against the sea certainly reduces society's vulnerability to coastal flooding it comes at a price to natural ecosystems. Building levees and sea walls along the coast disrupts or destroys the plant and animal communities located there. They also break the connection between marine and terrestrial ecosystems, and greatly alter the run-off patterns from inland that determine the rate and patterns of sedimentation along the shoreline, among other impacts.

A major challenge with improving current coastal defenses or building new ones is the sheer scale of the task. Such mega-scale investments not only require massive funding, but the time required for planning such huge structures is also very long since considerations needs to be made for issues such as property loss, livelihoods, nature habitats and of course investment costs. In many cases it takes about 30 years to plan and build such large infrastructures in modern democracies. This needs to be considered alongside the timing of

---

Box 6.3 Super adaptation with super dikes

"Super-dikes", or "super levees" are so named because of their extreme thickness (300–500 meters). One of the faults of conventional dikes is that they can be damaged or can fail under heavy water pressure by being overtopped by floodwaters or by being undermined by water seepage. Super dikes are built so wide that they can be overtopped without collapsing and are so broad that seepage and underflows are not expected to be a major problem. In Japan another motivation for building super wide levees is that they are resistant to the frequent earthquakes experienced in that country. Some super-dikes have such a gentle inner slope that parkland and buildings can be constructed on top of them, giving city residents access to the waterside (see picture below). These massive structures are already in place in Japanese cities such as Nagoya, Osaka, and Tokyo.[53]

The costs of these installations are also in the "super" category, reaching US$100 million per kilometer as compared to roughly US$1.5–7.5 million per kilometer for restoring or building new conventional dikes, sea walls and other sea protection structures in Europe.[54]

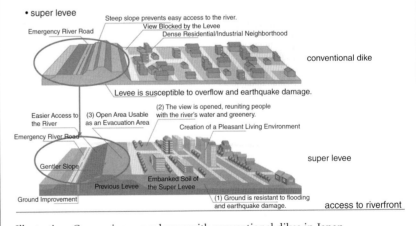

Illustration: Comparing super levees with conventional dikes in Japan.
Source: Stalenberg.[55] Reprinted with permission from Ministry of Land, Infrastructure, Transport and Tourism, Kanto Regional Development Bureau, Japan.

climate change, which under some of the scenarios will put several of the current European coastal defenses in jeopardy within such a time horizon. Planning therefore needs to start now to provide the necessary coastal defenses for the 21st century.

Accommodating to climate change

Thorough accommodation steps have been taken to bolster the resilience of structures to coastal flooding, for example by building coastal roads and bridges higher. More ambitious actions include constructing buildings on "flood-proof" concrete pillars or various flotation devices. The acceptance of sea level rise and recurrent coastal flooding in the Netherlands has prompted a host of innovative designs for floating homes and buildings. Among the best known examples are designs by the Dutch architectural firm Dura Vermeer which has constructed a group of homes on the Maas River in the Netherlands with foundations that float when the river reaches flood stage (Figure 6.4). Even while the homes float, they remain connected to sewage and electricity systems via flexible pipes.

Retreating inland

While the preferred strategies are to protect or accommodate coastlines to climate risks, under some circumstances these measures are too costly and the best

Figure 6.4 Accommodating to climate change on the coastline: "Floatable" houses in Maasbommel on the Maas River in the Netherlands designed by architects Dura Vermeer. The houses are set on foundations that float when the river reaches flood stage. Photo: Dura Vermeer. Reprinted with permission from Dura Vermeer Groep NV.

policy is to move people and their infrastructure safely inland. "Full retreat" – giving land back to the sea – is happening on a small scale in Great Britain and Denmark[56] but of course is a difficult option considering how densely most of Europe's coastline is populated. The Environment Agency in the UK calls this alternative "managed realignment" and lists various options for carrying it out including: retreating to higher ground, or building a back-up line of defense, such as a sea wall somewhat inland from the current shoreline.[57] An orderly retreat from the shoreline can also be achieved through various community policies such as:[58]

- *setbacks* which require that new developments along the coastline have to be located a minimum distance from the shore;
- *rolling easements*, which are provisional approvals of development under the stipulation that the building or other structure will eventually be removed to allow natural features to develop;
- *density restrictions* that place an upper limit on the number of people or types of activities permitted along the coastline.

"Retreating inland" has taken on a new and more positive meaning over the past few years because it is viewed as an opportunity for reversing the destruction of Europe's coastal ecosystems and a way of restoring its natural ecosystems (see Chapter 7).[59]

Integrated coastal management

While there are many options available to communities for adapting their shorelines to climate change, the question arises, how should they decide among the different options? Experience has shown that the least expensive and most efficient way of proceeding is to embed these adaptation options into a larger strategy for coastal development and conservation called "Integrated Coastal Zone Management" (ICZM). The European Environment Agency defines ICZM as:

> a dynamic, multidisciplinary and iterative process to promote sustainable management of coastal zones. It covers the full cycle of information collection, planning (in its broadest sense), decision making, management and monitoring of implementation. ICZM seeks, over the long-term, to balance environmental, economic, social, cultural and recreational objectives . . . 'Integrated' in ICZM refers to the integration of objectives and . . . the many instruments needed to meet these objectives . . . It [also] means integration of the terrestrial and marine components of the target territory, in both time and space.

From its definition and basic principles (Box 6.4), it is clear that ICZM is a noble effort of its proponents to reconcile the host of human and conservation interests

---

**Box 6.4 The principles of integrated coastal zone management (ICZM)
Source: European Commission[62]**

- A broad holistic perspective (thematic and geographic).
- A long-term perspective.
- Adaptive management.
- Local specificity.
- Working with natural processes and respecting the carrying capacity of ecosystems.
- Participatory planning.
- Support and involvement of relevant administrative bodies.
- Use of a combination of instruments.

---

that converge on the coastlines of Europe. On one hand, the natural features of the coast – the estuaries, wetlands, beaches and dunes – provide important ecosystem services such as the recycling of nutrients, the maintenance of the physical integrity of the coastline, and production of marine food resources. On the other hand, the harbors and beaches along Europe's coastline are among the most economically important and desirable locations for tourism, trade, and industrial development. With such a clash of interests it is not surprising that the consensus-type approach of ICZM has not always been successful. Reflecting on it's experience, the UK Department of Environment, Food, and Rural Affairs summarized those aspects of ICZM that have worked and those that have not worked so well:[60]

- mixed picture in implementing ICZM principles;
- current framework still sectoral;
- strong on local voluntary action;
- strong on involving stakeholders;
- weak on integrated long-term planning;
- weak on ICZM leadership and communication;
- weak on funding for ICZM initiatives.

Nevertheless, ICZM has been accepted as a good approach for reconciling interests along European coasts. Recognizing this, the European Union has forged ahead in making it a keystone of European policy. In 2002 the EU Council and Parliament adopted an earlier European Commission recommendation[61] that Member States set up national strategies for ICZM and involve all coastal stakeholders by March 2006.

In the background to these political developments, climate change will slowly but surely makes the clash of interests along Europe's coastlines one step more complicated since it will have a profound effect on both nature and humans.

Despite its weaknesses, ICZM will be needed more than ever as we adapt to life in Europe under climate change.

The longer perspective

Sea level rise caused by global warming will not cease at the end of the 21st Century. It will continue for many more centuries, even with a stabilization of greenhouse gas emissions. On one hand, the rise of sea level will be gradual, allowing time for society to adapt. On the other, even with the time available for adaptation, the challenges will be huge to protect Europe's coastlines.

Notes

1 EEA, 2006: *Vulnerability and adaptation to climate change in Europe EEA.* Technical report No 7/2005. 79 pp.

2 Nicholls, R., Klein, R., 2003: *Climate change and coastal management on Europe's Coast.* EVA Working Paper No. 3.

3 EEA, 2006.

4 Bindoff, N., Willebrand, J., Artale, V., Cazenave, A., Gregory, J., Gulev, S., Hanawa, K., le Quere, C., 2007: Observations: Oceanic climate change and sea level. In: *Climate Change 2007: The Physical Science Basis. Contributions of Working Group I to the Fourth Assessment Report of the Intergovernmental Panel on Climate Change,* Solomon, S., Quin, D., Manning, M. Chen, Z., Marquis, M. Averyt, K. Tignor, M., Miller, H. (eds) Cambridge University Press. Cambridge. UK, 385–482.

5 EEA, 2004: *Impacts of Europe's changing climate An indicator-based assessment.* EEA Report No 2/2004. 101 pp.

6 EEA, 2004.

7 EEA, 2004.

8 Meehl, G.A., Stocker, T.F., Collins, W.D., Friedlingstein, P., Gaye, A.T., Gregory, J.M., Kitoh, A., Knutti, R., Murphy, J.M., Noda, A., Raper, S.C.B., Watterson, I.G., Weaver, A.J., Zhao, Z.C., 2007: Global climate projections. In: *Climate change 2007: The physical science basis. Contribution of Working Group I to the Fourth Assessment Report of the Intergovernmental Panel on Climate Change,* Solomon, S., Qin, D., Manning, M., Chen, Z., Marquis, M., Averyt, K.B., Tignor, M., Miller, H.L. (eds). Cambridge University Press, Cambridge, United Kingdom and New York, NY, USA, p. 747–845.

9 EEA, 2004.

10 EEA, 2004.

11 Beaugrand, G., Ibañez, F., Lindley, J.A., Reid, P.C., 2002: Diversity of calanoid copepods in the North Atlantic and adjacent seas: species associations and biogeography. *Marine Ecology Progress Series* 232, 179–195.

12 Reid, P.C., Edwards, M., 2001: Long-term changes in the pelagos, benthos and fisheries of the North Sea. in: Kröncke, I., Türkay, M., Sündermann, J. (eds): Burning issues of North Sea ecology, Proceedings of the 14th International Senckenberg Conference 'North Sea 2000', *Senckenbergiana Maritima* 31, 107–115.

13 This effect will be compensated somewhat by the so-called "biological pump". As the ocean warms it will stimulate the production of phytoplankton, some of which also convert carbon dioxide into their carbonaceous shells. When the phytoplankton die, their carbonaceous shells sink to lower levels of the ocean, thereby effectively taking the carbon dioxide from the atmosphere with them. However, this "biological pump" will only partly compensate for the decline in carbon dioxide absorbing ability of the ocean.

14 Caldeira, K., Wickett, M., 2003: Anthropogenic carbon and ocean pH. *Nature* 425, 365.

15 Fischlin, A., Midgely, G., Price, J., Leemans, R., Gopal, B., Turley, C., Rounsevell, M., Dube, O., Tarazona, J., Velichko, A., 2007: Ecosystems, their properties, goods and services. In: *Climate Change 2007: Impacts, Adaptation and Vulnerability. Contribution of Working Group II to the Fourth Assessment Report of the Intergovernmental Panel on Climate Change*, Parry, M., Canziani, O., Palutikof, J., van der Linden, P., Hanson, C. (eds). Cambridge University Press. Cambridge. UK, 211–272.

16 In Scandinavia and some other locations, continents are slowly rising which compensates somewhat for the increase in the volume of seawater.

17 Bindoff, N. et al 2007. These estimates have an uncertainty of ±5 mm per decade.

18 Liebsch, G., Novotny, K., Dietrich, R., 2002: *Untersuchung von Pegelreihen zur Bestimmung der Änderung des mittleren Meeresspiegels an den europäischen Küsten*, Technische Universität Dresden (TUD), Germany. As quoted by: EEA (European Environment Agency). 2004: Impacts of Europe's changing climate An indicator-based assessment. EEA Report No. 2/2004. 101 pp.

19 Smith, D.E, Raper, S.B, Zerbini, S., Sánchez-Arcilla, A. (eds), 2000: *Sea level change and coastal processes: implications for Europe*. Office for Official Publications of the European Communities, Luxembourg, 247 pp.

20 Liebsch, G. et al., 2002. As cited in: EEA, 2004.

21 These are "best estimates" relative to the period 1980 to 1999 computed by models under climate scenarios based on the so-called "SRES" scenarios of the Intergovernmental Panel on Climate Change (see Chapter 1). These estimates are cited in Nicholls, R., Wong, P., Burkett, V., Codignotto, J., Hay, J., McLean, R., Ragoonaden, S., Woodroffe, C., 2007: Coastal systems and low-lying areas, *Climate Change 2007: Impacts, Adaptation and Vulnerability. Contribution of Working Group II to the Fourth Assessment Report of the Intergovernmental Panel on Climate Change*, Parry, M., Canziani, O., Palutikof, J., van der Linden, P., Hanson, C. (eds). Cambridge University Press. Cambridge. UK, 315–356.

22   Rahmstorf, S., 2007: A semi-empirical approach to projecting future sea-level rise. *Science* 19, 368–370.

23   Pfeffer, W.T., Harper, J.T., O'Neill, S., 2008: Kinematic constraints on glacier contributions to 21st century sea-level rise. *Science* 321, 1340–1343.

24   Woodworth, P.L., 2006: Some important issues to do with long-term sea level change. *Philosophical Transactions of The Royal Society A* 364, 787–803.

25   Nicholls, R., Wong, P., Burkett, V., Codignotto, J., Hay, J., McLean, R., Ragoonaden, S., Woodroffe, C., 2007: Coastal systems and low-lying areas. *Climate Change 2007: Impacts, Adaptation and Vulnerability. Contribution of Working Group II to the Fourth Assessment Report of the Intergovernmental Panel on Climate Change*, Parry, M., Canziani, O., Palutikof, J., van der Linden, P., Hanson, C. (eds). Cambridge University Press. Cambridge. UK, 315–356.

26   Gregory, J.M., Church, J.E., Boer, G.J., Dixon, K.W., Flato, G.M., Jackett, D.R., Lowe, J.A., O'Farrell, S.P., Roekner, E., Russell, G.L., Stouffer, R.J., Winton, M., 2001: Comparison of results from several AOGCMs for global and regional sea-level change 1900–2100. *Climate Dynamics* 18, 225–240.

27   Stone, G.W., Orford, J.D. (eds), 2004: Storms and their significance in coastal morpho-sedimentary dynamics. *Marine Geology* 210, 1–365.

28   Cooper, J.A.G., Pilkey, O.H., 2004: Sea-level rise and shoreline retreat: time to abandon the Bruun Rule. *Global and Planetary Change* 43, 157–171.

29   Taylor, J., Murdock, A., Pontee, N., 2004: A macro scale analysis of coastal steepening around the coast of England and Wales. *Geography Journal* 170, 179–188.

30   Nicholls, R. et al., 2007.

31   Kont, A., Jaagus, J., Aunap, R., Ratasa, U., Rivisa, R., 2008: Implications of sea-level rise for Estonia. *Journal of Coastal Research* 24, 423–431.

32   Jevregjeva, S, Drabkin,V., Kostjukov, J., Lebedev. A., Lepparanta, M., Mironomv, Y., Schmelzer, N., Sztobryn, M., 2004: Baltic Sea ice seasons in the twentieth century. *Climate Research* 25, 217–227.

33   Arnell, N.W., Livermore, M.J.L., Kovats, S., Levy, P.E., Nicholls, R., Parry, M.L., Gaffin, S.R., 2004: Climate and socio-economic scenarios for global-scale climate change impacts and assessments: characterising the SRES storylines. *Global Environmental Change* 14, 3–20.

34   Nicholls, R. et al., 2007.

35   EEA, 2006.

36   Devoy, R.J.N., 2008: Coastal vulnerability and the implications of sea-level rise for Ireland. *Journal of Coastal Research* 24, 325–342. Nicholls, R.J., 2004: Coastal flooding and wetland loss in the 21st century: changes under the SRES climate and socio-economic scenarios. *Global Environmental Change* 14, 3–20.

37   Nicholls, R. et al., 2007.

38   Cahoon, D.R., Hensel, P.F., Spencer, T., Reed, D.J., McKee, K.L., Saintilan, N., 2006: Coastal wetland vulnerability to relative sea-level rise: wetland eleva-

tion trends and process controls. Wetlands as a Natural Resource, Vol. 1: *Wetlands and Natural Resource Management*, Verhoeven, J., Whigham, D., Bobbink, R., Beltman, B. (eds), Springer Ecological Studies series, Chapter 12.

39 Kaufman, L.S., Dayton, P.J., 1997: Impacts of marine resource extraction on ecosystem services and sustainability. In: *Nature's Services: Societal Dependence on Natural Ecosystems*, G. Daily (ed.), Island Press, Washington, D.C. (USA), p. 275–293.

40 de Groot, Th.A.M., Orford, J.D., 2000: Implications for coastal zone Management. In: *Sea Level Change and Coastal Processes: Implications for Europe*, Smith, D.E, Raper, S.B., Zerbini, S., Sánchez-Arcilla, A. (eds). Office for Official Publications of the European Communities, Luxembourg, 214–242. Tol, R.S.J., 2002: Estimates of the damage costs of climate change, Parts 1 & 2. *Environmental Resource Economics* 21, 47–73 & 135–160.

41 Lozano, I., Devoy, R.J.N., May, W., Andersen, U., 2004: Storminess and vulnerability along the Atlantic coastlines of Europe: analysis of storm records and of a greenhouse gases induced climate scenario. *Marine Geology* 210, 205–225.

42 Tsimplis, M.N., Woolf, D.K., Osbourn, T.J., Wakelin, S., Wolf, J., Flather, R., Woodworth, P., Shaw, A.G.P., Challenor, P., Yan, Z., 2004: Future changes of sea level and wave heights at the northern European coasts. *Geophysical Research Abstracts* 6, 00332.

43 Busuioc, A., 2001: Large-scale mechanisms influencing the winter romanian climate variability. In: *Detecting and Modelling Regional Climate Change*, Bruner, M. and D. Lopez (eds), Spinger-Verlag, Berlin, 333–344.

44 Woth, K., Weisse, R., von Storch, H., 2006: Climate change and North Sea storm surge extremes: an ensemble study of storm surge extremes expected in a changed climate projected by four different regional climate models. *Ocean Dynamics* 56, 3–15.

45 Crichton, D., 2007: What can cities do increase resilience? *Philosophical Transactions of The Royal Society A* 365, 2731–2739.

46 Nicholls, R. et al., 2007.

47 van Goor, M., Zitman, T., Wang, Z., Stive, M., 2003: Impact of sea level rise on the morphological equilibrium state of tidal inlets. *Marine Geology* 202, 211.227.

48 IPCC – CZMS (Coastal Zone Management Subgroup). 1990: *Strategies for adaptation to sea level rise. Report of the Coastal Zone Management Subgroup, Response Strategies Working Group*. Ministry of Transport, Public Works, and Water Management, the Netherlands. 122 pp.

49 Nicholls, R. et al., 2007: Citations in Figure 6.3: IPCC CZMS, 1990: *Strategies for adaptation to sea-level rise. Report of the Coastal Zone Management Subgroup, Response Strategies Working Group of the Intergovernmental Panel on Climate Change. Ministry of Transport, Public Works and Water Management (the Netherlands)*, x+122 pp.; Klein, R.J.T., Tol, R.S.J., 1997: *Adaptation to*

climate change: Options and technologies – An overview paper. Technical Paper FCCC/TP/1997/3. UNFCC Secretariat; Cooper, N., Barber, P.C., Bray, M.C., Carter, D.J., 2002. *Shoreline management plans: a national review and an engineering perspective*. Proceedings of the Institution of Civil Engineers, Water and maritime engineering 154, 221–228.

50 EEA, 2007: *Climate change and water adaptation issues*. EEA Technical Report No 2/2007. 110 pp.

51 Quoted in EEA, 2006: *Vulnerability and adaptation to climate change in Europe*. EEA Technical report No. 7/2005. 79 pp.

52 Quoted in EEA, 2006.

53 Stalenberg, B., Vrijling, H., Kikumori, Y., 2008: Japanese lessons for Dutch urban flood management. Conference on "Water Down Under" (15–17 April). Adelaide, Australia

54 Estimates of costs for "super-dikes": Kundzewicz, Z. 2004: Floods and flood protection: Business as usual? In: Rodda, J., Ubbertini, L. (eds) *The basis of civilization – Water science?* IAHS (International Association of Hydrological Sciences) Publication 286. 201–209. Costs of new dikes along the Elbe River: 1.4 to 1.7 million euro per kilometer (2.1 to 2.6 million US\$, at 1 euro = US\$1.5): Meyerhoff, J., Dehnhardt, A., *The European Water Framework Directive and Economic Valuation of Wetlands*. Working Paper on Management in Environmental Planning 11/2004 Institute for Landscape Architecture and Environmental Planning. Technical University of Berlin. Franklinstraße 28/29. D- 10587 Berlin. Costs of improving dikes in Netherlands: Average: 5 million euro per km (7.5 million US\$, at 1 euro = 1.5 US\$) from: van der Vlies, A., Stoutjesdijk, K., Waals, H. Effects of climate change on water management in the Netherlands. Water authority Hollandse Delta www.riool.net/riool/binary/retrieve (retrieved 24 May, 2008). Costs of various coastal defenses in England and Wales (excluding replenishment of dunes and breakwaters): 970,000 to 4.7 million euros (US\$1.5–7.1 million, at 1 euro = US\$1.5).

55 Stalenberg, B. 2007: *River basin management – Japanese state of the art*. Presentation at Technical University, Delft.

56 Safe Coast, 2007: *Quick scan – climate change adaptation – with a focus on coastal defense policies in five North Sea countries*. www.safecoast.nl/editor/databank/File/Quick_scan_climate_change_adaptation_SC_template.pdf (retrieved 17 September, 2008).

57 UK Environment Agency. www.intertidalmanagement.co.uk/contents/index.htm

58 McLean, R., Tsyban, A., Burkett, V., Codignotto, J., Forbes, D., Mimura, N., Beamish, R., Ittekkot, V., 2001: Chapter 6. Coastal Zones and Marine Ecosystems. Climate Change 2001: Impacts, Adaptation, and Vulnerability. McCarthy, J., Canziani, O., Leary, N., Dokken, D., White, K. (eds) www.grida.no/publications/other/ipcc_tar/

59 DEFRA (Department of Environment, Food, and Rural Affairs). 2008: www.intertidalmanagement.co.uk/contents/index.htm

60   DEFRA, 2004: *ICZM in the UK: A Stocktake Management*. www.defra.gov.uk/
      environment/water/smarine/uk/iczm/stocktake/

61   European Commission. 2000: Commission proposal for a European Parlia-
      ment and Council Recommendation concerning the implementation of Inte-
      grated Coastal Zone Management in Europe (COM/2000/545), adopted 8
      September, 2000. http://ec.europa.eu/environment/iczm/proprec.htm

62   European Commission. 2000.

7 Ecosystems and biodiversity

Life in Europe Under Climate Change, First Edition. Joseph Alcamo and Jørgen E. Olesen.
© 2012 Joseph Alcamo and Jørgen E. Olesen. Published 2012 by John Wiley & Sons, Ltd.

Introduction

Biological diversity – or biodiversity – is one of the key terms in conservation, encompassing the richness of life and the diverse patterns it forms. The Convention on Biological Diversity (CBD) defines biological diversity as "the variability among living organisms from all sources including, inter alia, terrestrial, marine and other aquatic ecosystems and the ecological complexes of which they are part; this includes diversity within species, between species and of ecosystems".

The diversity is often understood in terms of the wide variety of plants, animals and micro-organisms. The number of species known is about 1.75 million, but the total number of species on earth may be as high as 13 million, most of which are small creatures such as micro-organisms and insects. Biodiversity also includes the genetic variation within species.[1] Yet another aspect of biodiversity is the variety of ecosystems such as those that occur in deserts, forests, wetlands, mountains, oceans, lakes, rivers and agricultural landscapes. In each ecosystem a variety of living creatures interact with each other and with the surrounding air, water and soil. It is this interaction that makes earth a habitable place, and therefore a primary outcome of biodiversity is to provide a large number of goods and services for sustaining human and other life on earth.

Biodiversity matters for ethical, environmental and economic reasons. Biodiversity has intrinsic value, and ecosystems provide spiritual and aesthetic experiences. Ecosystems offer outstanding opportunities for recreation, both in daily life and during vacations. They clean our water, purify our air and maintain our soils. They regulate the climate, recycle nutrients and provide food, fiber and energy. They provide raw materials and resources for medicines and other purposes. All of these services, products and values are difficult to express in common financial terms. Hence, when faced with threats to biodiversity, it is difficult (but not impossible) for scientists to argue for preserving biodiversity because of its economic worth.

Human well-being is dependent on ecosystem services provided by nature, such as water and air purification, fish production, timber production and nutrient cycling. Many of these "free" services have no markets and no prices, so their loss often is not detected by our current economic incentive system and can thus continue unabated. A variety of pressures resulting from population growth, changing diets, urbanization, climate change and many other factors is causing biodiversity to decline, and ecosystems are continuously being degraded. The world's poor are most at risk from the continuing loss of biodiversity, as they are the ones that are most reliant on the ecosystem services being degraded.

At the 1992 Earth Summit in Rio de Janeiro, world leaders agreed on a comprehensive strategy for "sustainable development". One of the key agreements adopted at Rio was the United Nations Framework Convention on Climate Change (UNFCCC); another one was the Convention on Biological Diversity (CBD). The CBD sets out commitments for maintaining the world's ecological

Box 7.1 Natura 2000

Natura 2000 is a network of protected nature areas in the EU. In 1992, the same year as the signing of the international Convention on Biodiversity, EU governments adopted legislation to protect the most seriously threatened habitats and species across Europe. The legislation is called the Habitats Directive and complements the Birds Directive adopted in 1979. These two directives form the basis for the Natura 2000 network. The Birds Directive requires the formation of special protected areas for birds, and the Habitats Directive requires special areas of conservation to be designated for other species and their habitats. Together these areas make up the Natura 2000 network of protected areas. The Natura 2000 network contributes to the Emerald Network of conservation areas set up under the Bern Convention on the Conservation of European wildlife and natural habitats.

As of 2008, the Birds Directive protects 10.5% of the terrestrial area and the Habitats Directive 13.3%. For marine areas, 66,084 km^2 is protected by the Birds Directive and 87,505 km^2 by the Habitats Directive.

In 2006, the EU launched an initiative to halt biodiversity loss by 2010 and beyond. A review in 2008 revealed that the EU is highly unlikely to meet this target by 2010, and that an intensive effort is needed to even come close to this target.[2] In 2011 the European Commission adopted a new strategy to halt the loss of biodiversity and ecosystem services in the EU by 2020. The review from 2006 showed that 50% of species and possibly 80% of habitat types of interest to conservation in Europe have unfavorable conservation status. Nevertheless, it is common belief that Natura 2000, and the Birds Directive in particular, have been very valuable in slowing the decline of species. One of the objectives of the EU Biodiversity Loss Initiative from 2006 is to help species in Europe adapt to climate change. To preserve biodiversity in the face of climate change will require climate change policies that are closely linked to actions for conserving and sustainably using ecosystems in Europe.

underpinnings as we continue the business of economic development. The biodiversity convention establishes three main goals: the conservation of biological diversity, the sustainable use of its components, and the fair and equitable sharing of the benefits from the use of genetic resources. In the EU nature protection is primarily handled through the Natura 2000 network (Box 7.1).

European ecosystems

Climate is the main determinant of the distribution of species and ecosystems. That means that dominant vegetation and associated biodiversity gradually changes along with changing climatic conditions as we move north- or southwards from the equator. The highest biodiversity occurs in the tropical rainforests found in the lower latitudes. A much lower diversity occurs in the large deserts girdling the equator, because few species thrive in areas that are both hot and dry. As we move northwards into the high latitudes with cooler temperatures, the evergreen forests around the Mediterranean are replaced by deciduous temperate forests. These are again replaced by extensive coniferous forests further north. The extreme cold and short growing season in the Arctic severely limit the number of species. The high north is often also very dry, which contributes to low biodiversity. A similar south–north pattern is observed in the sea. However, the pattern is also strongly influenced by the large oceanic currents which determine the distribution of temperature and nutrients and thus the algal growth that forms the basis for life in the sea.

Europe hosts a unique set of natural species. This includes hot spots such as the Mediterranean region which is home to many endemic[3] species, since this region has been spared from the severe consequences of glaciations during recurring ice ages. However, biodiversity loss has accelerated to an unprecedented level, both in Europe and worldwide. It has been estimated that the current global extinction rate is 1,000 to 10,000 times higher than the natural background extinction rate.[4] In Europe some 42% of European mammal species are endangered, together with 15% of bird species and 45% of the species of butterflies and reptiles.[5] The Arctic fox (*Alopex lagopus*) and the Iberian lynx (*Lynx pardinus*) are both under serious threat. For example, as of 2004 there were only a few hundred lynx left occupying four pockets of land in Spain. Cut off from one another, the big cat communities are being weakened by inbreeding.

With the exception of tropical ecosystems, Europe has most of the other major ecosystems represented, including sparsely vegetated deserts and shrublands in southeastern regions, various types of forests, grasslands and shrublands in temperate lowlands and mountains, and sparsely vegetated boreal and arctic areas (Figure 7.1). The land cover of Europe shows that agricultural croplands and grasslands dominate the European landscape, especially in the mild temperate parts of Europe. The fragmentation of the original landscape brought about by urban development, infrastructure and, in particular, agriculture has had large consequences for the native European ecosystems, and much of the biodiversity in the current European landscapes are dependent on continued agricultural management, especially with a low intensity of fertilizer and other inputs (Figure 7.2).

The five major threats to biodiversity in Europe are: habitat loss, overexploitation, pollution, invasive alien species, and climate change. Habitat loss is brought about by converting existing natural land to farmland or for infrastructure or settlements. Biodiversity is also threatened by air and water pollution coming from farmland and urban areas. Invasive alien species pose a long-term threat to ecosystems (Box 7.2). Climate change will act in parallel to the other four threats and magnify their impact. Hence policies and initiatives to conserve ecosystems must factor in the mitigation of, and adaptation to, climate change.

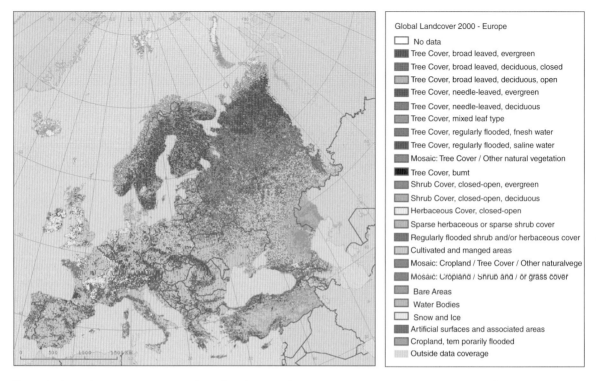

Global Landcover 2000 - Europe

- No data
- Tree Cover, broad leaved, evergreen
- Tree Cover, broad leaved, deciduous, closed
- Tree Cover, broad leaved, deciduous, open
- Tree Cover, needle-leaved, evergreen
- Tree Cover, needle-leaved, deciduous
- Tree Cover, mixed leaf type
- Tree Cover, regularly flooded, fnesh water
- Tree Cover, regularly flooded, saline water
- Mosaic: Tree Cover / Other natural vegetation
- Tree Cover, bumt
- Shrub Cover, closed-open, evergreen
- Shrub Cover, closed-open, deciduous
- Herbaceous Cover, closed-open
- Sparse herbaceous or sparse shrub cover
- Regularly flooded shrub and/or herbaceous cover
- Cultivated and manged areas
- Mosaic: Cropland / Tree Cover / Other naturalvege
- Mosaic: Cropland / Shrub and / or grass cover
- Bare Areas
- Water Bodies
- Snow and Ice
- Artificial surfaces and associated areas
- Cropland, tem porarily flooded
- Outside data coverage

Figure 7.1 Land cover in Europe based on classification of satellite imagery. Source: Bartholome et al.[6] Reprinted with permission from European Commission. (See color plate.)

Ecosystem services

Natural ecosystems provide a multitude of resources and processes that benefit humankind. These are generally known as ecosystem services and include products or goods like clean drinking water, food, feed and fiber (Chapter 4) and processes such as decomposition of wastes and cleaning of air. The services can be divided into five categories:[9] (1) provisioning services that include production of food, feed, fiber and energy; (2) regulating services such as carbon sequestration, climate regulation and nutrient cycling; (3) supporting services such as water and air purification and crop pollination; (4) cultural services such as recreational benefits; and (5) preservation, which includes guarding against extremes through the maintenance of overall system functioning and preservation of gene pools.

To some extent these services have traditionally been considered as free, invulnerable and infinitely available. However, developments over recent decades have clearly demonstrated this to be untrue. All of these services are likely to be profoundly affected by climate change through processes that affect ecosystem structure and functioning. However, these ecosystem services are not only being

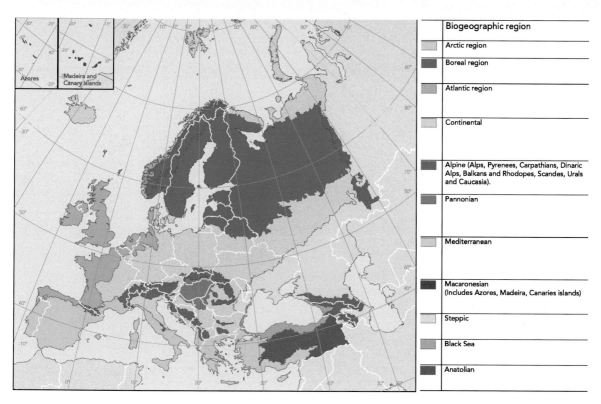

Figure 2.1 Biogeographic regions of Europe. Source: EEA.[4] Reprinted with permission from European Environmental Agency, Copenhagen.

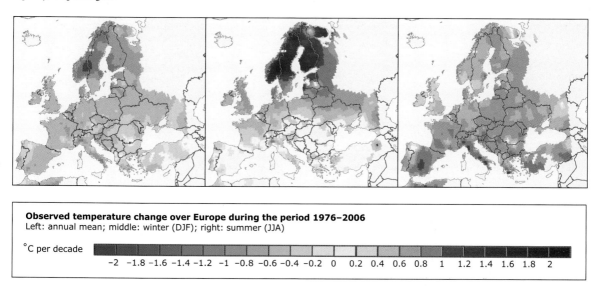

Observed temperature change over Europe during the period 1976–2006
Left: annual mean; middle: winter (DJF); right: summer (JJA)

°C per decade

| -2 | -1.8 | -1.6 | -1.4 | -1.2 | -1 | -0.8 | -0.6 | -0.4 | -0.2 | 0 | 0.2 | 0.4 | 0.6 | 0.8 | 1 | 1.2 | 1.4 | 1.6 | 1.8 | 2 |

Figure 2.2 Observed temperature change over Europe during the period 1976–2006. Left: annual mean; middle: winter (December to January); right: summer (June to August). Source: EEA.[8] Reprinted with permission from European Environmental Agency, Copenhagen, and KNMI, Netherlands.

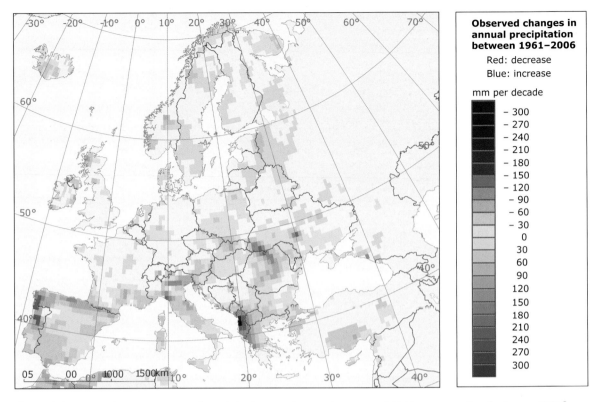

Figure 2.3 Observed changes in annual precipitation over Europe for 1961 to 2006 in mm per decade. Source: EEA.[9] Reprinted with permission from European Environmental Agency, Copenhagen, and KNMI, Netherlands.

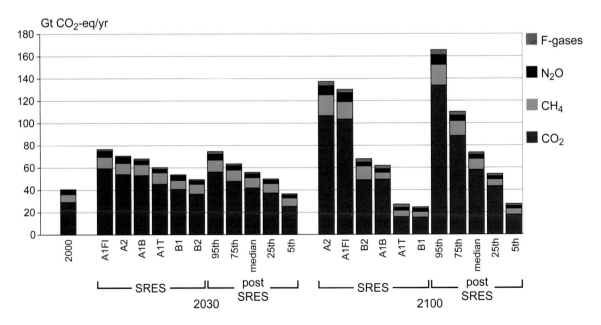

Figure 2.4 Annual global greenhouse gas emissions, expressed in Gt CO_2 equivalents, for 2000 and projected emissions for 2030 and 2100. Shown are six IPCC-SRES scenarios (Box 2.5) and a frequency distribution of emissions in post-SRES scenarios. The "F-gases" depict halocarbons. Source: Metz et al.[16] Reprinted with permission from Intergovernmental Panel on Climate Change, Geneva.

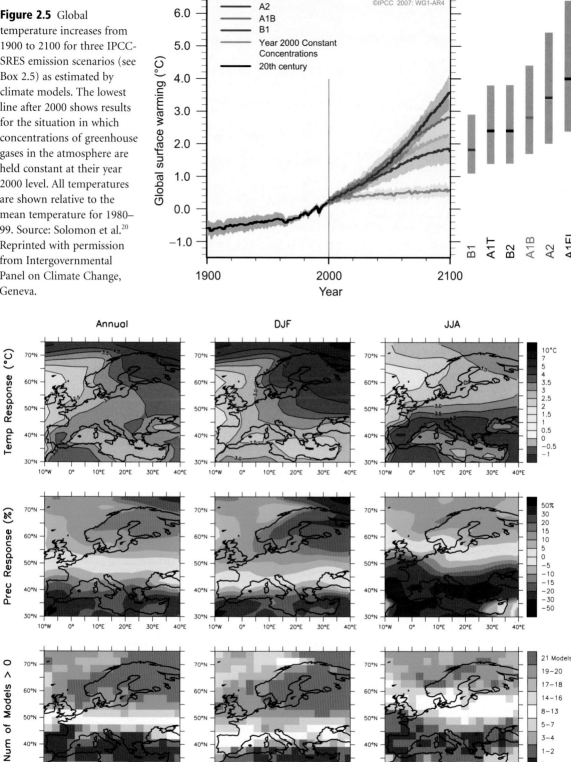

Figure 2.5 Global temperature increases from 1900 to 2100 for three IPCC-SRES emission scenarios (see Box 2.5) as estimated by climate models. The lowest line after 2000 shows results for the situation in which concentrations of greenhouse gases in the atmosphere are held constant at their year 2000 level. All temperatures are shown relative to the mean temperature for 1980–99. Source: Solomon et al.[20] Reprinted with permission from Intergovernmental Panel on Climate Change, Geneva.

Figure 2.6 Modeled change in annual, winter and summer mean temperature (upper maps, from left to right) and precipitation (middle row maps) over Europe from 1980–1999 and 2080–2099 averaged for 21 global climate models, all using the A1B emission scenario. The bottom row maps show the number of models that out of 21 project increases in precipitation. Source: Christensen et al.[22] Reprinted with permission from Intergovernmental Panel on Climate Change, Geneva.

Figure 3.1 Schematic representation of how an increase in average annual temperature would affect annual total of temperature-related deaths, by shifting distribution of daily temperatures to the right under climate change as indicated for 2050 compared with 2005. Some acclimation occurs to the new climate regime resulting in a change in curve shape. Additional heat-related deaths in summer would outweigh the extra winter deaths averted (as may happen in some northern European countries). Average daily temperature range in temperate countries would be about 5–30°C. Source: McMichael et al.[4] Reprinted with permission from Elsevier Limited, Oxford.

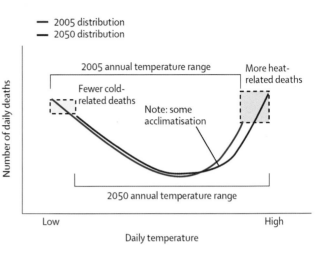

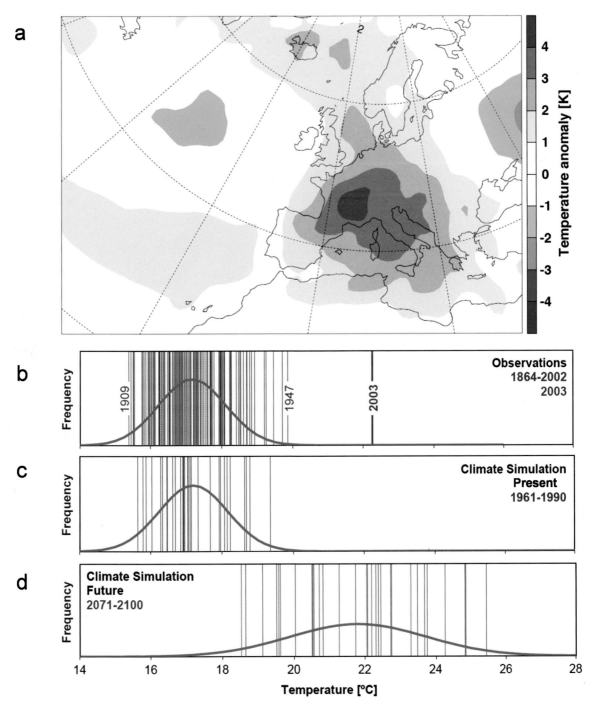

Figure 3.2 Characteristics of the summer 2003 heat wave in terms of summer (June to August) temperatures:
(a) Temperature anomaly with respect to 1961–90 (b) Temperatures for Switzerland observed during 1864–2003
(c) Simulated using a regional climate model for the period 1961–1990 (d) Simulated for 2071–2100 under the A2
emission scenario. Source: Schär et al.[7] Reprinted with permission from Macmillian Publishers Limited: Nature.

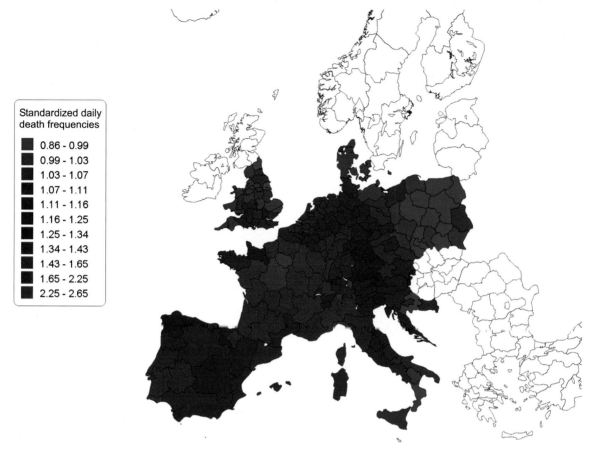

Figure 3.3 Standardized daily death frequencies during the summer heatwave from 3 to 16 August 2003 in 16 European countries (1 means equal to normal and 2 means twice the normal death rate). Source: Robine et al.[8] Reprinted with permission from Elsevier Masson SAS.

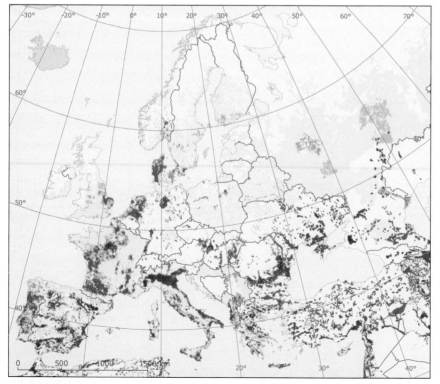

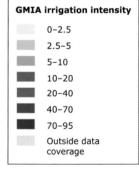

Figure 4.4 Irrigation intensity across Europe as illustrated by the percentage of area equipped with irrigation. Source: EEA.[44] Reprinted with permission from European Environmental Agency, Copenhagen.

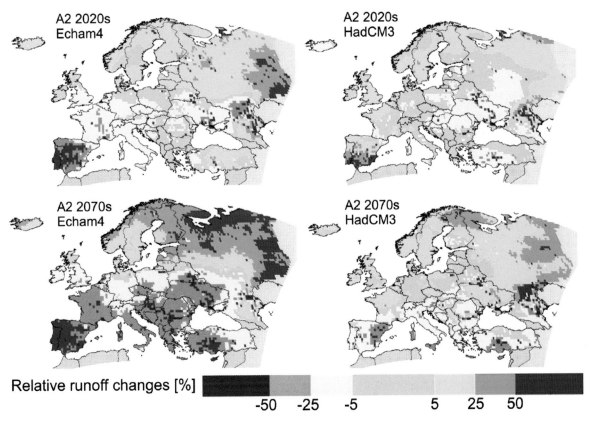

Relative runoff changes [%]

-50 -25 -5   5   25   50

Figure 5.1 Change in annual river run-off in Europe between a 1961–90 baseline period and two future periods (2020s and 2070s). Reprinted with permission from Intergovernmental Panel on Climate Change, Geneva. (See text for full caption.)

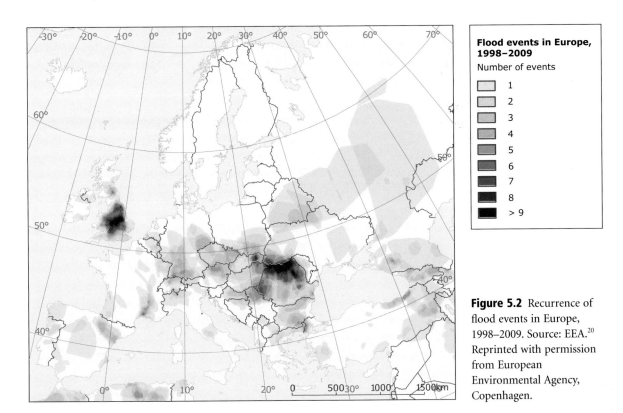

Flood events in Europe, 1998–2009

Number of events

1
2
3
4
5
6
7
8
> 9

Figure 5.2 Recurrence of flood events in Europe, 1998–2009. Source: EEA.[20] Reprinted with permission from European Environmental Agency, Copenhagen.

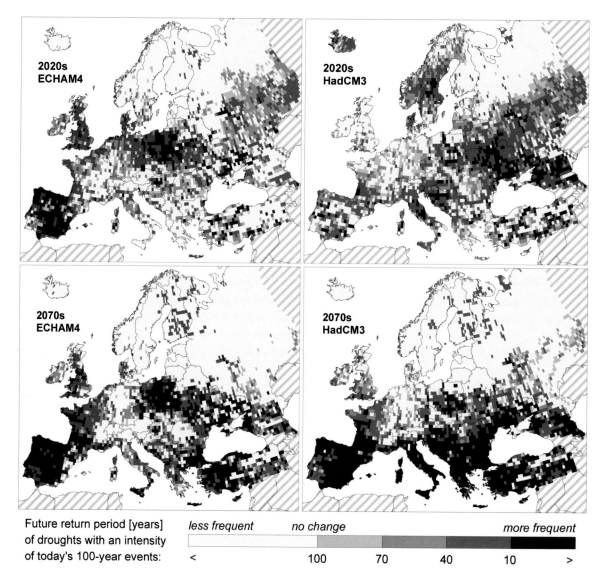

Future return period [years]	*less frequent*	*no change*				*more frequent*

Figure 5.4 Change in the recurrence interval of 100-year droughts in Europe between a 1961–90 baseline period and two future periods (2020s and 2070s). Calculations are from the WaterGAP model using climate data from two different climate models (left and right side of figure). Source: Lehner et al.[30] Reprinted with permission from Springer Science + Business Media.

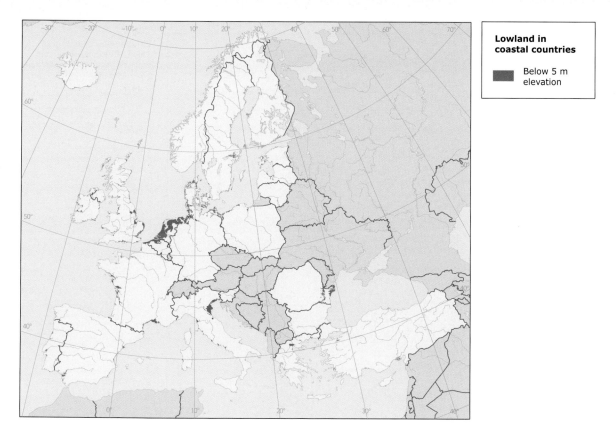

Figure 6.2 Coastal lowlands (elevation below 5 meters) in Europe. Source: EEA.[35] Reprinted with permission from European Environmental Agency, Copenhagen.

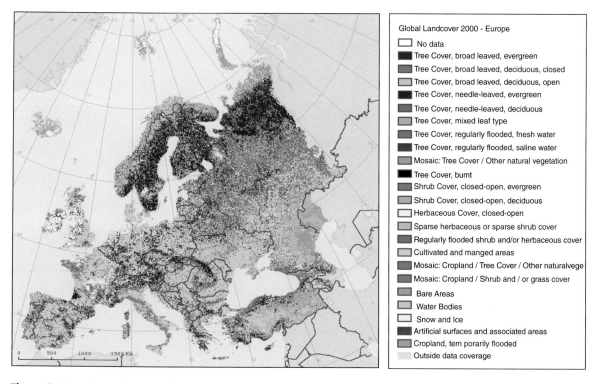

Global Landcover 2000 - Europe
- No data
- Tree Cover, broad leaved, evergreen
- Tree Cover, broad leaved, deciduous, closed
- Tree Cover, broad leaved, deciduous, open
- Tree Cover, needle-leaved, evergreen
- Tree Cover, needle-leaved, deciduous
- Tree Cover, mixed leaf type
- Tree Cover, regularly flooded, fnesh water
- Tree Cover, regularly flooded, saline water
- Mosaic: Tree Cover / Other natural vegetation
- Tree Cover, bumt
- Shrub Cover, closed-open, evergreen
- Shrub Cover, closed-open, deciduous
- Herbaceous Cover, closed-open
- Sparse herbaceous or sparse shrub cover
- Regularly flooded shrub and/or herbaceous cover
- Cultivated and manged areas
- Mosaic: Cropland / Tree Cover / Other naturalvege
- Mosaic: Cropland / Shrub and / or grass cover
- Bare Areas
- Water Bodies
- Snow and Ice
- Artificial surfaces and associated areas
- Cropland, tem porarily flooded
- Outside data coverage

Figure 7.1 Land cover in Europe based on classification of satellite imagery. Source: Bartholome et al.[6] Reprinted with permission from European Commission.

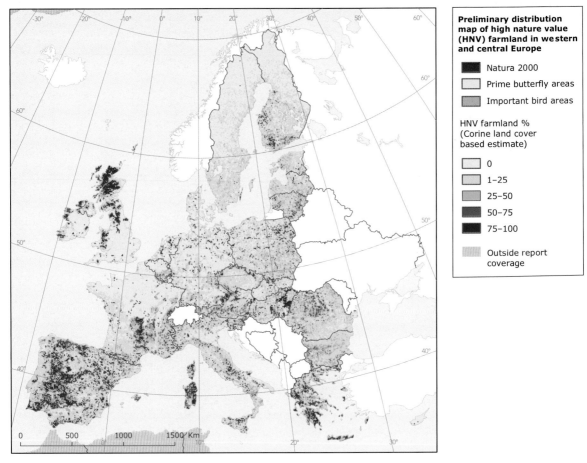

Figure 7.2 Distribution of high nature value farmland in Western and Central Europe. Source: EEA.[7] Reprinted with permission from European Environmental Agency, Copenhagen.

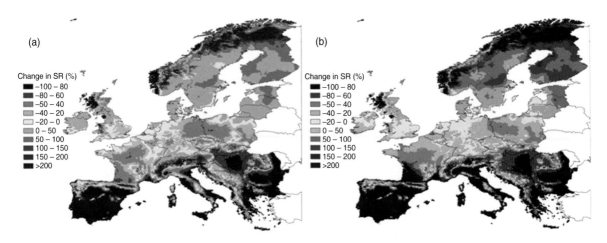

Figure 7.4 Modeled changes in mammalian species richness by 2080 in percentage, under (a) a low emission scenario (B1) and (b) a high emission scenario (A2). Source: Levinsky et al.[26] Reprinted with permission from Springer Science + Business Media.

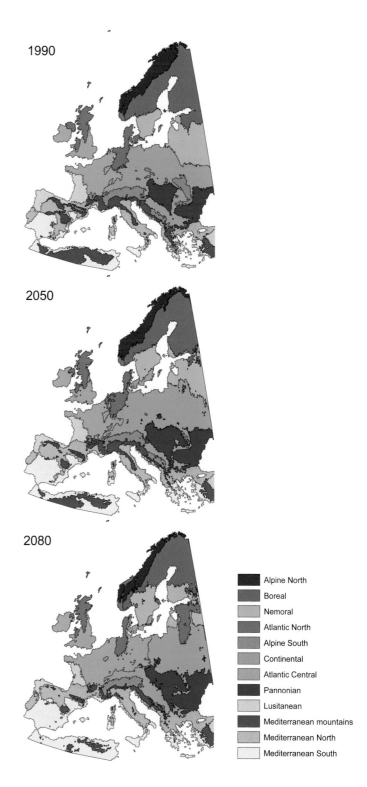

1990

2050

2080

Figure 7.3 Shifting environmental zones for Europe under one climate change scenario (A2 emission scenario, CGCM2 GCM). Source: Metzger et al.[24] Reprinted with permission from Cambridge University Press.

Alpine North
Boreal
Nemoral
Atlantic North
Alpine South
Continental
Atlantic Central
Pannonian
Lusitanean
Mediterranean mountains
Mediterranean North
Mediterranean South

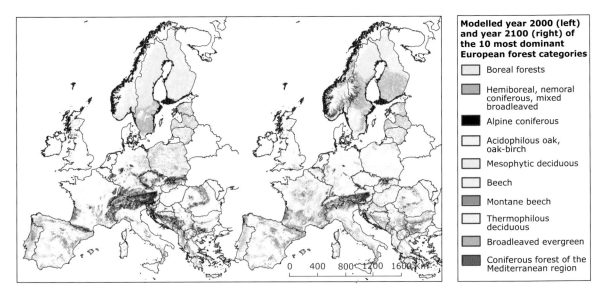

Modelled year 2000 (left) and year 2100 (right) of the 10 most dominant European forest categories

- Boreal forests
- Hemiboreal, nemoral coniferous, mixed broadleaved
- Alpine coniferous
- Acidophilous oak, oak-birch
- Mesophytic deciduous
- Beech
- Montane beech
- Thermophilous deciduous
- Broadleaved evergreen
- Coniferous forest of the Mediterranean region

0   400   800   1200   1600 KM

Figure 7.6 Modeled forest cover types under current (2000) climate (left) and future (2100 for A1B emission scenario and NCAR CCM3 GCM) (right) climate. Source: EEA.[42] Reprinted with permission from European Environmental Agency, Copenhagen.

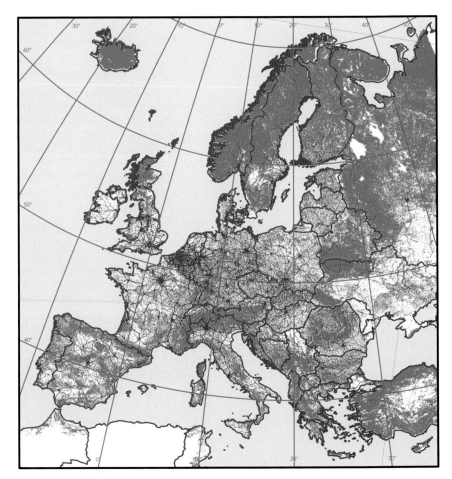

Figure 7.7 Pressures from urbanization and transport (red) on semi-natural areas in Europe. Source: EEA.[78] Reprinted with permission from European Environmental Agency, Copenhagen.

Current Arctic Conditions

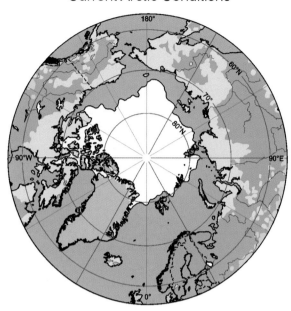

Projected Arctic Conditions

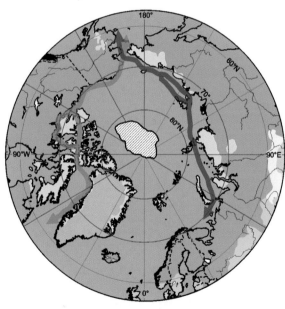

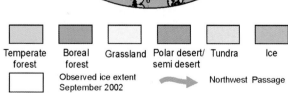

Figure 8.1 Melting ice cap will open up new shipping lanes in the far north. Shown here is the observed minimum sea-ice extent for September 2002 (upper picture) compared to the projected extent for the period 2080–100 (lower picture). Also shown in the lower picture are possible new or improved sea shipping routes. Source: Anisimov et al.,[37] based on Instanes et al.[38] and Walsh et al.[39] Although the melting of the Arctic ice cap might be an advantage to shipping it will also disrupt important biological and physical processes with likely negative impacts on society and the rest of nature (Chapter 7). The depicted change in vegetation is based on simulations of the LPJ Dynamic Vegetation model. For further information about vegetation changes see explanation in Anisimov et al.[40] Reprinted with permission from Intergovernmental Panel on Climate Change, Geneva.

Figure 9.1 Key vulnerabilities of European systems and sectors to climate change during the 21st century for the main biogeographic regions of Europe: TU: Tundra, pale turquoise. BO: Boreal, dark blue. AT: Atlantic, light blue. CE: Central, green; includes the Pannonian Region. MT: Montane, purple. ME: Mediterranean, orange; includes the Black Sea region. ST: Steppe, cream. SLR: sea-level rise. NAO: North Atlantic Oscillation. Source: Alcamo et al.[1] Source of map of biogeographic regions: European Environment Agency, copyright EEA, Copenhagen. www.eea. europa.eu. Reprinted with permission from European Environmental Agency, Copenhagen.

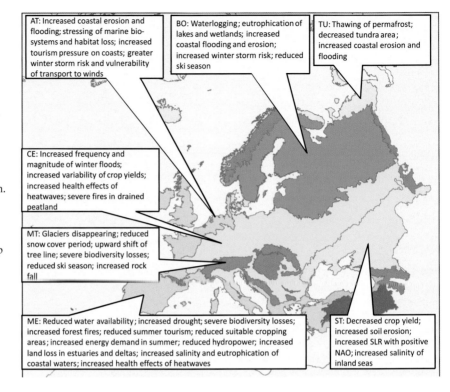

Figure 9.2 Some expected negative economic impacts of climate change according to biogeographic regions of Europe. See caption of Figure 9.1 for explanation of biogeographic regions. For details, see text, especially Chapter 8. Source of map of biogeographic regions: European Environment Agency, copyright EEA, Copenhagen. www.eea. europa.eu

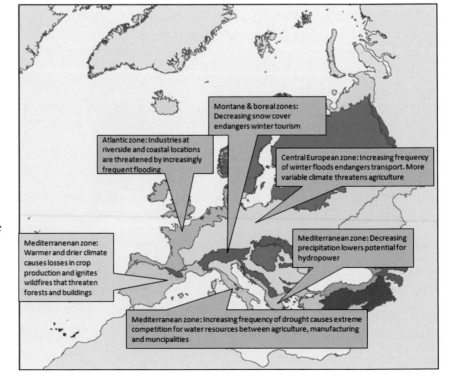

Figure 10.1 "Burning embers" diagram showing vulnerability of different parts of the European and global environment to global warming. The "X" axis uses the increase in average global surface temperature (°C) (relative to pre-industrial climate) as a proxy for the specific changes in climate occurring at the regional and global levels. The redder the color, the more widespread and or greater magnitude of negative impacts or risks. Note that the impacts of climate change on potential food production may be positive in the first phase of climate change because of warmer and moister conditions, but they eventually turn negative because of the impact of higher temperatures. (See Chapter 4). Sources: MNP,[1] redrawn by EEA.[2] Reprinted with permission from Netherlands Environmental Assessment Agency.

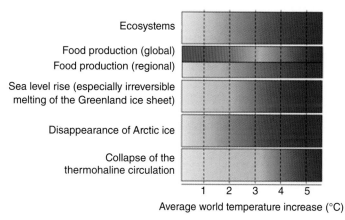

Figure 10.2 "Burning embers" diagram showing increasing threat to ecosystems as climate changes. Change in global surface temperature relative to 1900 is used as a proxy for specific changes in climate occurring at the regional and global levels. The redder the color, the more widespread and or greater magnitude of negative impacts or risks. Source: IPCC.[3] Reprinted with permission from Intergovernmental Panel on Climate Change, Geneva.

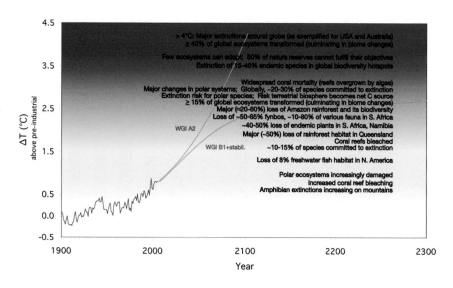

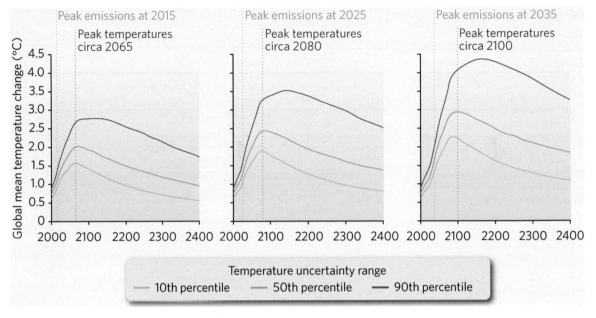

Figure 10.3 Increase in average global surface temperature for scenarios of peak greenhouse gas emissions in either 2015, 2025 or 2035 with 3%-per-year reductions in greenhouse gas emissions. Note: the later the peak in global emissions, the later and higher the peak in temperature. Source: Parry et al.[7] Reprinted with permission from Macmillian Publishers Limited: Nature.

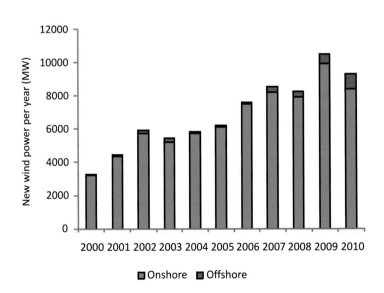

Figure 10.7 Newly installed capacity of onshore and offshore wind power in the European Union. Source: EWEA.[25] Redrawn with permission from European Wind Energy Association.

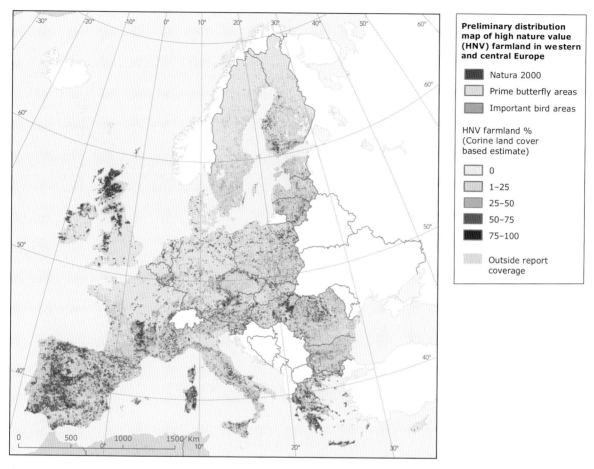

Figure 7.2 Distribution of high nature value farmland in Western and Central Europe. Source: EEA.[7] Reprinted with permission from European Environmental Agency, Copenhagen. (See color plate.)

affected by climate change – many of them are also under serious pressure by other human-related factors. Air and water quality are increasingly being compromised globally, oceans are being overfished, deforestation is eliminating flood control around human settlements, species are being lost at alarming rates, and pest and diseases are spreading beyond historical ranges. Since ecosystems also perform critical services in terms of maintaining greenhouse gas concentrations and climate in general, their functioning is critical to avoiding negative climate change impacts.

Scenario-based analysis of ecosystem service supply in Europe during the 21st century in response to different socio-economic and climate change scenarios have shown that large changes in climate and land use typically result in large changes in ecosystem supply.[10] In these scenarios trends in the technological effects on crop yields often tend to outweigh climate effects on land use, which is

Box 7.2 Invasive alien species

Invasive alien species constitute a threat to biodiversity by outcompeting native species. This not only threatens biodiversity, but can also cause considerable economic damage. Species including the zebra mussel (*Dreissena polymorpha*), American red swamp crayfish (*Procambarus clarkii*), American mink (*Mustela vison*), water hyacinth (*Elchornia crassipes*), and giant hogweed (*Heracleum mantegazzianum*) currently cause hundreds of millions of euros of damage every year. Unless action is taken to eradicate or control these invasive species the damage will only increase – and the trend is almost certain to be exacerbated by climate change.[8]

Europeans today are more mobile than ever before. This has many benefits, but it also increases the number of points of entry for new species. Some highly invasive species are intentionally imported as pets or ornamental plants, while others arrive as "hitchhikers" or contaminant organisms through trade. Common causes include species escaping from gardens or aquariums, from captivity, or from fish farms. The deliberate stocking by anglers of freshwater alien fauna is another cause. For example, the introduced North American brook trout (*Salvelinus fontenalis*) is threatening European freshwater fish and could replace brown trout (*Salmo trutta*) in river mountain ecosystems. In the marine environment, harmful aquatic organisms are often introduced via ballast water in ships, taken on in one part of the world and released at some distant location.

The environmental consequences are considerable, and range from whole-scale ecosystem changes and the near extinction of native species to more subtle ecological changes and decreased biodiversity. An example is the European mink (*Mustela lutreola*) which is threatened by the American mink (*Mustela vison*). Another example are the ecosystems of the Mediterranean sea which have suffered extensive disruption by the toxic algae, *Caulerpa taxifolia*. Much of western Europe has suffered serious environmental and economic damage due to the zebra mussel, which clogs power plant intakes and competes with native freshwater mussel populations. The Asian topmouth gudgeon (*Pseudorasbora parva*) has spread rapidly throughout Europe since being introduced into Romanian ponds close to the Danube in the 1960s, with serious consequences for native species because of the parasites plaguing the gudgeon.

Climate change will in many cases exacerbate problems with invasive alien species, because it may provide them with a competitive advantage in some regions. Also, climatic warming will mean that invasive alien species may expand their range to areas not previously suitable, as in the case of ragweed (*Ambrosia artemisiifolia*), which is spreading to northern Europe (Box 3.4).

a major driver of changes in European landscapes. Some of the modeled trends were positive (e.g., increases in forested area and productivity) or offer opportunities (e.g. more land for agricultural extensification and bioenergy production). However, many changes increase vulnerability as a result of a decreasing supply of ecosystem services, especially in the Mediterranean and mountain regions (e.g. declining soil fertility, declining water availability, increasing risk of forest fires).

Climate regulation

Not only does climate affect ecosystems, but conversely, changes in ecosystems have a strong influence on global climate system. Land use changes can both increase and decrease greenhouse gas emissions.[11] Intensively cultivated countries like Netherlands and Denmark are sources of CO_2, whereas countries like Slovenia and Slovakia are carbon sinks. Moreover, forests are a major storehouse of carbon. Carbon dioxide is released during forest fires, or whenever trees are cut down and allowed to decompose. European forests were almost completely cleared during the expansion of agriculture during the 16th to 18th centuries. This released considerable amounts of CO_2 from vegetation and soil. The recent regrowth of forests in Europe meant that the European terrestrial biosphere absorbed 7–12% of the annual anthropogenic CO_2 emissions from Europe in the 1990s.

Peatlands provide many environmental services, such as improving water quality, and they are important biodiversity havens and stopover points for migratory birds. They also hold roughly one third of the soil carbon stored worldwide.[12] In Europe, peat extraction for horticulture, agriculture and energy is greatly reducing amounts of stored carbon. In Russia, the warming climate is causing permafrost peatlands to melt and dry out causing more frequent fires, resulting in large sources of CO_2 emissions.

There are many complex feedback mechanisms between ecosystems and climate (Box 2.4). Higher temperatures and increased atmospheric CO_2 can stimulate forest growth and hence CO_2 uptake in central and northern European forests. In high latitudes, replacement of shrub/tundra vegetation with trees will affect the radiation balance, since the new vegetation will tend to absorb more solar radiation than the tundra it replaces.

The ability of ecosystems to regulate climate is in itself influenced by climate and climate change. This was clearly illustrated with the European heatwave in 2003 (Box 3.2) which was accompanied by annual precipitation deficits up to 300 mm. The heatwave was associated with a prolonged drought, which reduced gross primary production of terrestrial ecosystems over Europe by 30%.[13] This resulted in carbon losses equivalent to three years of carbon gain in normal years. The projected climate change is likely to increase the frequency of such heatwaves (Chapter 2), which could turn European ecosystems into a net carbon source rather than a sink.

Direct effects of climate change

Climate change will impact on ecosystems in multiple ways through the specific responses of individual species to changing environmental conditions and through the interactions with other species in the trophic network (food-web) (Box 7.3). Plant and animal species have different environmental requirements. Temperature changes affect the timing for most plant and many animal species and also the rate and duration of growth processes. Rainfall affects wetness of the soil and provides water for many ecosystem processes, including plant growth. Abundance of water is essential for many ecosystems (e.g. lakes and peatlands). It is not only climatic variables such as temperature and precipitation that will affect ecosystems under climate change. The increasing concentration of CO_2 will enhance the photosynthesis of many species. But the response of plants to higher CO_2 levels will vary from species to species, and will also depend on the nutritional status of the ecosystem.[14]

In ecosystems there are many interactions between constituent species and the physical and chemical environment. Such interactions are seldom fully understood and are therefore also difficult to describe and model (Box 7.4). Hence, the models used for describing and projecting climate change impacts on ecosystems and biodiversity tend to be relatively simple and many uncertainties remain. An important one is the effect of extreme climate events (e.g., droughts and floods) on ecosystems. It is known that such events can sometimes trigger substantial and even permanent shifts in the structure and functioning of ecosystems (so-called regime shifts). European species and ecosystems are reported to have responded to climate changes in a wide variety of ways, such as an upward shift of species (e.g., tree line and alpine species), phenological changes (e.g., advanced timing of flowering, breeding and migration), increasing productivity and forest carbon sinks, invasion of evergreen broad-leaved species in Alpine forests, disappearance of wetlands and changes in vegetation composition.[18]

Ecosystem and species range shifts

The most obvious impacts are the effects of temperature and rainfall changes on species ranges and ecosystem boundaries. Any particular ecosystem consists of an assemblage of species, some of which will be near the edge of their ranges and others which are not. Those at the edge of their ranges may need to move owing to climate change. Species that are highly mobile and opportunistic are likely to benefit at the expense of those that are not. This may even facilitate spreading and establishment of invasive alien species (Box 7.2).

A warming of 1°C in Europe corresponds approximately to a 150–200 km shift northwards, with larger shifts in continental Europe than along the Atlantic coast, or a shift upwards in altitude by 150–180 m. In temperate and colder areas of Europe, range shifts have thus often involved the northwards or upslope extension of climate space – and thus of geographic range – for many species, which may

Box 7.3 Trophic dynamics and trophic mismatch

Trophic dynamics in ecology refers to the system of trophic levels which describe the positions of various organisms in the food chain (i.e. what the organism eats and by whom it is eaten). Primary producers such as algae and plants harvest the sunlight and turn it into biomass, which is in turn consumed by other organisms (herbivores), which then are eaten by carnivores. In terrestrial ecosystems plants form the first trophic level. Next are the herbivores (primary consumers), followed by carnivores (secondary consumers). In the ocean phytoplankton are usually the primary producers. Phytoplankton are consumed by microscopic animals called zooplankton, and these in turn are consumed by larger zooplankters and by fish. Fish that eat zooplankton constitute the fourth trophic level; mammals, such as seals, that consume fish make up the fifth trophic level. In the Arctic, polar bears that eat seals constitute the sixth and final trophic level. Because more and more energy is required to harvest food at higher trophic levels, ecosystems seldom have more than six levels. Only very small populations can be supported at the top levels.

In nature, species in one trophic level are commonly dependent on species in other trophic levels. These dependencies can be disrupted by climate change. If the population of one species is reduced by climate change this could reduce the food supply of another species at a higher trophic level, which could disrupt the food supply of species in the next higher trophic level, and so on. This is called "trophic mismatch". Such a disruption of food supply is particularly harmful to "specialist" species that depend on a single or small number of other species in the food web. "Generalist" species that live on a much wider range of food sources have fewer problems in adapting to climate change.

Over time, the characteristics of different species become synchronized with one another. For example, some migrant birds arrive in a particular area just as insects make their first appearance for the year. If climate change causes either the behavior of migrant birds or the hatching of insects to change, this could cause a trophic mismatch in time, and disrupt the life cycles of both birds and insects.[15] A similar mismatch in time could occur if climate change disrupts the synchronized bursting of oak buds and hatching of winter moth eggs.[16] Species with low adaptability to phenology changes, such as plants responding to day length rather than temperature, may suffer greater stress or even the risk of extinction under rapid climate change.

A "spatial trophic mismatch" occurs when climate change reduces the overlap in the suitability ranges of two interdependent species. Current niche spaces of the butterfly *Boloria titania* and its host plant American bistort (*Polygonum bistorta*) are already beginning to show some degree of mismatch.[17] Under climate change, the mismatch is expected to be even larger, leading to an uncertain future for this "specialist" butterfly which currently lives in the Alps, the British highlands and Scandinavia.

Box 7.4 Models of biodiversity change

Projections of changes in biodiversity in response to future climatic changes are frequently based on modeling techniques such as bioclimatic envelope modeling and dynamic vegetation modeling.[19] Results from such models can be used to identify regions and taxa or ecosystems that are most threatened by climate change. Results can then be applied for conservation planning, although such model projections are still subject to considerable uncertainty.

The most popular current approach for projecting climate change impacts on biodiversity are "bioclimatic envelope" models, also referred to as niche-based models or habitat models. These models relate current species' distributions (either presence/absence or abundance) with current climate variables and thereby define the climatic "envelope" of each species. By doing so, they are strictly empirical, include all biotic interactions constraining a species' distribution and are thus based on the realized climatic niche of species. By applying changing climate variables to the model, the potential future space of a species can be projected. This technique has been used to quantify potential changes for a large number of species. Such models, however, do not take into account several processes that affect species distributions such as population dynamics and competition, land use, dispersal and the direct physiological effects of CO_2. The dispersal capabilities of species is typically addressed by two extreme options, allowing either unlimited or no migration at all for the species. However, tests of the validity of the approach have shown remarkable predictive accuracy at large spatial scales.

Dynamic vegetation models explicitly simulate the population dynamics of shrubs, trees and other vegetation types, but they tend to be either parameterized for particular study sites or else vegetation is represented by a few "plant functional types" (PFTs) such as broad-leaved deciduous trees, not specifically accounting for changes in species richness and implicitly assuming unlimited dispersal of species or PFTs. Furthermore, this approach is generally only applied to woody plants, summarizing all herbaceous species in one or two types. These models focus on ecosystem processes, such as net primary productivity, water cycling of terrestrial ecosystems, and carbon sequestration.

---

or may not be accompanied by a contraction of climate space at the southern or lower altitude range limit.

There is plenty of evidence for range shifts caused by recent climate change.[20] Trees and shrubs have replaced many Arctic and tundra communities. The tree line and the level at which alpine plants are found in Europe is moving towards higher altitudes. Range shifts northwards have often been reported for well-studied species groups such as birds and butterflies, but also increasingly for plants.[21] In European seas, temperate species have migrated about 250 km northward per decade,[22] whereas sub-Arctic and Arctic species have declined in numbers. More temperate species are now found in the North Sea driven by increasing sea temperatures.

European climate zones are generally projected to move from south-west towards north-east. This can be illustrated by an analysis where Europe is divided into 84 environmental strata (e.g., the Po valley or the British uplands), and where each of these strata is well defined in terms of climatic space.[23] Applying this approach to a wide change of climate change scenarios shows that Europe is likely to experience major environmental shifts northwards, with pronounced regional variations (Figure 7.3). In particular, the southern Mediterranean strata are projected to expand, whereas Atlantic environments remain much more stable. Alpine and Mediterranean mountain environments decline dramatically, whereas the Scandinavian zones show inconsistent patterns of change.

Plants are very likely to expand their ranges northwards while contracting in southern European mountains and in the Mediterranean region, resulting in a net increase of species richness in northern Europe and a probable decrease in Mediterranean countries. Similar patterns are expected for amphibians and reptiles, although their limited dispersal abilities might prevent northward expansions.[25] For these taxa (in particular amphibians), southwestern Europe is the region most likely to be negatively affected, essentially due to increased aridity. Other European studies of plants, insects, birds and mammals indicate that a general range shift from the south-west to the north-east is projected, but with possible differences among species. The potential mammalian species richness is projected to become drastically reduced in the Mediterranean region, but increase towards the north-east and in higher elevations (Figure 7.4).

Changes in phenology

Climate change is also causing a shift in the reproductive cycles and growing seasons of certain species. Most biological events in nature follow a cyclical timing called "phenology". Many of these events are annual in nature because they are triggered by the annual variation in temperature, precipitation and day length. Changes in climate are thus expected to cause changes in the timing of many of these events as well as interactions with other events such as changing day length.

There are many reports of alterations in phenology in response to climate change in Europe.[27] Overall, 62% of the observed variability of timing of life-cycle events can be explained by climate. However, variability differs between events, with those occurring earlier (i.e. spring) being more variable than later events. Of all leaf unfolding, flowering, and fruiting events across Europe, 78% show an advancing trend, and only 3% a significant delay.[28] On average, spring/summer phenological events are advancing by 2.5 days per decade.[28] One of the effects is

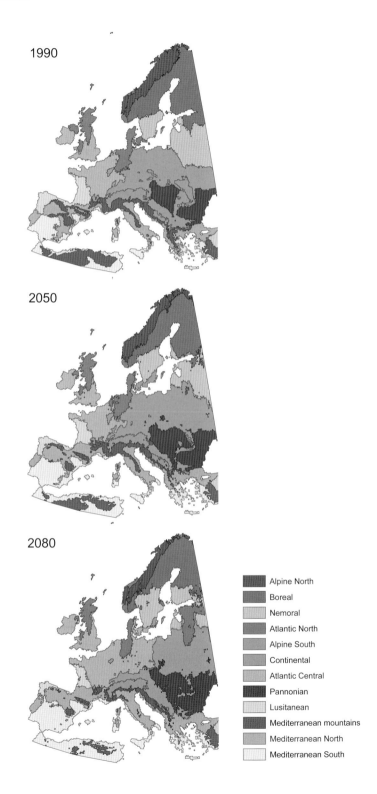

Figure 7.3 Shifting environmental zones for Europe under one climate change scenario (A2 emission scenario, CGCM2 GCM). Source: Metzger et al.[24] Reprinted with permission from Cambridge University Press. (See color plate.)

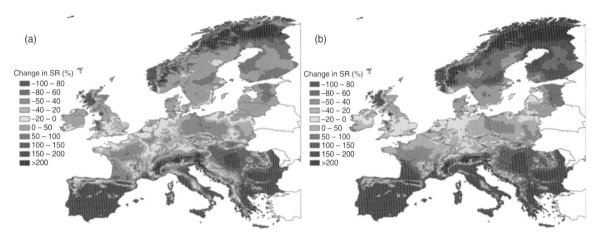

Figure 7.4 Modeled changes in mammalian species richness by 2080 in percentage, under (a) a low emission scenario (B1) and (b) a high emission scenario (A2). Source: Levinsky et al.[26] Reprinted with permission from Springer Science + Business Media. (See color plate.)

that the pollen season now starts on average 10 days earlier and is of a longer duration than 50 years ago[29] (Chapter 3). In particular, rapid climate-induced changes in spring phenomena have been observed over the Arctic over the last 10 years.[30]

Spring migration of birds typically follows weather variations. Short-distance migrants, spending the winter in Europe, have been found to respond more strongly to climate change than long-distance migrants,[31] though the latter group has also have been observed to respond to climate change in their wintering grounds and along their migration routes.

Phenological changes under climate change will alter growing seasons with resulting effects on ecosystem production, species interactions and community dynamics.[32] Different species show different phenological responses, for example, annuals and insect-pollinated species are more likely to flower early than perennials and wind-pollinated species. While advancing trends in seasonal events will continue as climate warming increases in the decades to come, it is uncertain how different species will respond when temperature thresholds are reached and whether linear relationships between temperature and growing season duration will be realized in the future.

The future impacts of climate change on animal phenology are poorly understood, but could include increasing trophic mismatch and disturbance to ecosystem functioning (Box 7.3). The trend towards warmer springs may continue to induce earlier breeding and migration activity. Unpredictable cold snaps are likely to cause high mortality among early movers. Meanwhile, species whose life cycles are calibrated according to day length, and which do not respond so readily to changing temperatures, will not be able to exploit earlier spring resources unless they can adapt.

Effects of extreme events

Climate change related extreme events such as heatwaves, droughts, floods, wind storms, fire or diseases are likely to increase (Chapter 2), and this can seriously affect the state of ecosystems and biodiversity. Such extreme events may affect organisms, populations and ecosystems more than gradual changes in temperature and rainfall.[33] Such changes are typically associated with changes in disturbance regimes[34] (Box 7.5).

Heatwaves such as the European event in 2003 (Box 3.2) have both short and long-term impacts, especially if accompanied by drought conditions. High temperatures and longer dry spells increase the flammability of vegetation, and during the 2003 heatwave a record-breaking incidence of extensive wildfires was observed

---

Box 7.5 Ecosystem disturbance regimes

Ecosystem structure is not only affected by soils and average climatic conditions. Often disturbances in the form of extreme events such as fire or pest outbreaks can have substantial short and long-term effects, and this is sometimes independent of the influence of average climate.[36]

Climate change may not introduce new disturbance regimes, but rather change the extent and impacts of existing ones. However, in some cases changes in disturbance regimes may have profound effects on ecosystem structure and functioning leading to major and permanent ecosystem changes (regime shifts). Some of the disturbance regimes affected by climate change include the following:

- Fire is an important natural (and in some cases human managed) disturbance factor in many ecosystems, including forests, shrublands and grasslands. Under prolonged drought peatlands may become susceptible to devastating fires.
- The impacts of insect outbreaks can be huge in some ecosystems, especially in boreal forests. Such insect outbreaks can be caused by the effect of long-term air temperature changes on insect phenology and reproduction, and indirectly through effects of windfall of trees following intense storms that leaves wood available for insect reproduction (e.g. for bark beetles (*Ips typographus*) in spruce).[37]
- Greater storminess and higher return rates of extreme sea water levels increases disturbances to coastal ecosystems, leading to changes in biodiversity and hence ecosystem functioning.
- Invasive alien species (Box 7.2) can also change disturbance regimes (e.g. by increasing vegetation flammability). Also, ecosystems are often more susceptible to invasion of alien species following extreme events.

in European countries. More frequent heatwaves are very likely to cause changes in biome[35] type, particularly by promoting highly flammable, shrubby vegetation that burns more frequently rather than less flammable vegetation types such as forests, and as seen in the tendency of burned woodlands to reburn at shorter intervals. The conversion of vegetation structure in this way on a large enough scale may even accelerate climate change by releasing stocks of carbon from the soil and vegetation to the atmosphere. Projections for Europe suggest significant reductions in species richness even under "average" scenarios of climate change, and an increased frequency of such extremes is likely to exacerbate overall biodiversity losses.

Species extinctions

Extinction occurs when there has been a global loss of all individuals of a species. The International Union for Conservation of Nature (IUCN) Red List of threatened species categorizes a species as extinct "when there is no reasonable doubt that the last individual has died" (Box 7.6). Some scientists have already linked the extinction of species of butterflies and amphibians to the shrinking of their range by climate change.[38] But climate change was not thought to be the only factor causing the extinctions. Furthermore, species extinctions often lag behind climate change. Long-lived species, in particular, can survive beyond the time their habitat has been disturbed. Hence, extinctions may be delayed until some event or disturbance finally removes the species from the landscape.

Species extinctions resulting from climate change are often caused by trophic mismatches (Box 7.3), where a mismatch between food resource and predator as a result of climate change can lead to critical population declines. While there is relatively little evidence that recent climate change has caused extinctions, a number of projections raise serious concern for the future of many species. A continental-scale study indicated that more than 50% of 1,350 modeled European plant species might become either vulnerable, endangered, critically endangered or committed to extinction by 2080, under the assumption that they are unable to migrate. The actual extinction rates therefore critically depend on the capability of species to shift their range. Studies have shown that mountain plants are particularly susceptible to extinction, primarily because they frequently have little or no ability to shift their range.

Some species and ecosystems are more vulnerable to extinction than others, particularly small populations or those restricted to small areas. In mountainous regions of Europe, endemic tree species have been replaced by other species (such as spruce and pine), which have migrated upwards owing to a number of factors, including climate change. These endemic mountain species can also migrate upwards, but only until they run out of mountain. In the Arctic, tundra areas have virtually no possibility for adapting under global warming since they cannot migrate beyond the Arctic Ocean.

Box 7.6 IUCN and the Red List

The International Union for Conservation of Nature (IUCN), is the world's oldest and largest global environmental network of more than 1,000 government and NGO membership organizations, and involving almost 11,000 scientists in more than 160 countries. Its headquarters are located in Gland, near Geneva, in Switzerland. IUCN is concerned with developing knowledge and management to conserve nature and biodiversity. The issue of climate change is becoming an increasing part of the concern for IUCN, which works to include biodiversity concerns in adaptation and mitigation polices and practice, as well as furthering natural resource management strategies that help species and ecosystems adapt to the impacts of climate change.

IUCN maintains the IUCN Red List of Threatened Species. This list was first conceived in 1963 and sets the standard for conservation assessment efforts. Over time, the extinction risk of more and more plant and animal species have been evaluated using a set of standardized criteria. There are nine categories in the Red List system: extinct, extinct in the wild, critically endangered, endangered, vulnerable, near threatened, least concern, data deficient, and not evaluated. Classification into the categories for species threatened with extinction (vulnerable, endangered, and critically endangered) is through a set of five quantitative criteria, which are based on biological factors related to extinction risk and include: rate of decline, population size, area of geographic distribution, and degree of population and distribution fragmentation. The recent 2008 update of the Red List shows that climate change is increasingly becoming a driver for species extinction.

Indirect effects of climate change

It is not just climate change itself that can impact biodiversity. In some cases, the strategies that are taken to mitigate or adapt to climate change can affect biodiversity, sometimes by improving conditions for vulnerable species and ecosystems and in other cases by enhancing pressures on biodiversity (Figure 7.5).

Mitigation and adaptation can be seen as complementary strategies for dealing with climate change, and both are necessary to cope with climate change. The relationship between mitigation and adaptation is complex, as the costs and benefits are not equally experienced between different actors or spatial and temporal scales. There is one primary difference between mitigation and adaptation efforts, in that mitigation benefits are essentially global, whereas adaptation benefits are more localized. However, both efforts might compromise biodiversity, if not properly applied.

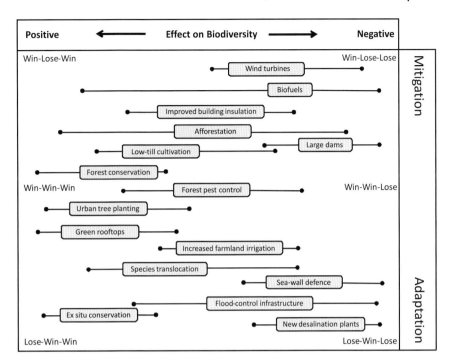

Figure 7.5 Relationships between adaptation and mitigations measures and impacts on biodiversity. Source: Paterson et al.[39] Reprinted with permission from John Wiley & Sons.

Mitigation

Climate change mitigation seeks a net reduction of greenhouse gas emissions, which also concerns the protection and promotion of carbon sinks through land use and habitat management. For example, two important habitats which can make a potentially significant contribution to carbon sequestration and storage are forest and wetlands/peatlands. Mitigation also involves the encouragement of the use of non-carbon or carbon-neutral energy sources, and the improvement of energy efficiency.

The installation of infrastructures required to support the necessary rise in renewable energy could have detrimental impacts on biodiversity. Wind energy has undergone a tremendous growth in recent years in Europe. These facilities were initially installed on land, but are now increasingly being deployed offshore. Wind farms potentially cause various problems for species and habitats, including: (1) collision of birds and bats with moving turbine blades; (2) disturbance of breeding birds cause by presence of the turbines and their noise; and (3) barriers to movement, leading to some disruption of ecological links.

The large projected increases in bioenergy production pose additional environmental pressures on biodiversity and on soil and water resources, if not managed properly.[40] These negative impacts of bioenergy can be minimized by growing low-impact bioenergy crops and not allowing the ploughing of permanent grasslands, or by adapting the intensity of residue extraction to the local soil conditions, so that soil fertility is preserved. Care must also be taken in converting high nature value farmland to bioenergy crop production, since these bioenergy crops are most often highly productive monocultures, where biodiversity is intrinsically low.

Adaptation

Adaptation to climate change in other sectors of society will in many cases have unintended consequences for ecosystems and biodiversity. Effects may be both favorable and detrimental to biodiversity.

One of the major consequences of climate change is rising sea levels, which in many areas calls for improvement and extension of large-scale physical coastal defenses against flooding, which often take the form of sea walls (Chapter 6). Such coastal defenses often negatively affect biodiversity, since they limit the formation of natural shorelines and the often rich coastal wetlands. Alternative adaptation options, such as strategic placement of artificial wetlands can benefit biodiversity, but with sea-level rise there is a compromise to be made with the loss of land (and property) against the protection of coastal ecosystems and biodiversity. In this context some of the most vulnerable ecosystems are located along areas with relatively little tidal impacts (e.g., the Baltic Sea and the Mediterranean).

Adaptation practices in agriculture and forestry that affect land use will also have serious effects on biodiversity. In some areas of Northern Europe, the wetter winter climate and more intense rainfall events will make some low-lying areas along rivers and lakes unsuitable for agricultural cultivation and use. Restoring such cultivated wetland to pristine conditions will have positive consequences not only for biodiversity of the wetlands but also for reducing nutrient leakages to freshwater ecosystems and thus improving their ecological condition.[41] However, adaptation in agriculture can also have negative consequences for biodiversity, for example, through increasing use of pesticides for preventing pests and diseases and through the need for expanding irrigation (Chapter 4). Some of the high nature value in Europe is associated with extensive agricultural systems (Figure 7.2), which may suffer from increasing heatwaves and droughts in a future climate.

How will individual ecosystems be affected?

The impacts of climate change on ecosystems and biodiversity will vary between regions. The most rapid changes in climate and associated effects on ecosystems are expected in the far north (Arctic), in the Mediterranean and in mountainous regions. Unfortunately, these regions are also home to many species that have no alternative habitats to which they can migrate in order to survive. The following

Table 7.1 Positive and negative impacts of climate change on forest growth and structure. Source: EEA.[43] Reprinted with permission from European Environmental Agency, Copenhagen.

Direction	Influencing factor
Improved growth	Warmer winters will extend the growing season and increase productivity.
	Elevated atmospheric CO_2 increases photosynthesis and growth rates.
	Cold and snow related damages will become less common.
Reduced growth	Reduced summer rainfall will reduce growth and prolonged and recurrent droughts may kill trees.
	Elevated atmospheric ozone levels may negatively affect growth.
	Increased storm intensity may lead to more storm fall and damage.
	Possible increase in spring frost damage may occur, due to increased susceptibility during leafing.
	Increased damage may result from forest fires and insect pest outbreaks.

briefly describes how climate change will most likely affect some of the major ecosystems of Europe.

Forests

Forest ecosystems in Europe are very likely to be strongly influenced by climate change and other global changes (Table 7.1). The forested area is expected to expand in the north, along with a decrease in the current tundra area, but contract in the south. Native conifers are likely to be replaced by deciduous trees in western and central Europe (Figure 7.6), and the distribution of a number of typical tree species is likely to decrease in the Mediterranean.

In the northern and maritime temperate zones of Europe, and at higher elevations in the Alps, forest productivity is likely to increase throughout the century. However, increasing droughts and high temperatures in continental central Europe and in southern Europe is likely to decrease forest productivity by the end of the century. Some of these reductions in productivity from increasing droughts may be partly offset by the higher atmospheric CO_2 concentrations.

Abiotic hazards and related disturbance regimes (Box 7.5) for forests are likely to increase, although expected impacts are regionally specific and depend on forest management systems used. In northern Europe, snow cover will decrease, and soil-frost-free periods and winter rainfall increase, leading to increased soil waterlogging and winter floods. Fire danger, length of the fire season, and fire frequency and severity will increase in the Mediterranean,[44] and lead to increased dominance of shrubs over trees. Fire danger is also likely to increase in central,

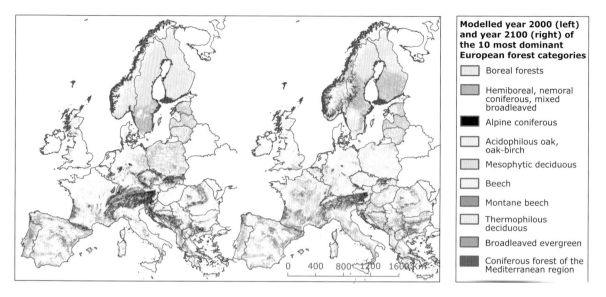

Figure 7.6 Modeled forest cover types under current (2000) climate (left) and future (2100 for A1B emission scenario and NCAR CCM3 GCM) (right) climate. Source: EEA.[42] Reprinted with permission from European Environmental Agency, Copenhagen. (See color plate.)

eastern and northern Europe. This, however, does not translate directly into increased fire occurrence or changes in vegetation. The range of important forest insect pests will expand northwards, but the net impact on forest growth and biodiversity in combination with other anthropogenic influences are difficult to predict.

Shrublands

The area of shrubland has increased in Europe over the last decades, particularly in the south. Climate change is likely to affect shrubland ecosystem services such as carbon storage, nutrient cycling, and biodiversity. The response to warming and drought will depend on the current conditions, with cold-moist sites being more responsive to temperature changes, and warm-dry sites being more responsive to changes in rainfall.[45] In northern Europe, warming will increase growth and productivity, hence enabling higher grazing intensities, partly associated with increased dominance of grasses. In southern Europe, warming and, particularly, increased drought are likely to lead to reduced plant growth and changes in community structure. Shrubland fires will increase owing their high propensity to burn. With increased rainfall intensity under climate change (Chapter 2) these more frequent fires can lead to increased erosion risk, due to reduced ground cover and slow plant regeneration.[46]

Grasslands

Climate change will alter the community structure of grasslands in ways specific to their location and type. Management and species richness may increase resilience to change. Fertile, early succession grasslands have been found to be more responsive to climate change than more mature and/or less fertile grasslands, indicating that a higher biodiversity increases the resilience to climate change. In general, intensively managed and nutrient-rich grasslands are expected to respond positively to both increased CO_2 concentration and temperature, given that water and nutrient supply is sufficient. Nitrogen-poor and species-rich grasslands on the other hand may respond to climate change with small changes in productivity in the short term. Overall, productivity of temperate European grasslands is expected to increase, but warming alone is likely to have negative effects on productivity and biodiversity.[47] In the Mediterranean, changes in precipitation patterns are likely to negatively affect productivity and biodiversity of grasslands.

Mountain regions

The duration of snow cover is expected to decrease by several weeks for each degree Celcius increase in temperature in the Alpine region at middle elevations. An upward shift of the glacier equilibrium line is expected from 60 to 140 m per 1°C, and glaciers will experience a substantial retreat during the 21st century. Small glaciers will disappear, while larger glaciers will suffer a volume reduction between 30% and 70% by 2050.[48] The lower elevation of permafrost is likely to rise by several hundred meters. Rising temperatures and melting permafrost will destabilize mountain walls and increase the frequency of rock falls, threatening mountain valleys. In northern Europe, lowland permafrost will eventually disappear.

These climate-induced changes will severely affect European mountain flora. Change in snow cover duration and growing season length will have much more pronounced effects than direct effects of temperature changes on plant growth. Overall trends are towards increased growing season, earlier phenology and shifts of both plant and animal species distributions towards higher elevations. The treeline will shift upward by several hundred meters. There is evidence that this process has already begun in Scandinavia, the Ural Mountains, West Carpathians and the Mediterranean.[49]

These changes, together with the effect of abandonment of traditional alpine pastures, will restrict the alpine zone to higher elevations, severely threatening nival flora.[50] The composition and structure of alpine and nival communities are very likely to change. Local plant species losses of up to 62% are projected for Mediterranean and Lusitanian mountains by the 2080s under the A1 emission scenario.[51] Mountain regions may additionally experience a loss of endemism, owing to invasive species.[52] Similar extreme impacts are expected for habitat and animal diversity as well, making mountain ecosystems among the most threatened in Europe.

Arctic regions

The Arctic covers the areas north of the tree line (i.e. areas covered by tundra). Here the mean temperature of the warmest month does not exceed 12°C, which limits vegetation to small bushes, grasses and herbs. The part of the Arctic furthest north is called the High Arctic in comparison to the warmer Low Arctic. Bushes in the Low Arctic grow up to half a meter high, whereas plants in the High Arctic seldom exceed 5–10 cm. At the far north the High Arctic is actually desert, and the precipitation from snow and rain does not exceed that of precipitation in the Sahara.

Plants and animals in the Arctic have during millennia been adapted to the extreme conditions, and many species are directly dependent on ice and snow. Polar bears feed almost solely on seals that they hunt on sea ice (Box 7.7). Many more species that depend on sea ice will experience problems as the sea ice melts and is reduced in extent, in particular during summer. Similar problems will occur on land as bushes and trees take over the tundra. Vegetation models based on climate change scenarios show that more than half of the Arctic tundra could be replaced by shrublands or forests by the end of this century.[53] This will displace the plants and animals that currently depend on the low vegetation of the tundra.

Increasingly vigorous vegetation in the tundra and the disappearance of snow and ice will, of course, benefit species that can expand northwards, but this is often species that are common and widely distributed in the areas south of the Arctic. As they spread further north, they will replace Arctic species, which will, in turn, spread further northwards when possible. But here lies the core of the problem, since plant and animals cannot spread further than to the rim of the Arctic Ocean, and the species that depend on sea ice have nowhere to move if the ice disappears during summer.[54] The High Arctic zone will be particularly hard hit, since it in most cases occupies only a narrow rim between the Low Artic zone to the south and the Arctic Ocean to the north. This means that species that live under the extreme conditions of the High Arctic will disappear or be reduced considerably in number and extent. The warmer winter climate in High Arctic also means that there will be more periods with thaw in winter. The result is ice encasement on the snow making it difficult for musk ox (*Ovibos moschatus*) and other herbivores to reach the vegetation beneath the snow.[55] They may therefore also face extinction over vast territories.

The warmer climates could lead to progressive drying of the tundra wetland habitats for migrant birds, and many species of Arctic-breeding shorebirds and waterfowl are projected to undergo major population declines of up to 50% as tundra habitat shrinks with global temperature increases of up to 2°C.[56] Many are migrant species that overwinter in coastal areas further south in Europe, where habitats are vulnerable to sea-level rise.

Box 7.7 Polar bears – icons of climate warming

There are an estimated 20,000–25,000 polar bears (*Ursus maritimus*) world-wide, mostly inhabiting the sea ice of the circumpolar Arctic from Russia to Alaska, from Canada to Greenland and Norway's Svalbard archipelago. Here they may wander for thousands of kilometers per year. They are specialized in eating ice-breeding seals and are therefore dependent on sea ice for survival. Female bears require nourishment after emerging in spring from 5 to 7 months of fasting in nursing dens. They are therefore very dependent on close proximity between land and sea ice before it breaks up. Continuous access to sea ice allows the bears to hunt throughout the year, but in areas where the sea ice melts completely each summer, they are forced to spend several months in tundra fasting on stored fat reserves until freeze-up. Already now several hundred polar bears every summer are stranded on the Svaldbard islands north of Norway as the ice melts away so that there is several hundred kilometers to the rim of the sea ice. On the islands they live a miserable life, and may starve or eat each other.

Polar bears face great challenges from the consequences of global warming, primarily because reductions in sea ice will drastically shrink the marine habitat for them, ice-inhabiting seals and other animals.[57] Polar bears depend on frozen ice from where to hunt seals. Without ice the bears are unable to reach their prey. In a warmer Arctic the bears may not be able to swim the distances required to reach solid ice when reduced ice coverage leads to more open water. In a future warmer world, bears may be leaving the sea ice to den on land. In Russia, large number of bears have already been stranded on land by long summers which prevent the advance of a permanent ice pack. Although the Arctic has experienced warm periods before, the present shrinking of the Arctic's sea ice is rapid and unprecedented. The IUCN Polar Bear Specialist Group in 2005 decided to upgrade the IUCN Red List classification of the polar bear to vulnerable, the primary reason being the concern for reductions in population size due to loss of habitat. If sea ice continues to decline according to some projections, polar bears will face a high risk of extinction with warming of 2.8°C above the pre-industrial level.[58]

Wetlands and freshwater ecosystems

The predicted climate change may significantly impact northern peatlands. The common hypothesis is that elevated temperature will increase productivity of wetlands and intensify peat decomposition, which will accelerate carbon and nitrogen emissions to the atmosphere.[59] This in itself feeds back to the climate system, enhancing warming (Chapter 2). Loss of permafrost in the Arctic is likely to cause a reduction of some types of wetlands in the current permafrost zone. During dry years, catastrophic fires are expected on drained peatlands in European Russia.

European freshwater systems are already being affected by many human activities, which makes it difficult to disentangle effects of climatic factors from other pressures. There is therefore limited data to demonstrate unequivocally the impact of climate change on water quality and freshwater ecology. On the other hand, there are many indications that freshwaters are highly susceptible to climate change, and that climate change will severely hinder attempts to restore many polluted water bodies to good ecological status.[60]

Throughout Europe, in lakes and rivers that freeze in the winter, warmer temperatures may result in earlier ice melt and in longer growing seasons. A consequence of these changes could be a higher risk of algal blooms and increased growth of toxic cyanobacteria in lakes.[61] Higher precipitation and reduced frost may enhance nutrient loss from cultivated fields. These factors may result in higher nutrient loadings, which will intensify the eutrophication of lakes and wetlands.[62] Streams in catchments with impermeable soils may have increased run-off in winter and deposition of organic matter in summer, which could reduce invertebrate diversity.

The reductions of rainfall in southern Europe will reduce the water volume of lakes and rivers and lead to increased salinization. Many ephemeral[63] ecosystems may disappear, and permanent ones shrink. Although an overall drier climate may decrease the external loading of nutrients to inland waters, the concentration of nutrients may increase because of the lower volume of inland waters. An increased frequency of high rainfall events could also increase nutrient discharge to wetlands.

Warming will also affect the physical properties of inland waters, in particular through changing water temperatures. In stratified lakes the temperature of the bottom water and the duration of stratification will increase in summer, leading to higher risk of oxygen depletion at the lake bottom. Higher temperatures will also reduce dissolved oxygen saturation levels and increase the risk of oxygen depletion.[64] These increased risks of hypoxia[65] will have large detrimental effects on lake biodiversity and may lead to regime shifts.

Marine ecosystems

The oceans play an important role in regulating climate and transporting heat and nutrients across the globe, not least in Europe, where the Gulf Stream and its extensions influence European weather patterns and will greatly modify climate change regionally (Chapter 2). However, the oceans themselves are also impacted by climatic conditions, and the resulting changes in temperature, salinity and pH affect marine ecosystems.

Climate change impacts are observed in all European seas[66] with consequent effects on fisheries (Chapter 4). The primary physical impact of climate change on European regional seas is increased sea surface temperature. However, because of different geographical constraints, climate change is expected to affect physical conditions differently in different seas, and consequently biological impacts also vary depending on the region.

In the Northeast Atlantic, sea surface temperature changes have already resulted in a lengthening of the marine growing season and a northward movement of marine zooplankton. Some fish species are shifting their distributions northward in response to increased temperatures.

In the Baltic Sea climate models project a mean increase of 2–4°C in the sea surface temperature in the 21st century, and increasing run-off and decreasing frequency of Atlantic inflows, both of which will decrease the salinity of the sea.[67] Consequently, the extent of sea-ice is expected to decrease by 50–80% over the same period and stratification is expected to become stronger, increasing the probability of a deficiency of oxygen (hypoxia) that kills marine life in the region. Changes in stratification are expected to affect commercially important regional cod fisheries, because stratification appears to be an important parameter for the reproductive success of cod in the Baltic Sea.

In the Mediterranean temperature is projected to increase and run-off to decrease. In contrast to the Baltic Sea, the combination of these two effects is not expected to change stratification conditions greatly because of the compensating effects of increasing temperature and increasing salinity on the density of sea water.[68] The invasion and survival of alien marine species in the Mediterranean is correlated with the general sea surface temperature increase, resulting in the replacement of local fauna with new species affecting ecosystem composition.

Coastal ecosystems

Coastal ecosystems are not only affected by changes in temperature and rainfall. They will also be highly impacted by sea-level rise (Chapter 6). In many cases this will lead to "coastal squeeze", where the coastal habitats will increasingly be restricted in area extent between the rising seas and the coastal defenses. This is particularly the case for intertidal coastal habitats that will decline every-where in Europe if the policy of "holding the line" of existing sea defenses continues.

The most vulnerable intertidal habitats are around the Black Sea, Mediterra-nean and the Baltic. Salt marsh and mudflats on these coasts are likely to shrink as sea-level rises, reducing their coastal protection functions. The threat to salt marsh and mudflats throughout Europe will increase during this century, particu-larly under high emissions scenarios. The length of coastline in northwest Europe that has a high vulnerability to sea-level rise is predicted to increase by 46% under the 2080s high emissions scenario.[69]

How will ecosystems adapt to climate change?

Adaptation is vital to avoid unwanted impacts of climate change, especially for ecosystems that are vulnerable to even moderate levels of warming. It is also seen as a means for maintaining or restoring ecosystem resistance/resilience to single or multiple stresses. In ecology a distinction is made between *resistance* and *resilience* of a system to change in the environment. Resistance is the ability to withstand climatic extremes, whereas resilience is the ability to recover from such extremes. Such a distinction between resistance and resilience is often not made in economic and social sciences. Ideally, adaptation should promote both resistance and resilience of systems to climate change, but in practice it may be impossible to ensure that ecosystems can withstand climate change, in particular where more frequent extremes occur. Therefore, it becomes increasingly relevant that ecosystems are resilient and can recover from such extremes.

The IPCC considers two types of adaptation: autonomous (or spontaneous) adaptation and planned (or societal) adaptation. In the case of biodiversity, the former occurs at the species and habitat levels and includes the various responses to climate change as have already been observed, and the latter includes human management and policy actions aimed at facilitating cost-effective and timely adaptation that preserves the ecosystem services associated with biodiversity.

The role of autonomous adaptation

When climatic conditions for individual species change beyond their tolerance levels, they may be forced to respond by shifting the timing of life-cycle events (phenology), shifting their geographical boundaries (range shift), or changing their morphology, behavior or genetic constitution. When neither adaptation nor shifting range is possible, extinction becomes a likely scenario.[70] The cause behind the response can be either plastic (changes within individuals during their lifetimes) or genetic (changes in genotypes between generations and among populations).

Climate on earth and the conditions for life have constantly changed over millions of years. For the last two million years changes have been dominated by intermittent ice ages. This led to massive extinctions – in particular at the onset of ice ages. But new species have also evolved during the millennia, and their distribution on earth has constantly changed as consequence of environmental changes and competition from other species. History thus shows that life on earth – in general terms – is quite robust. However, the climate changes we are now facing are expected to yield a warmer climate than the earth has experienced for more than a million years – with the major warming occurring over just several decades. Ecosystems are thus facing a new type of climate change that will proceed with unprecedented speed.

Even though we cannot draw any direct parallels between current climate change and the changes that have occurred previously, we can draw some conclusions. For example, we must expect considerable changes in biological systems, including loss of species. The projected climate changes will happen so fast that most species are unlikely to be able to adapt through evolutionary pathways (i.e. changes in genetic make-up). Therefore, among their few possibilities to avoid extinction are to change habitat range or shift seasonal patterns.

The situation is complicated by the fact that ecosystems are not only affected by climate change. The great growth in human population over the last 100 years and the increasing need for food, feed, fiber and fuel have so far been the main causes of extinction in Europe and worldwide. Large areas have been allocated for agriculture, urbanization and infrastructures, resulting in massive fragmentation of natural habitats, and many species will have great difficulty in spreading between these isolated pockets. Furthermore, many of these "natural" areas have been affected by pollution and other forms of human activity and are therefore less suited habitats for wild animals and plants than previously. This greatly reduces the efficiency of natural (autonomous) adaptation processes such as range shifts and changing species interactions.

Management of forests, shrublands and grasslands

Since forests are usually intensely managed in Europe there is a wide range of available management options that can be employed for adapting forests to climate change. General strategies include changing the species composition of forest stands, and planting forests with seedlings genetically suited to a new climate. Adaptive forest management could substantially decrease the risk of forest destruction by wind and other extreme weather events. Strategies for coniferous forests include planting of deciduous trees better adapted to the new climate as appropriate, and the introduction of multi-species planting into currently mono-species coniferous plantations. These are complex issues and new long-term planning tools that account for the changing environment over the coming decades are therefore needed.[71]

Adaptation strategies need to be specific to different parts of Europe. The range of alternatives is constrained, among other factors, by the type of forest. Forests that are already moisture-limited (Mediterranean forests) or temperature-limited (boreal forests) will have greater difficulty in adapting to climate change than other forests (e.g. in central Europe). Fire protection will be important in Mediterranean and boreal forests and includes replacement of highly flammable species, regulation of age class distributions, and widespread management of accumulated fuel, eventually through prescribed burning. Public education and forest health monitoring are important prerequisites for such adaptation policies to work effectively.

Productive grasslands are closely linked to livestock production. Dairy and cattle farming may become less viable because of climate risks to fodder production and grasslands could therefore be converted to cropland or abandoned, leading to conversion to shrubland or forest.[72] A range of grassland management measures are available to regulate the grassland community structure and productivity, including regulating grazing pressure, revising cutting strategies, fertilization and irrigation. Shrublands will in many cases need some grazing to maintain ecological structure and functioning – also under climate change. The management changes for grasslands and shrublands under climate change are – like forests – going to be locally specific, depending on the community structure, the direction and magnitude of climate change and extent of other pressures on these ecosystems. There is therefore a substantial need for improved local knowledge and research on improved management of these ecosystems under climate change.

Management of wetlands and aquatic ecosystems

Better management practices are needed to compensate for possible climate-related increases in nutrient loading to aquatic ecosystems from cultivated agricultural areas in northern Europe.[73] These practices include improved fertiliser management and improved crop rotations, such as growing cover crops to reduce nutrient losses. A higher level of treatment of domestic and industrial sewage and reduction in farmland areas can further reduce nutrient loadings to surface waters and also compensate for climate-related increases in these loadings. To further reduce the load of nutrients to surface waters and transport to the sea, several methods are available. A key measure is to enhance the retention time of water by the re-meandering of channelized streams, and (re)establishment of wetlands, including constructed wetlands in cropland areas.

Freshwater wetlands are rare and are threatened by many factors, including climate change. Therefore, protection and restoration of wetlands have high priority in many European countries. Conversion and restoration, however, are not easy tasks, because the type of wetland that can exist at a given spot depends on water regime, water- and soil chemistry and fluctuations therein. In practice this implies that minor adjustments in the hydrological system often result in complete conversion of one wetland type into another. Local restoration has little or no effect in such situations, which becomes even more complex under changing climatic conditions.

Practical possibilities of adaptation of northern wetlands and peatlands to climate change are limited and could be realized only as part of integrated landscape management including minimizing other human pressures, avoiding physical destruction of surfaces and applying appropriate technologies of infrastructure development on permafrost. Protection of drained peatlands against fire in European Russia is an important regional problem, which requires restoration of drainage systems and regulating water regimes in such territories.

In southern Europe, to compensate for increased climate-related risks (lowering of the water table, salinization, eutrophication, and species loss), a lessening of the overall human burden on water resources is needed. This would involve stimulating water saving in agriculture, relocating intensive farming to less environmentally sensitive areas and reducing diffuse pollution, increasing the recycling of water, increasing the efficiency of water allocation among different users, facilitating the recharge of aquifers, and restoring riparian vegetation.

Management of coastal areas

Adaptation in coastal systems can be summarized in a variety of ways (Chapter 6). A common approach is to consider the three planned approaches of: (1) protection, (2) accommodation, and (3) retreat.[74] "Protection" uses coastal defenses to reduce the risk of flooding and erosion but inadvertently prevents the onshore migration of coastal ecosystems (resulting in "coastal squeeze"). "Accommodation" reduces the impacts of flooding and erosion by changing land use and building design for this purpose. Planned "retreat" reduces risk by removing assets from the areas threatened by flooding and erosion. Both retreat and, to a lesser extent, accommodation allow wetland habitats to migrate inland countering coastal squeeze, and in the case of retreat, allows an increase in intertidal area.

The retreat option (also sometimes called "managed realignment") is attracting increasing interest in Europe.[75, 76] As described in Chapter 6, it is seen as an opportunity for sustainable coastal development, including the restoration of coastal ecosystems.

Safeguarding ecosystems and their services

Biodiversity loss and ecosystem degradation continue in Europe and globally, despite the fact that policy makers, administrators, NGOs and businesses around the world have been seeking ways to halt these processes. There are many reasons for this, but perverse economic drivers as well as failures in markets, information and policy are significant factors.[77] Markets tend not to assign economic values to the largely public benefits of conservation, but do assign value to private goods and services which may result in ecosystem damage.

Market failure can cover anything from the lack of markets for public goods and services (e.g., absence of "markets" for species conservation or for most of the regulating and supporting services of ecosystems) to imperfections in structure or process around markets which cause inefficiency and distortions (e.g., by not effectively including and pricing all carbon emissions). The size of the challenge of market failure should not be underestimated, and for some services (e.g., scenic beauty, hydrological functions and nutrient cycling) it is difficult even to obtain a profile of demand and supply.

Failure to provide appropriate and adequate information on the values of the services is often one of the contributors to market failure. Decision makers often have insufficient facts, tools, arguments or support for taking different decisions and avoid biodiversity loss. There are many cases of local economy and local societal losses from ecosystem degradation in the interests of short-term private gain.

Policy failures arise when incentives encourage harmful action. Taxes and subsidies can sometimes encourage ecosystem degradation, even where nature offers a sustainable flow of services to the economy and to society. Environmentally harmful subsidies discriminate against sound environmental practices while encouraging other, less desirable activities. Subsidy policies within agriculture and fisheries have often had these effects, also in Europe. Policy reforms within the EU have, over the past decades, sought to rectify some of these incentive failures. Many agricultural practices can support high-value biodiversity. But without appropriate recognition, for example through payments for environmental services, some good practices risk disappearing.

Therefore policy measures and incentives have to be put into place to support protection of biodiversity and ecosystems services under climate change, and must try to avoid the above mentioned deficiencies and failures within markets and policy. This requires not only specific measures to protect existing nature reserves and high-value habitats, but a concerted effort across a wide range of policy areas (Table 7.2). Indeed, the many ways in which climate and biodiversity interact suggest a need for more integrated policy responses. This would mean mainstreaming biodiversity protection and the associated climate

Table 7.2 Policy areas of concern and associated measures and instruments for ensuring adaptation of European ecosystems to climate change

Policy area	Policy measures and aims	Policy instruments
Nature conservation	Protection of natural habitats	Nature reserves, national parks
	Protection of threatened species and germplasm	Move species to new locations, maintain species in botanical gardens and zoos
Spatial planning	Ecological corridors in the landscape	Restrict urban sprawl and promote connectivity of habitats
Pollution	Remove both point source and diffuse pollution of ecosystems	Legislation, monitoring and control
Agriculture	Maintenance of high nature farmland	Subsidies or specific support schemes
	Reduce nutrient loading	Nutrient quotas or taxation
Fisheries	Avoid overfishing of threatened species	Fishing quota or protected areas
Water	Avoid water overuse (irrigation) in dry regions	Water planning and pricing
Energy	Promote environmentally friendly bioenergy crops	Standards for sustainable bioenergy production
Research	Studies on species interactions and efficient ecosystem management	Public funding and tax exemptions
	Monitoring networks	Funding
Information	Increase awareness of climate change and biodiversity issues.	Information campaigns and educational policies.

change issues into a range of policy areas needed to support effective adaptation measures. To do so requires effective policy incentives that in many cases also need to be supported by efficient planning, financing, administration and control. Species range shifts will not respect political or administrative boundaries, so effective conservation and biodiversity policies require regional and international collaboration.

Land and water policies

There are many pressures besides climate change that impact upon the environment and on biodiversity, in both urban and rural environments, not least related to the use of land and water. These pressures may interact among themselves and interact with climate change, to produce further indirect effects. Many of these pressures are related to demographic changes with expanding urban boundaries and increasing infrastructure, not least related to traffic. In addition, the changes towards more intensive agriculture also impacts on biodiversity.

Urban sprawl may encroach on sites that are important for wildlife directly (protected sites) or which contribute to the quality of wildlife sites elsewhere, for example, water gathering grounds where precipitation can infiltrate and reach groundwater and river systems. This urban sprawl and the associated traffic infrastructure are greatly degrading natural habitats in western and central Europe (Figure 7.7). The massive fragmentation of natural habitats in the landscape greatly reduces the possibilities for migration of plants and animals and is a major obstacle to the natural adaptation of ecosystems to climate change. It is questionable whether even much stricter land use and planning policies would be able to ensure efficient corridors for migration in the most populated parts of western and central Europe.

There are also many pressures on natural habitats and biodiversity in rural areas as a result of habitat destruction, fragmentation or over-exploitation. The rapid post-war transformation and intensification of agriculture in many parts of Europe brought with it a decline in many important species and habitats found on agricultural land.[79] The former included many arable weeds, insects, gamebirds, field birds and a variety of other species. The reasons for this decline relate to new agricultural practices ranging from changes in cultivation techniques (e.g. altered sowing dates and greater use of fertilizers and pesticides), to the loss of hedgerows and field margins. Within the EU these changes are greatly affected by the Common Agricultural Policy (CAP) (Chapter 4), and some of the agri-environmental schemes in the CAP seek to support biodiversity in the agricultural landscape, which forms a large part of the European area (Figure 7.1). However, so far the climate change issue and the need for new farming and landscape-based practices to preserve biodiversity has not to any significant extent been included in these agri-environmental schemes.

Policy development in other areas may affect land available to biodiversity (e.g., policy to increase generation of renewable energy). Climate change mitigation

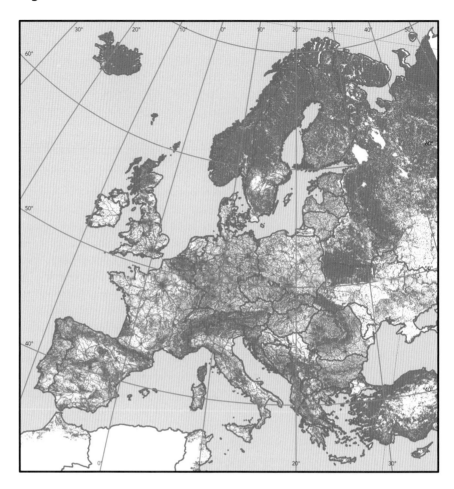

Figure 7.7 Pressures from urbanization and transport (red) on semi-natural areas in Europe. Source: EEA.[78] Reprinted with permission from European Environmental Agency, Copenhagen. (See color plate.)

through intensified forestry and land allocation to biofuel production could lead to increased pressure on biodiversity. Land use and management clearly fulfill various, often conflicting goals, such as food, timber and biofuel production, carbon sequestration, soil erosion protection, maintenance of biodiversity and recreation. Economic incentives and legislation should be used to balance strategies for fulfilling the different goals.

Coastal areas in particular demonstrate an array of competing land uses (e.g., housing, tourism, ports and energy), all of which may interact further, affecting biodiversity. However, there is increasing interest in managed coastal realignment across Europe, especially in northern Europe, offering opportunities for intertidal habitat creation that might serve adaptation purposes to preserve coastal ecosystems.

Activities in the water supply and drainage industry (Chapter 5), such as inter-catchment transfer, reservoir building and water abstraction may have significant consequences for aquatic, wetland and terrestrial biodiversity, particularly where species may be under stress as a consequence of direct impact of climate change, such as drought conditions. The promotion of infiltration across catchments may require restrictions on some activities within flood plains. The increasing replacement of permeable surfaces in residential areas by impermeable surfaces caused by, house extension, lawn replacement and so on, may fall outside planning powers.[80]

Nature conservation and reserves

Climate change threatens the assumption that the ranges of species will remain unchanged which underpins current conservation policy. The ability of countries to meet the requirements of EU Directives and other international conventions (Box 7.1) is likely to be compromised by climate change, and a more dynamic strategy for conservation is required for sustaining biodiversity.[81]

Conservation strategies relevant to climate change can take three forms: (1) selection, design and management of conservation areas (protected areas, nature reserves, Natura 2000 sites, wider countryside); (2) conservation of germplasm in botanical gardens, museums and zoos; and (3) translocation of species into new regions or habitats.[82] In Europe, such conservation measures for mitigating climate change impacts have not yet been put in place. However, an expansion of reserve areas will be necessary to conserve species in Europe. An assessment thus shows that European protected areas need to be increased by 18% to meet the EU goal of providing conditions by which 1,200 European plant species can continue thriving in at least 100 km^2 of habitat.[83] To meet this goal under climate change by 2050 it was estimated that the current reserve area must be increased by 41%. The study showed that it would be more cost-effective to expand protected areas proactively rather than waiting for climate change impacts to occur and then responding to changes.

Another important adaptation tool is dispersal corridors for species, although large heterogeneous reserves that maximize microclimate variability might sometimes be a suitable alternative. Despite the importance of modifying reserve areas, some migratory species are vulnerable to loss of habitat outside Europe.[84] For these migratory species transcontinental conservation policies need to put in place.

The existing network of protected nature areas in Europe form a very patchy and fragmented pattern (Figure 7.2), which, to a great extent, are separated by intensive farmland (Figure 7.1) or urban and traffic infrastructures (Figure 7.7). Adaptation measures to link networks or isolated patches under these conditions are a great challenge and require a combination of the following measures:[85] (1) creation of new habitat patches; (2) enlarging existing habitat patches: (3) creation of robust corridors; (4) improvement of permeability of the landscape matrix; and (5) mitigation of barriers caused by, for instance, roads. Ensuring such a major change in connectivity of ecological networks in Europe requires substantial efforts in spatial planning linked with a range of other policies such as agri-environmental schemes and all policies affecting urban development and traffic planning.

Strategies based on providing opportunities for species and ecosystems to change in many different ways will be increasingly important compared to ones targeting individual species. Ensuring that a high diversity of habitat types are available for native species seems likely to provide the best opportunity for ensuring the persistence of species diversity at local, regional and European scales. It will help buffer against changes in bioclimatic niches, changes in trophic interactions and disturbance regimes. This suggests that conservation policies should have the following priorities under climate change:[86]

- Conservation managers should develop a broad understanding of biodiversity responses to climate change and the implications for conservation, before committing to solutions based solely on range shift models.
- There should be a renewed emphasis on protecting the diversity of habitat types at local, regional, national and European scales to provide as much opportunity as possible for as many species as possible to survive. This will provide future habitats and buffering against changed species interactions and threats under climate change.
- Management of habitats should reduce stresses from invasive species, land use intensification, altered hydrology, excessive fires and other threats.
- Connectivity in the landscape should be tactically managed (sometimes increasing it and sometimes decreasing it) to allow migration and to protect climatic refuges.

Many conservation practitioners in Europe are now taking these messages on board and arguing for a landscape approach to reduce the negative impacts of climate change on biodiversity.[87] The focus thus shifts from protecting individual sites to ensuring an overall protection of the landscape, which also involves consideration of the interaction among landscape components, which may also change under climatic warming. This calls for larger protected areas or landscapes that cover a range of elevations, microclimates and ecosystems, and this will allow improved ecosystem interactions and ease the migration of species across the landscape. Such concepts are also embedded in the Natura 2000 network, where Article 10 of the EU Habitats Directive requires Member States to strengthen the coherence of the network of habitats. If this is indeed effectively implemented, we will see a substantially different European landscape under climate change, but it will still be rich in nature values and continue to support the many services that ecosystems deliver for the benefit of individual Europeans and the European society.

Notes

1   Sala, O.E., Chapin III, F.S, Armesto, J.J., Berlow, J., Bloomfield, E., Dirzo, R., Huber-Sanwald, E., Huenneke, L.F., Jackson, R.B., Kinzig, A., Leemans, R., Lodge, D.M., Mooney, H.A., Oesterheld, M., Poff, N.L., Sykes, M.T., Walker, B.H., Walker, M., Wall, D.H., 2000: Global biodiversity scenarios for the year 2100. *Science* 287, 1770–1774.

2   European Commission (EC), 2008: *A mid-term assessment of implementing the EC biodiversity action plan.* European Commission.

3   Endemic species are those species unique to a specific geographical location. Physical, climatic, biological and historic factors can contribute to endemism.

4   Millennium Ecosystem Assessment, 2005: *Ecosystems and Human Well-being. Biodiversity synthesis.* World Resources Institute, Washington DC.

5   EC, 2008: Saving the natural world – it's a big deal. European Commission. *Environment for Europeans* No. 16.

6   Bartholome, E.M., Belward, A.S., Beuchle, R., Eva, H., Fritz, S., Hartley, A., Mayaux, P., Stibig, H.-J., 2004: Global land cover 2000 map. (Catalogue No. LB-55–03.099–EN-C). Copyright European Commission.

7   EEA, 2007: *Europe's environment – The fourth assessment.* State of the environment report no. 1/2007. European Environment Agency, Copenhagen.

8   EC, 2008: *Invasive alien species.* Natura 2000. European Commission and DG ENV Nature Conservation Newsletter No. 25.

9   Millennium Ecosystem Assessment, 2005.

10 Schröter, D., Cramer, W., Leemans, R., Prentice, I.C., Araújo, M.B., Arnell, N.W., Bondeau, A., Bugmann, H., Carter, T.R., Gracia, C.A., de la Vega-Leinert, A.C., Erhard, M., Ewert, F., Glendining, M., House, J.I., Kankaapää, S., Klein, R.J.T., Lavorel, S., Lindner, M., Metzger, M.J., Meyer, J., Mitchell, T.D., Reginster, I., Rounsevell, M., Sabaté, S., Sitch, S., Smith, B., Smith, J., Smith, P., Sykes, M.T., Thonicke, K., Thuiller, W., Tuck, G., Zaehle, S., Zierl, B., 2005: Ecosystem service supply and vulnerability to global change in Europe. *Science* 310, 1333–1337.

11 Janssen, I.A., Freibauer, A., Ciais, P, Smith, P., Nabuurs, G.J., Folberth, G., Schlamadinger, B., Hutjes, R.W., Ceulemans, R., Schulze, E.D., Valentini, R., Dolman, A.J., 2003: Europe's terrestrial biosphere absorbs 7 to 12% of European anthropgenic CO2 emissions. *Science* 300, 1538–1542.

12 Reid, H., 2006: Climate change and biodiversity in Europe. *Conservation and Society* 4, 84–101.

13 Ciais, P., Reichstein, M., Viovy, N., Granier, A., Ogée, J., Allard, V., Aubinet, M., Buchmann, N., Bernhofer, C., Carrara, A., Chevallier, F., De Noblet, N., Friend, A.D., Friedlingstein, P., Grünwald, T., Heinesch, B., Keronen, P., Knohl, A., Krinner, G., Loustau, D., Manca, G., Matteucci, G., Miglietta, F., Ourcival, J.M., Papale, D., Pilegaard, K., Rambal, S., Seufert, G., Soussana, J.F., Sanz, M.J., Schulze, E.D., Vesala, T., Valentini, R., 2005: Europe-wide reduction in primary productivity caused by the heat and drought in 2003. *Nature* 437, 529–533.

14 Ainsworth, E.A., Long, S.P., 2005: What have we learned from 15 years of free-air CO2 enrichment (FACE)? A meta-analytic review of the responses of photosynthesis, canopy properties and plant production to rising CO2. *New Phytologist* 165, 351–372.

15 Parmesan, C., 2007: Influences of species, latitudes and methodologies on estimates of phenological response to global warming. *Global Change Biology* 13, 1860–1872.

16 Visser, M.E., Holleman, L.J.M., 2001: Warmer springs disrupt the synchrony of oak and winter moth phenology. *Proceedings of the Royal Society B: Biological Sciences* 268, 289–294.

17 Schweiger, O., Settele, J., Kudrna, O., Klotz, S., Kühn, I., 2008: Climate change can cause spatial mismatch of trophically interacting species. *Ecology* 89, 3472–3479.

18 Alcamo, J., Moreno, J.M., Novaky, B., Bindi, M., Corobov, R., Devoy, R.J.N., Giannakopoulos, C., Martin, E., Olesen, J.E., Shvidenko, A., 2007: *Europe. Chapter 12 in: Climate Change 2007: Impacts, Adaptation and Vulnerability. Contribution of Working Group II to the Fourth Assessment Report of the Intergovernmental Panel on Climate Change (IPCC)*, Parry, M.L., Canziani, O.F., Palutikof, J.P., van der Linden, P.J., Hanson, C.E. (eds). Cambridge University Press, Cambridge, UK.

19 Thuiller, W., Albert, C., Araújo, M.B., Berry, P.M., Cabeza, M., Guisan, A., Hickler, T., Midgley, G.F., Paterson, J., Schurr, F.M., Sykes, M.T., Zimmermann, N.F., 2008: Predicting global change impacts on plant species' distributions: Future challenges. *Perspectives in Plant Ecology, Evolution and Systematics* 9, 137–152.

20 Parmesan C., 2006: Ecological and evolutionary responses to recent climate change. *Annual Review of Ecology, Evolution, and Systematics* 37, 637–669.

21 Olofsson, J., Hickler, T., Sykes, M.T., Araujo, M.B., Baletto, E., Berry, P.M., Bonelli, S., Cabeza, M., Dubuis, A., Guisan, A., Kühn, I, Kujala, H., Piper, J., Rounsevell, M., Settele, J., Thuiller, W., 2008: Climate change impacts on European biodiversity – observations and future projections. MACIS deliverable 1.1. report. www.macis-project.net.

22 Parmesan, C., Yohe, G., 2003: A globally coherent fingerprint of climate change impacts across natural systems. *Nature* 421, 37–42.

23 Metzger, M.J., Bunce, R.G.H., Jongman, R.H.G., Mücher, C.A., Watkins, J.W., 2005: A climatic stratification of the environment of Europe. *Global Ecology and Biogeography* 14, 549–563.

24 Metzger, M.J., Bunce, R.G.H., Leemans, R., Viner, D., 2008: Projected environmental shifts under climate change: European trends and regional impacts. *Environmental Conservation* 35, 64–75.

25 Araújo, M.B., Thuiller, W., Pearson, R.G., 2006: Climate warming and the decline of amphibians and reptiles in Europe. *Journal of Biogeography* 33, 1712–1728.

26 Levinski, I., Skov, F., Svenning, J.-C., Rahbek, C., 2007: Potential impacts of climate change on the distributions and diversity patterns of European mammals. *Biodiversity and Conservation* 16, 3803–3816.

27 Menzel A., Sparks, T.H., Estrella, N., Koch, E., Aasa, A., Ahas, R., Alm-Kübler, K., Bissolli, P., Braslavská, O., Briede, A., Chmielewski, F.M., Crepinsek, Z., Curnel, Y., Dahl, Å., Defila, C., Donnelly, A., Filella, Y., Jatczak, K., Måge, F., Mestre, A., Nordli, Ø., Peñuelas, J., Pirinen, P., Remišová, V., Schneifinger, H., Striz, M., Susnik, A., van Vliet, A.J.H., Wielgolaski, F.-E., Zach, S., Zust, A.,

2006: European phenological response to climate change matches the warming pattern. *Global Change Biology* 12, 1969–1976.

28   Menzel, A. et al., 2006.

29   EEA, 2008: Impacts of Europe's changing climate – 2008 indicator-based assessment. EEA Report 2008/4. European Environmental Agency, Copenhagen.

30   Høye, T.T., Post, E., Meltofte, H., Schmidt, N.M., Forchhammer, M.C., 2007: Rapid advancement of spring in the high Arctic. *Current Biology* 17, 449–451.

31   Rubolini, D., Møller, A.P., Rainio, K., Lehikoinen, E., 2007: Intraspecific consistency and geographic variability in temporal trends of spring migration phenology among European bird species. *Climate Research* 35, 135–146.

32   Fitter, A.H., Fitter, R.S.R., 2002: Rapid changes in flowering time in British plants. *Science* 296, 1689–1691.

33   Walther, G.-R., Post, E., Convey, P., Menzel, A., Parmesan, C., Beebee, T., Fromentin, J.-M., Hoegh-Guldberg, O., Bairlain, F., 2002: Ecological responses to recent climate change. *Nature* 416, 389–395.

34   Disturbance regime covers various ways by which vegetation is being widespread replaced, e.g. by fire, floods, pests or diseases.

35   Biomes are climatically and geographically defined areas of ecologically similar conditions defining communities of plants, animals and soil organisms, and they are also referred to as ecosystems.

36   Bond, W.J., Woodward, F.I., Midgley, G.F., 2005: The global distribution of ecosystems in a world without fire. *New Phytologist* 165, 525–537.

37   Wermelinger, B., 2004: Ecology and management of the spruce bark beetle *Ips typographus* – a review of recent research. *Forest Ecology and Management* 202, 67–82.

38   Olofsson, J. et al., 2008.

39   Paterson, J.S., Araujo, M.B., Berry, P.M., Piper, J.M., Rounsevell, M.A., 2008: Mitigation, adaptation, and the threat to biodiversity. *Conservation Biology* 22, 1352–1355.

40   Searchinger, T., Heimlich, R., Houghton, R.A., Dong, E.A., Fabiosa, J., Tokgoz, S., Hayes, D., Tun-Hsiang, Y., 2008: Use of U.S. Croplands for biofuels increases greenhouse gases through emissions from land-use change. *Science* 319, 1238–1242.

41   Jeppesen, E., Kronvang, B., Meerhoff, M., Søndergaard, M., Hansen, K.M., Andersen, H.E., Lauridsen, T.L., Bekioglu, M., Ozen, A., Olesen, J.E., 2009: Climate change effects on runoff, phosphorus loading and lake ecological state, and potential adaptations. *Journal of Environmental Quality* 38, 1930–1941.

42   EEA, 2008.

43   EEA, 2008.

44   Moriondo, M., Good, P., Durao, R., Bindi, M., Gianakopoulos, C., Corte-Real, J., 2006: Potential impact of climate change on fire risk in the Mediterranean area. *Climate Research* 31, 85–95.

45 Peñuelas, J., Gordon, C., Llorens, L., Nielsen, T., Tietema, A., Beier, C., Emmett, B., Estiarte, M., Gorissen, A., 2004: Nonintrusive field experiments show different plant responses to warming and drought among sites, seasons, and species in a north-south European gradient. *Ecosystem* 7, 598–612.

46 Delitti, W., Ferran, A., Trabaud, L., Vallejo, V.R., 2005: Effects of fire recurrence in *Quercus coccifera* L. shrublands of the Valencia Region (Spain): I. plant composition and productivity. *Plant Ecology*, 177, 57–70.

47 de Boeck, H.J., Lemmens, C.M.H.M., Bossuyt, H., Malchair, S., Carnol, M., Merckx, R., Nijs, I., Ceulemans, R., 2006: How do climate warming and plant species richness affect water use in experimental grasslands? *Plant and Soil*, 10.1007/s11104–006–9112–5.

48 Paul, F., Kääb, A., Maish, M., Kellengerger, T., Haeberli, W., 2004: Rapid disintegration of Alpine glaciers observed with satellite data. *Geophysical Research Letters* 31, L21402, doi:10.1029/2004GL020816.

49 Alcamo, J. et al., 2007.

50 Nival flora refers to plants growing in snow.

51 Thuiller W., Lavorel, S., Araújo, M.B., Sykes, M.T., Prentice, I.C., 2005: Climate change threats to plant diversity in Europe. *Proceedings of the National Academy of Sciences of the United States of America* 102, 8245–8250.

52 Viner, D., Sayer, M., Uyarra, M., Hodgson, N., 2006: *Climate change and the European countryside: Impacts on land management and response strategies.* Report Prepared for the Country Land and Business Association, UK.

53 Callaghan, T.V., Björn, L.O., Chapin III, F.S., Chernov, Y., Chapin, T., Christensen, T.R., Huntley, B., Ims, R., Johansson, M., Riedlinger, D.J., Jonasson, S., Matveyeva, N, Panikov, N., Oechel, W., Panikov, N., Shaver, G., 2005: Arctic tundra and polar desert ecosystems. In Symon, C., Arris, L., Heal, B. (eds). *Arctic Climate Impact Assessment (ACIA): Scientific Report.* Cambridge University Press, p. 243–252.

54 Post, E., Forchhammer, M.C., Bret-Harte, S., Callaghan, T.V., Christensen, T.R., Elberling, B., Fox, A.D., Gilg, O., Hik, D.S., Høye, T.T., Ims, R.A., Jeppesen, E., Klein, D.R., Madsen, J., McGuire, A.D., Rysgaard, S., Schindler, D.E., Stirling, I., Tamstorf, M.P., Tyler, N.J.C., van der Wal, R., Welker, J., Wookey, P.A., Schmidt, N.M., Aastrup, P., 2009: Ecological dynamics across the Arctic associated with recent climate change. *Science* 325, 1355–1358.

55 Aanes, R., Saether, B.E., Smith, F.M., Cooper, E.J., Wookey, P.A., Oritsland, N.A., 2002: The Arctic Oscillation predicts effects of climate change in two trophic levels in a high-arctic system. *Ecological Letters* 5, 445–453.

56 Folkestad, T., New, M., Kaplan, J.O., Comiso, J.C., Watt-Cloutier, S., Fenge, T., Crowley, P., Rosentrater, L.D., 2005: Evidence and implications of dangerous climatic change in the Arctic. In: Schellnhuber, H.J., Cramer, W., Nakicenovic, N., Wigley, T.M.L., Yohe, G. (eds). *Avoiding dangerous climate change.* Cambridge University Press, UK.

57 Fischlin, A., Midgley, G.F., Price, J.T., Leemans, R., Gopal, B., Turley, C., Rounsevell, M.D.A., Dube, O.P., Tarazona, J., Velichko, A.A., 2007: Ecosys-

tems, their properties, goods, and services. Chapter 4 in: *Climate Change 2007: Impacts, Adaptation and Vulnerability. Contribution of Working Group II to the Fourth Assessment Report of the Intergovernmental Panel on Climate Change (IPCC)*, Parry, M.L., Canziani, O.F., Palutikof, J.P., van der Linden, P.J., Hanson, C.E. (eds). Cambridge University Press, Cambridge, UK.

58 Symon, C., Arris, L., Heal, B., 2005: *Arctic Climate Impact Assessment (ACIA): Scientific Report*. Cambridge University Press. Cambridge, UK.

59 Weltzin, J.F., Bridgham, S.D., Pastor, J., Chen, J.Q., Harth, C., 2003: Potential effects of warming and drying on peatland plant community composition. *Global Change Biology* 9, 141–151.

60 EEA, 2008.

61 Eisenreich, S.J., 2005: *Climate change and the European Water Dimension*. Report to the European Water Directors. European Commission – Joint Research Centre, Ispra, Italy. EUR 21553.

62 Jeppesen, E. et al., 2009.

63 An ephemeral waterbody is a wetland, spring, stream, river, pond or lake that only exists for a short period following rain or snowmelt.

64 Sand-Jensen, K., Pedersen, N.L., 2005: Broad-scale differences in temperature, organic carbon and oxygen consumption among lowland streams. *Freshwater Biology* 50, 1927–1937.

65 Hypoxia (less than 2 mg/l dissolved oxygen) occurs in aquatic environments when the concentration of dissolved oxygen becomes reduced to a harmful low level for aquatic organisms.

66 Halpern, B.S., Walbridge, S., Selkoe, K.A., Kappel, C.V., Micheli, F., D'Agrosa, C., Bruno, J.F., Casey, K.S., Ebert, C., Fox, H.E., Fujita, R., Heinemann, D., Lenihan, H.S., Madin, E.M.P., Perry, M.T., Selig, E.R., Spalding, M., Steneck, R., Watson, R., 2008: A global map of human impact on marine ecosystems. *Science* 319, 948–952.

67 Meier, H.E.M., Kjelström, E., Graham, L.P., 2004: Estimating uncertainties of projected Baltic Sea salinity in the late 21st century. *Geophysical Research Letters* 33, L15705, DOI:10.1029/2006GL026488.

68 EEA, 2008.

69 BRANCH, 2007: *Planning for biodiversity in a changing climate – BRANCH project final report*. Natural England, UK.

70 Rosenzweig, C., Casassa, G., Karoly, D.J., Imeson, A., Liu, C., Menzel, A., Rawlins, S., Root, T.L., Seguin, B., Tryjanowski, P., 2007: Assessment of observed changes and responses in natural and managed systems. In: *Climate Change 2007: Impacts, Adaptation and Vulnerability. Contribution of Working Group II to the Fourth Assessment Report of the Intergovernmental Panel on Climate Change*, Parry, M.L., Canziani, O.F., Palutikof, J.P., van der Linden, P.J., Hanson, C.E. (eds). Cambridge University Press, Cambridge, 79–131.

71 Pretzsch, H., Grote, R., Reineking, B., Rötzer, T., Seifert, S., 2008: Models for forest ecosystem management: A European perspective. *Annals of Botany* 101, 1065–1087.

72 Holman, I.P., Rounsevell, M.D.A., Shackley, S., Harrison, P.A., Nicholls, R.J., Berry, P.M., Audsley, E., 2005: A regional, multi-sectoral and integrated assessment of the impacts of climate and socio-economic change in the UK: Part I Methodology. *Climatic Change* 70, 9–41.

73 Jeppesen, E., Kronvang, B., Olesen, J.E., Audet, J., Søndergaard, M., Hoffmann, C.C., Andersen, H.E., Lauridsen, T.L., Liboriussen, L., Beklioglu, M., Meerhoff, M., Özen, A., Özkan, K., 2011: Climate change effects on nitrogen loading from catchment: implications for nitrogen retention, ecological state of lakes and adaptation. *Hydrobiologia* 663, 1–21.

74 Berry, P.M., Jones, A.P., Nicholls, R.J., Vos, C.C., 2007: *Assessment of the vulnerability of terrestrial and coastal habitats and species in Europe to climate change.* Annex 2 of Planning for biodiversity in a changing climate – BRANCH project Final Report, Natural England, UK.

75 Managed realignment (or managed retreat) allows areas that were not previously exposed to flooding by the sea to become flooded by removing coastal protection.

76 Rupp, S., Nicholls, R.J., 2007: Coastal and estuarine retreat: A comparison of the application of managed realignment in England and Germany. *Journal of Coastal Research* 23, 1418–1430.

77 EC, 2008: *The economics of ecosystems and biodiversity. An interim report.* Cambridge, UK.

78 EEA, 2002: *Environmental signals 2002 – Benchmarking the millennium.* European Environmental Agency, Copenhagen.

79 Berger, G., Kaechele, H., Pfeffer, H., 2006: The greening of the European common agricultural policy by linking the European-wide obligation of set-aside with voluntary agri-environmental measures on a regional scale. *Environmental Science and Policy* 9, 509–524.

80 Piper, J.M., Wilson, E.B., Weston, J., Thompson, S., Glasson, J., 2006: *Spatial planning for biodiversity in our changing climate.* Annex 1 of Planning for biodiversity in a changing climate – BRANCH project Final Report. Natural England, UK.

81 Harrison, P.A., Berry, P.M., Butt, N., New, M., 2006: Modelling climate change impacts on species' distributions at the European scale: implications for conservation policy. *Environmental Science & Policy* 9, 116–128.

82 Edgar, P.W., Griffiths, R.A., Foster, J.P., 2005: Evaluation of translocation as a tool for mitigating development threats to great crested newts (*Triturus cristatus*) in England, 1990–2001. *Biological Conservation* 122, 45–52.

83 Hannah, L., Midgley, G.F., Andelman, S., Araújo, M.B., Hughes, G., Martinez-Meyer, F., Pearson, R.G., Williams, P.H., 2007: Protected area needs in a changing climate. *Frontiers in Ecology and the Environment* 5, 131–138.

84 Viner, D. et al., 2006.

85 Vos, C.C., Opdam, P., Steingröver, E.G., Reijnen, R., 2007: Transferring ecological knowledge to landscape planning: a design method for robust corridors. In: Wu, J., Hubbs, R.J. (eds). *Key topics in landscape ecology.* Cambridge University Press, UK, 227–245.

86 Dunlop, M., Brown, P.: 2009. Conserving biodiversity in the face of uncertain ecological responses to climate change. *IOP Conf. Series: Earth and Environmental Science* 6, 312016, doi:10.1088/1755-1307/6/1/312016

87 Hannah, L., Midgley, G.F., LoveJoy, T., Bond, W.J., Bush, M., Lovett, J.C., Scott, D., Woodward, F.I., 2002: Conservation of biodiversity in a changing climate. *Conservation Biology* 16, 264–268.

8 Impacts on Europe's industries

Introduction

In previous chapters we have seen how global warming will alter everyday life in Europe from the air temperatures we feel, to the blossoms we smell, and the shapes of coastlines we perceive. But climate change will alter Europe in another, perhaps less perceptible, but still important way, namely by impacting its industry. Under climate change the structure, location and mode of industry will undergo changes. In the same way that nature will have to adapt to climate change or face the consequences, so too will Europe's manufacturing, forestry, agricultural and other industries have to adjust to new conditions. The facilities of water-intensive manufacturers may disappear south of the Alps because of diminishing water supply. Modes of producing energy will undergo changes, not only because global warming will make some types of energy production more economic than others, but because the motivation to reduce greenhouse gases will make some fuels more

Life in Europe Under Climate Change, First Edition. Joseph Alcamo and Jørgen E. Olesen.
© 2012 Joseph Alcamo and Jørgen E. Olesen. Published 2012 by John Wiley & Sons, Ltd.

desirable than others. As summers becoming steadily hotter, tourism will be pushed northward, and warmer winters will force many ski resorts in montane Europe to abandon their slopes. Neglected ski areas will eventually revert to a more pastoral alpine character.

Yet climate change may not disrupt industry as much as it disturbs other facets of society and the rest of nature. Many, perhaps most, industries have substantial financial and other resources for adapting to climate change, especially if the change is not too abrupt.[1] Some of their adaptation options are discussed later. Moreover, the future vitality of industry in Europe will depend on many factors other than climate change. The local and global demand for products, the style of management, the magnitude of governmental support, the size of a market, the costs and availability of labor and resources – these are all factors that will strongly influence economic success. Climate change will only be a complicating factor among them.

Another point to note is that climate impacts on industry will not be felt equally everywhere, and severe impacts will be concentrated in particular geographic areas, such as along coastlines and in river basins,[2] especially where climate change scenarios suggest an increasing frequency of floods and other extreme events (Box 8.1). Coastal flooding will pose a threat to the viability of beach tourism, and river flooding the location of manufacturers. More frequent droughts will threaten the water supply needed at thermal power plants to cool their turbines. Chapters 4 and 7 show that climate change threatens the agriculture and the forest industries in some parts of Europe, and one overriding lesson we can learn is that some industries will be more vulnerable than others, either because they depend on current climate conditions, or because they will find it hard to cope with a new climate regime. The same industry will be affected differently in different regions – changing the competition between European regions.

Box 8.1 The high costs of high temperatures

In Chapter 3 we noted that the great European summer heatwave of 2003 was a possible harbinger of climate change and cost the lives of about 70,000 European citizens while affecting many others. This event had a price not only in lives but also in economic resources.[3] In France the productivity of construction workers fell off, and cold storage systems in food-related businesses were found to be inadequate for their products. Electricity shortages were caused not only by the obvious elevated demand for air conditioning and cooling, but also because many power plants could not operate correctly during the heatwave because water withdrawn from rivers to cool turbines was too warm or river levels too low.[4] The record low flows of Central European rivers also affected inland navigation and restricted transport capacity on rivers like the Rhine.

Manufacturing, retail and service industries

Manufacturers and industrial producers are considered to be relatively less vulnerable overall to climate change than other industries.[5] The conventional wisdom is that they have the capital for adaptation measures and are not as reliant on climate factors as agriculture or tourism, for instance. Of course, this is again under the assumption that climate change does not happen too rapidly. Some changes in the weather may actually benefit manufacturers: for example, fewer severe winters with large burdens of ice and snow mean fewer disruptions to the entire supply chain of industry from extraction of raw materials to the marketing of finished goods.[6] A decrease in their frequency will be one less disturbance business will have to face.

Among the more vulnerable industries is the food processing sector because it depends on inputs from agriculture, a climate-sensitive sector. Also, industries with longer-term assets such as water suppliers with reservoirs or energy producers with power plants, are more exposed to global warming. Since their infrastructure is built to last for decades, it will be too entrenched to be easily replaced, even if it is found to be sensitive to climate change. Industries with extended supply chains, including the retail-distribution industry, are also vulnerable because the disruption of one link can upset the entire chain of production.[7] Furthermore, there are many different links that can be disrupted. Also vulnerable are the chemical, leather, paper and beverage industries because they rely on large quantities of high quality water,[8] which will be declining in supply south of the Alps.

Retail and service industries will also be affected by climate change (the "tourism" service industry is discussed below). It is likely that global warming will have some effect on the entire chain of service provision including the distribution network of services, the comfort of the workforce and the patterns of consumption.[9] For example, the distribution of food and household products to various retail centers could be disrupted by more frequent river and coastal flooding. Stronger winds are likely to threaten freight vehicles on roads and bridges and delay ship transport.[10] But there are actually very few studies about climate impacts on the retail and service industries, and this is one area that requires more research.

Energy demand and production

In contrast to many industries, the energy sector is likely to be especially sensitive to global warming because climate-related resources play an important role here. By energy sector we mean both its demand side (e.g. illumination, heating, cooling, industrial production, transportation), and its supply and production side (e.g. power plants and transmission systems).

Energy demand

Because of increasing temperatures throughout Europe, the demand for heating will decrease while cooling will increase.[11] One study estimated that under the plausible assumption of a 2°C warming by 2050, space heating requirements in the UK and Russia will substantially decline, thereby decreasing their fossil fuel use by around 5–10% and electricity demand by a few percent.[12] Other studies examined a range of plausible climate scenarios and projected that heating demand in winter could decrease by 6–8% in Hungary and Romania[13] and by 10% in Finland[14] up to the period 2021–50 compared to current average demands. Over a longer time horizon (up to 2100), the energy needed for winter heating in winter could drop by 20–30% in Finland,[15] and by about 40% in the residential sector of Switzerland, relative to current heating demand.[16] By 2030, under an "average" climate scenario, the southeast Mediterranean might require 10% less energy for heating.[17] Another study found that, by 2050, communities around the Mediterranean might need to heat their homes for two to three fewer weeks each year.[18]

On the other hand, these same studies found a strong increase in cooling requirements. In the southeastern part of the Mediterranean 28% more energy will be needed for cooling by 2030.[19] Around the whole of the Mediterranean, cooling will be required for an additional two to three weeks each year along the coast and five weeks inland,[20] assuming a continuation of current patterns of heating and cooling. Cooling requirements are expected to grow in central, southern[21] and other parts of Europe. The increase in summer season space cooling will have a spin-off effect on electricity demand. All other factors remaining the same, the demand for more cooling might increase electricity requirements up to 50% in Italy and Spain by the 2080s.[22]

So the demand for heating is likely to decrease, while the demand for cooling will increase, but how will this balance out? For much of Europe, increases in cooling energy demand due to climate change will be outweighed by reductions in the need for heating energy. Studies for the entire Europe show that this would reduce the energy need for heating and cooling in the residential sector by about 2000 PJ per year,[23] which corresponds to about 3% of current total energy consumption in the EU. However, these effects vary strongly both by season and region, and in southern and eastern Europe the consequences are a much greater demand for electricty for cooling during the summer period. This will in some regions put increasing stresses on electricity supply during the part of the year where cooling water for thermal power generation is less available.

Energy supply and production

To assess the impact of climate change on the supply side of Europe's energy system, we have to examine both current and future energy carriers. In 2006, 89% of electricity in Europe was produced at thermal power stations,[24] driven mostly by nuclear and fossil fuels. Except for a small number of air-cooled facilities, power plants require large volumes of relatively low temperature water for cooling their turbines. But as air temperature increases, so too will the temperature of intake water and this will reduce the effectiveness of cooling water. In the southern part of the continent, power plant operators will confront the additional problem of declining river run-off, which will reduce the supply of water to power plants.[25]

Another type of impact will be disruption to the transport of fuels around the world. An increasing frequency of extreme weather events could pose a threat to energy infrastructure located on or near the coast (e.g. oil rigs, electrical transmission towers, wind generators). Likewise, the melting of permafrost will cause soil to subside and this could threaten transmission towers, oil and gas pipelines, natural gas processing plants and other energy infrastructure in the Arctic.[26] This is already happening in northern Russia.[27]

Chapter 5 discussed the likely impact of climate change on Europe's potential for producing hydroelectricity, so here we only revisit the main points. Since the potential of hydroelectric plants obviously depends on the amount of water flowing through its turbines, it should also be expected that a change in water supply will change the output of these facilities. It follows that a wetter climate and higher river flows by the 2070s could increase potential hydroelectric production over many parts of northern Europe.[28] Meanwhile, less precipitation and lower run-off may lower the potential at existing facilities in southern and southeastern Europe by 25% or more. In total, 22 out of 40 European countries examined in one study were projected to have a net decline of 25% or more in potential hydroelectric production.[29] It seems that just as Europeans are trying to reduce CO_2 emissions by expanding the capacity of hydroelectric facilities and other renewable energy sources, the potential for hydroelectricity will be diminishing in the south.

Changes in climate will also have implications for other renewable energy sources. As the output from wind generators claim a steadily larger fraction of total electricity production in Europe (Chapter 10), the effect of climate change on their production will become an important issue. One study found that more frequent heavy winds over the Atlantic and northern regions of Europe could substantially increase the potential output of wind generators during the winter season, but increase it only slightly when averaged over the entire year.[30] But few studies of this kind have been conducted, and more are needed.

The production of biofuels is another renewable energy source likely to be affected by global warming. Since biofuel crops grow like any other crops, they also depend on local climatic conditions, and changes in temperature, precipitation and other climate variables will likewise affect their yield. Higher temperatures in northern Europe will substantially increase the productivity of perennial bioenergy crops such as willow and miscanthus.[31] This may cause a shift from growing annual bioenergy crops such as rapeseed to these perennial crops.

As the patterns of average cloudiness and air temperature changes, so too will the output of different kinds of solar energy devices such as photovoltaic arrays and solar thermal collectors. One study showed that climate change on the whole may increase the potential for exploiting solar energy in the Mediterranean region.[32]

Not only will the use and production of energy be affected by climate change, but also its distribution systems. Various factors could decrease the efficiency of transmitting electricity over electrical lines. For example, the resistance in electricity transmission lines increases with rising temperature[33] (and likewise its losses), and the efficiency of gas pipeline compressors decreases. But researchers have barely begun to study climate impacts on these aspects of the energy system.

The future energy system will look different from the one we know now, not least because in order to mitigate climate change we have to change the source of energy from existing fossil fuel sources to renewable energy sources (Chapter 10). Many of these renewable energy sources produce electricity (hydro, wind, wave and solar), and it is likely that the future European energy supply will be much more electricity-based than it is now. This also increases the reliance on the electricity grid for a constant supply of energy. Since these renewable energy sources strongly depend on climate, there is a great interest also in the energy sector on how climate change will affect energy production.

Transport

Changing climate is likely to have many subtle but important effects on the transport of goods and people in the future. One place where these effects are likely to be felt is the Arctic. Although Arctic sea ice has hindered ocean transport of cargo up to now, the minimum extent of sea ice (which occurs at the end of the summer melt season) fell by the substantial rate of 7.7 percent per decade between 1978 and 2005.[34] If this trend continues, the Northwest passage and Northern Sea Route will eventually become navigable for shipping[35] and it may be economically feasible to establish new permanent shipping lanes over Europe and other northern continents. Indeed, computer simulations under climate change show a significant shrinking of permanent Arctic sea ice in the course of this century (Figure 8.1). There is already now an increasing use of the Arctic shipping routes although these involve considerable risk, not only for the ships and their cargo, but also for the environment.[36]

On the negative side, temperatures have been rising over the land surface of the Arctic and this has caused permafrost to melt, roads to sag and ice road seasons to shorten.[41] But warmer temperatures over Europe could also benefit land transportation, because fewer days with frost and snow will reduce maintenance and service costs on rail and roads. Nevertheless, an increasing intensity of rainfall over many parts of Europe (Chapter 2) could also reduce safety on roadways.[42]

Chapter 2 showed that a wetter climate in northern Europe, and intensified winter rainfall throughout much of Europe, are likely to lead to more frequent flooding. This will especially threaten underground rail systems and roads with inadequate drainage.[43] Warmer temperatures expected throughout Europe will

Current Arctic Conditions

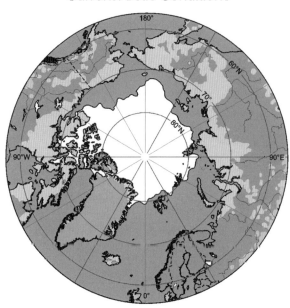

Figure 8.1 Melting ice cap will open up new shipping lanes in the far north. Shown here is the observed minimum sea-ice extent for September 2002 (upper picture) compared to the projected extent for the period 2080–2100 (lower picture). Also shown in the lower picture are possible new or improved sea shipping routes. Source: Anisimov et al.,[37] based on Instanes et al.[38] and Walsh et al.[39] Although the melting of the Arctic ice cap might be an advantage to shipping it will also disrupt important biological and physical processes with likely negative impacts on society and the rest of nature (Chapter 7). The depicted change in vegetation is based on simulations of the LPJ Dynamic Vegetation model. For further information about vegetation changes see explanation in Anisimov et al.[40] Reprinted with permission from Intergovernmental Panel on Climate Change, Geneva. (See color plate.)

Projected Arctic Conditions

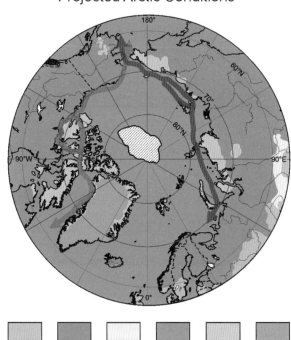

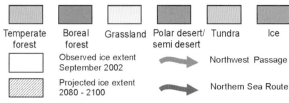

also damage the surfaces of road and rail systems.[44] Furthermore, higher wind speeds in some areas could pose an additional threat to air, land and sea transport.[45]

The rivers in Europe provide an important route for transporting goods as the large barges on the Rhine and other waterways attest to. Within the boundaries of the EU, water transport accounts for 6% of the total freight transported.[46] But its importance is regional; while it makes up a vital proportion of transport in the northwest of Europe, where rivers run full, it is relatively unimportant in the drier south. In some ways, climate change may have a positive impact on inland water transport, because higher temperatures will reduce the ice coverage of rivers. In sum, however, the increasing frequency of climate extremes is likely to disrupt river traffic by causing more unstable navigation conditions.[47]

Tourism

Europeans enjoy their vacations and travel widely throughout the year to various destinations within and outside of their countries. The European continent itself is the world's most important tourist destination; in 2005 it accounted for about 60% of all "international tourist arrivals", the metric used by the UN World Tourism Organization to report on the state of the tourist industry.[48] This is by far the largest percentage of any region. It is no surprise that tourism is an important part of the European economy, accounting for 4% of the EU's Gross Domestic Product (GDP) in 2005 and 24.3 million jobs.[49] If the various spin-offs of tourism are taken into account (automobile rentals, airline flights, and the like) tourism accounts for 11% of the EU's GDP.[50] The trend is also upwards according to estimates cited by the European Commission.[51]

Since tourism has so much to do with outdoor activities such as swimming, skiing and visiting outdoor cultural and historical sites, the link between climate and this part of the economy is particularly strong, and global warming will have a particularly disruptive effect. The ski industry is among the more vulnerable tourist sectors. Warmer temperatures in central Europe will substantially reduce snow cover, especially at the beginning and end of the ski season.[52] Under a very plausible, if not likely, temperature increase of 1°C, the winter ski season in the Austrian Alps will shrink by four weeks and the spring season by six weeks[53] (assuming no additional artificial snowmaking, which could prolong the ski season). At one Swiss alpine site, it was estimated that snow cover will decline by 30 days, assuming a 2°C warming and 50% increase in precipitation, and by 50 days with a 2°C warming but no precipitation change.[54]

Various studies also suggest a geographic and seasonal shift of tourism in Europe. As temperatures become warmer throughout Europe, summer tourism will become more uncomfortable in southern Europe but likewise more attractive in cooler northern and mountainous regions.[55] Changing temperature is expected to improve the so-called Tourism Comfort Index[56] in northern and western Europe, serving as a magnet for tourists to these regions.[57] Mountainous areas of France, Italy and Spain could become more appealing because of their relative coolness.

While high temperatures may reduce the desirability of visiting the Mediterranean region in summer, tourism may expand here during spring and autumn.[58] Greece and Spain, as examples, could experience fewer tourists in summer, but a longer overall tourist season.[59] One advantage of a lengthening and "flattening" of the tourist season is a reduction of electricity demand associated with tourism during summer months and a lessening of the visitor pressure on water supply during the dry summer season.[60]

Economic consequences in coastal regions

It was shown in Chapter 6 that as the seas and oceans bordering Europe warm and expand, and are swollen by melting glaciers and ice caps, their volume will grow and they will force their way inland. The frequency of storms will also increase along some stretches of coastline and the combination of sea-level rise and increasing storminess will cause increased coastal flooding. Apart from the direct threat to the safety of people and ecosystems along this coastline, coastal flooding also brings about costly property damage (Chapter 6).

Additional damage comes from the erosion of the coastline, which is slowly chipped away by long-term sea-level rise and intermittently by severe coastal storms. One estimate is that about one-fifth of the entire coastline of the EU was severely impacted by erosion in 2004.[61] The annual expenditures on coastal protection were estimated to be 3.2 billion euros in 2001.[62]

It should also be remembered that a large proportion of the settlements and thus also the industries in Europe are located along the coastline (Chapter 6). These industries are vulnerable not only to changes in weather events such a storms, but also to long-term changes in sea level. This is particularly true for the ports, and for industry located in the ports. However, sea-level rise is a relatively slow process, and with proper accounting of these changes investments in appropriate infrastructure will in most cases be able to limit damages to acceptable levels.

Adaptation by industry

While Europe's industry may be affected in many ways by climate change, it also has many adaptation options. Adapting, however, is likely to change the face of industry in Europe. According to the European Commission, it could "provoke significant restructuring in some economic sectors that are particularly weather dependent, e.g. agriculture, forestry, renewable energy, water, fisheries and tourism, or specifically exposed to climate change, e.g. ports, industrial infrastructure and urban settlements in coastal areas, floodplains and mountains . . ."[63] Some of the many adaptation options open to industry are examined below.

Manufacturing, retail and service industries

It was noted above that some industries, besides agriculture, particularly the chemical, leather, paper and beverage sectors, require large volumes of water and are therefore particularly sensitive to the expected decrease in water availability in southern Europe. One obvious option for these industries is to relocate to a more water-rich area, if they can afford to do so. Otherwise, they will have to sharply reduce their water requirements through water conservation, or secure new water supplies by building desalination plants, for example.

Energy demand and production

The public at large can help the energy sector to cope with climate change simply by consuming less energy. Less energy *consumption* usually means less energy *production* and fuel transport, which means fewer oil rigs to be exposed to heavy winds, fewer electrical transmission towers to sink into soils when permafrost melts, and fewer hydroelectric facilities that need to cope with reduced river run-off.

Although reducing energy use is a sensible all-around strategy, the size of the future energy sector of Europe will still be large because the continent will continue to have substantial energy needs, not least due to continued economic growth. With a substantial increase in renewable energies, it is likely that the share of energy consumption in the form of electricity will increase. Also, the number of thermal power plants powered by nuclear or fossil fuels will likely remain large (see Chapter 10 for a discussion of Europe's energy futures). As mentioned above, this type of power plant is particularly vulnerable to climate change because it relies on large volumes of water for cooling. But the increasing water temperature of rivers, and a reduction in run-off in some parts of Europe, could disrupt the ability of power plants to cool their turbines. One way for the energy sector to cope with this problem is to shift to a type of energy production that does not rely on large quantities of water. Included here are wind power generation, most solar energy applications, and hydropower facilities, although the potential for hydroelectricity generation is likely to be declining in southern Europe because of decreasing river run-off. These energy facilities have the additional advantage of producing very small quantities of greenhouse gases. Therefore, they are useful for both adapting to climate change as well as for mitigating its effects. There are, of course, physical, technical and economic limitations to the use of these energy carriers, which are discussed in Chapter 10. Some of the other options available to the energy industry are given in Table 8.1.

Transport

It has been pointed out that an increase in the rate of climate extremes is likely to disrupt river freight traffic by causing more unstable navigation conditions. A

Table 8.1 Adaptation measures for the energy industry. Source: Feenstra et al.[64] Reprinted with permission from Free University of Amsterdam.

Measure	Impact addressed	Type of Adaptation	Comments
Air conditioning efficiency standards	Increased cooling electricity costs	Threat modification	Increased air conditioning efficiency will reduce electricity expenditures, but will make initial costs higher. Standards will also reduce greenhouse gas emissions.
Thermal shell standards	Increased cooling electricity costs	Threat modification	Increased ceiling insulation and reduced shading coefficient requirements are often the most cost-effective measures. Standards will also reduce greenhouse gas emissions.
River-front power plant siting regulation	Power plant flooding and cooling system problems	Effect prevention	Approval for permits for power plants along river fronts should consider effects of reduction or increase in river flow.
Coastal production facility siting regulation	Plant flooding and extreme weather damage	Effect prevention	Approval for permits for coastal power plants and oil and gas production plants should consider effects of sea-level rise and increase in extreme weather.
Change approach to water management vis-à-vis hydropower generation	Loss of hydropower generation capability	Loss sharing/ use change	Reductions in or changes in patterns of river and stream flows may require changes in approach to water management. There is a potential for interregional conflicts.
Consider demand and hydropower generation changes in integrated resource planning	Changes in generation capacity requirements	Threat modification	Changes in electricity demand and hydropower generation may require a change in the generation capacity portfolio.
Information programs	Increased space cooling costs	Threat modification	Government agencies can provide information about energy efficiency measures that can reduce energy costs (e.g., appliance labeling programs).
Reduce/eliminate energy subsidies	Increased national electricity costs	Loss sharing	Subsidies to energy prices distort market signals and can result in wasteful consumption. Impacts on low-income groups can be ameliorated through targeted programs.

range of options are available for helping this segment of the transport industry adapt to climate change:[65] For example, dams and reservoirs could be added to waterways in order to smooth out the impacts of floods and droughts. Of course, damming navigation waterways will have the same disruptive effect on aquatic and riparian ecosystems as impounding other watercourses (Chapter 5). Alternatively, ship owners could build smaller freighters with smaller draughts (less weight), which would have a better chance of dealing with lower river flows. Another option is for managers of navigation routes, in cooperation with regional authorities, to implement or improve warning systems for droughts and floods and provide more precise forecasts of water levels for ship operators.

The increasing frequency of extreme climate events, in particular river and coastal flooding, will also affect land transport of people and freight. A straightforward strategy to protect against flooding is to increase the elevation of roadways and rail lines, although this will entail high costs. The planners of the Copenhagen Metro took such an approach by designing the system's entrance ways so that they would be above flood levels – even considering a sea-level rise of 52 cm by 2100.[66]

While the range of impacts looks extensive, transportation engineers and planners in Europe are accustomed to dealing with even greater challenges than climate change. Nevertheless, a changing climate will pose new problems to be solved and added complexity for routine transportation planning. Much of the transport infrastructure is highly dependent on extreme events (e.g., storms, floods or heatwaves), and changes in frequency of these events will therefore greatly influence the design criteria of transport networks.[67]

Tourism

Among the many different types of industry, tourism is notable for its ability to adapt to changing demographic and economic conditions.[68] Tourists themselves are likely to change their destinations as climate conditions change.[69] Therefore it would be no surprise if tourist operators perceive at an early stage that the summer heat in southern Europe is becoming unpleasant for their customers and respond by promoting alternative and cooler vacation sites. Alternatively, they might promote tourism along the Mediterranean during the cooler months of the year.[70]

One type of tourism is locked into its location, namely, the winter ski industry. Resort owners are already adapting to warmer temperatures and thinner snow cover by producing artificial snow. Indeed snow-making machines are a common feature at many ski resorts in Europe. Austria spent about 800 million euros between 1995 and 2003 on artificial snow-making installations and France nearly 500 million euros between 1990 and 2004.[71] But not all of these investments can be attributed to regional warming since artificial snow is also used to extend the traditional snow season or to expand the snow coverage of the resort beyond its previous natural limits. It is also a fact that snowmaking is far from being an environmentally-friendly practice since it requires large amounts of water and energy and produces air pollution. It may also turn out to be just a temporary solution, since it will be difficult to keep up with the steadily increasing air temperatures, and ski tourists may not be satisfied with completely artificial snow in ski resorts. In the end, ski resort owners may have to offer their customers alternatives such as hiking or grass-skiing.[72]

Moving from the mountains to the coasts, tourism here will be threatened by rising sea level and, along some stretches of coastline, by more frequent storm surges. The tourist industry can adapt by building or strengthening coastal dikes or other protection structures, or by moving their facilities further inland (Chapter 6).

The tourist industry in general can adapt to changing climate by promoting alternative types of activity such as eco-tourism or cultural tourism that could take into account changing weather conditions.[73] Yet adaptation options in the tourist industry will be limited by the increasingly high costs incurred by tourists in getting to their destination because of rising fuel prices. This may lead to a greater focus on local or regional tourism, or alternative modes of travel that rely on renewable energy sources.

Coastal flooding and erosion

Options for coastal flooding and erosion fall into three categories. *Protecting the shoreline* by building dikes and other structures, *accommodating the increased risk of flooding* by strengthening buildings and other property, and *retreating from the shoreline*, by moving people and property inland. Many different measures fall into these categories and are discussed in Chapter 6.

Property damage and the insurance industry

One of the methods used by modern society to cope with infrequent but damaging natural hazards is to purchase insurance for homes, vehicles and businesses. It is logical, then, that insurance should play a role as society learns to cope with climate change. At present, insurance coverage against natural hazards varies widely in Europe. For example, the availability of insurance as a hedge against flood damage varies from good in some parts of Europe to very bad in others. This implies that there is also a large variation in the vulnerability of property to an increasing frequency of extreme climate events. Expanding and standardizing this coverage, if affordable, would be an obvious and effective way for individuals and businesses to protect themselves against the risks of a changing climate.

But large uncertainties about future climate and future vulnerable property make it difficult to estimate how much insurance coverage will be needed. As an example, a British government agency projected UK damages in the 2080s due to river flooding to range from 1.7 to 21.8 times current average damages.[74]

Uncertainty notwithstanding, a consistent message coming from projections of climate-related property damage in Europe is that the risk of damage is sharply increasing. Another important point is that the damage from infrequent climate events such as catastrophic floods or windstorms, have much higher costs than more frequent events. Costs of an extreme climate event occurring once every 1000 years is 2.5 times larger than a 100-year event.[75] Similarly, reinsurance companies in Germany have estimated that storm damage increases roughly as the cube of maximum wind gusts,[76] implying that a storm with a maximum wind speed of 45 meters per second causes more than three times the damage of a storm with maximum winds of 30 meters per second.

In responding to climate change, insurance companies are beginning to factor in increasing risks to property, which is due to an increasing risk of floods, droughts, wind storms and other extremes. Their options to respond include raising insurance premiums, removing or restricting insurance coverage, or reinsurance.[77] With "reinsurance" an insurance company reduces the chance that it will have to make a huge payout because of a catastrophe by sharing a part of its liability with another insurer. Insurance companies are also beginning to take a more active role in public discussions about policies such as flood plain restrictions, which could reduce the exposure of property to floods and other hazards.[78]

Among the more costly climatic hazards in Europe are winter storms with heavy winds that occur over wide areas. Storms in Germany, for instance, accounted for more than 50% of economic losses related to natural hazards between 1970 and 1999.[79] Despite the severity of these storms it seems possible to build structures strong enough to withstand their impact. Consider the case of Germany, where insurance companies usually pay out damages when wind strength exceeds 20 meters per second because damage often occurs at these speeds. Apparently, since these wind speeds are seldom (occurring less than 2% of the time in Germany as a whole) many homeowners do not build their homes sturdily enough to withstand them when they do occur. But the situation is different where these winds happen more often. In the community of List on the windy island of Sylt off the coast of Germany, wind speeds above 20 meters per second occur about 20% of the time. Yet damage rarely occurs here, indicating that building owners have adapted their structures to the strong winds.[80]

Indeed, upgrading existing buildings to make them more climate-resistant with sturdy structural braces, stronger foundations, and more robust roofs, is an effective strategy for avoiding or reducing damage in the face of a changing climate. But retrofitting entails high costs, and this argues for another approach, namely for introducing climate-resistant building standards for new rather than existing buildings. The advantage here is that the additional costs of making a structure climate-resistant can be spread out over the long-term financing of the structure. Also, the additional costs of climate-proofing new buildings are quite small since much of the construction needs to be done anyway, and just requires slightly different materials and construction methods. On the other hand, although it is more affordable, this policy will take several years to have a noticeable effect on the building sector because homes, public edifices and other buildings in Europe have a very slow turnover rate.

This means that new, more climate-resistant building standards will only be used in a small fraction of the total building stock for many years. Governments may therefore have to enforce higher building standards not only for new buildings, but also when buildings are renovated. Alternatively, subsidies can be applied that only benefit climate-proofed renovation. The insurance sector also has an interest in improved building standards, and future insurance policies may require particular building standards to achieve low insurance premiums. The message here is that policy makers will have to grapple with some difficult issues in determining the best policies for adapting buildings to new climate hazards.

Notes

1   Wilibanks, T., Lankao, R., Bao, M., Berkhout, F., Cairncross, S., Ceron, J.-P., Kapshe, M., Muir-Wood, R., Zapata-Marti, R., 2007: Industry, settlement, and society. In: *Climate Change 2007: Impacts, Adaptation and Vulnerability. Contribution of Working Group II to the Fourth Assessment Report of the Inter-governmental Panel on Climate Change*, Parry, M.L., Canziani, O., Palutikof, J., van der Linden, P., Hanson, C. (eds). Cambridge University Press,. Cambridge, UK, 357–390.
2   Wilibanks, T. et al., 2007.
3   Wilibanks, T. et al., 2007.
4   Wilibanks, T. et al., 2007.
5   Wilibanks, T. et al., 2007.
6   Parry, M.L. (ed.), 2000: *Assessment of potential effects and adaptations for climate change in Europe: The Europe ACACIA project.* Jackson Environment Institute. University of East Anglia. Norwich, UK. 320 pp.
7   Wilibanks, T. et al., 2007.
8   Parry, M.L, 2000.
9   Wilibanks, T. et al., 2007.
10 Wilibanks, T. et al., 2007.
11 Aebischer, B., Henderson, G., Jakob, M., Catenazzi, G., 2007: *Impact of climate change on thermal comfort, heating and cooling energy demand in Europe.* ECEEE Summer Study 2007 – Saving Energy – Just do it. 859–870.
12 Kirkinen, J., Matrikainen, A., Holttinen, H., Savolainen, I., Auvinen, O., Syri, S., 2005: *Impacts on the energy sector and adaptation of the electricity network under a changing climate in Finland.* FINADAPT, working paper 10, Finnish Environment Institute.
13 Vajda, A., Venalainen, A., Tuomenvirta, H., Jylha, K., 2004: An estimate of the influence of climate change on heating energy demand on regions of Hungary, Romania and Finland. *Quarterny Journal of the Hungarian Meteorological Service* 108, 123–140.
14 Venalainen, A., Tammelin, B., Tuomenvirta, H., Jylha, K., Koskela, J., Turunen, M.A., Vehvilainen, B., Forsius, J., Jarvinen, P., 2004: The influence of climate

change on energy production and heating energy demand in Finland. *Energy and the Environment* 15, 93–109.

15 Kirkinen, J. et al., 2005.

16 Frank, T., 2005: Climate change impacts on building heating and cooling energy demand in Switzerland. *Energy and Buildings* 37, 1175–1185. Christenson, M., Manz, H., Gyalistras, D., 2006: Climate warming impact on degree-days and building energy demand in Switzerland. *Energy Conversion and Management* 47, 671–686.

17 Cartalis, C., Synodinou, A., Proedrou, M., Tsangrassoulis, A., Santamouris, M., 2001: Modifications in energy demand in urban areas as a result of climate changes: an assessment for the southeast Mediterranean region. *Energy Conversion and Management* 42, 1647–1656.

18 Giannakopoulos, C., Bindi, M., Moriondo, M., LeSager, P., Tin, T., 2005: *Climate change impacts in the Mediterranean resulting from a 2°C global temperature rise*, WWF report, Gland Switzerland, http://assets.panda.org/downloads/medreportfinal8july05.pdf. Retrieved February 2011.

19 Cartalis, C. et al., 2001.

20 Giannakopoulos, C. et al., 2005.

21 Fronzek, S., Carter, T.R., 2007: Assessing uncertainties in climate change impacts on resource potential for Europe based on projections from RCMs and GCMs. *Climatic Change* 81 (suppl. 1), 357–371.

22 Valor, E., Meneu, V., Caselles, V., 2001: Daily air temperature and electricity load in Spain. *Journal of Applied Meteorology* 40, 1413–1421. Giannakopoulos, C., Psiloglou, B.E., 2006: Trends in energy load demand for Athens, Greece: Weather and non-weather related factors. *Climate Research* 13, 97–108.

23 Isaac, M., van Vuuren, D.P., 2008: Modeling global residential sector energy demand for heating and air conditioning in the context of climate change. *Energy Policy* 37, 507–521.

24 For year 2006, for the 25 countries of the European Union. This figure is the total gross electricity generation minus 8.8% for hydroelectric power generation and minus 2.5% for wind energy generation (i.e. 88.7%). Source: European Commission. 2008. EU Energy in Figures 2007/2008. http://ec.europa.eu/dgs/energy_transport/figures/pocketbook/doc/2007/2007_energy_en.pdf (retrieved November 2008).

25 Arnell, N., Tomkins, E., Adger, N., Delaney, K., 2005: *Vulnerability to abrupt climate change in Europe*. ESRC/Tyndall Centre Technical Report No 20, Tyndall Centre for Climate Change Research, University of East Anglia, Norwich, UK, 63 pp.

26 Wilibanks, T. et al., 2007.

27 ACIA, 2004: *Impacts of a warming Arctic. Arctic Climate Impact Assessment.* Cambridge University Press, Cambridge, UK, 144 pp.

28 Lehner, B., Czisch, G., Vassolo, S., 2005: The impact of global change on the hydropower potential of Europe: a model-based analysis. *Energy Policy* 33, 839–855.

29 Lehner, B. et al., 2005.

30 Pryor, S.C, Barthelmie, R.J., Kjellström, E., 2005: Potential climate change impact on wind energy resources in northern Europe: analyses using a regional climate model. *Climate Dynamics* 25, 815–835.

31 Hastings, A., Clifton-Brown, J., Wattenbach, M., Mitchell, C.P., Stampfl, P., Smith, P., 2009: Future energy potential of *Miscanthus* in Europe. *Global Change Biology Bioenergy* 1, 180–196.

32 Santos, F.D., Forbes, K., Moita, R. (eds), 2002: *Climate change in Portugal: Scenarios, Impacts and Adaptation Measures*. SIAM project report, Gradiva, Lisbon, Portugal, 456 pp.

33 Santos, F.D. et al., 2002.

34 Lemke, P., Ren, J., Alley, R., Allison, I., Carrasco, J., Flato, G., Fujii, Y., Kaser, G., Mote, P., Thomas, R., Zhang, T., 2007: Observations: change in snow, ice and frozen ground. *Climate Change 2007: The Physical Science Basis. Contribution of Working Group I to the Fourth Assessment Report of the Intergovernmental Panel on Climate Change*, Solomon, S., Qin, D., Manning, M., Chen, Z., Marquis, M., Averyt, K.B., Tignor, M., Miller, H.L. (eds). Cambridge University Press, Cambridge, 337–384.

35 Anisimov, O.A., Vaughan, D.G., Callaghan, T.V., Furgal, C., Marchant, H., Prowse, T.D., Vilhjálmsson, H., Walsh, J.E., 2007: Polar regions (Arctic and Antarctic). *Climate Change 2007: Impacts, Adaptation and Vulnerability. Contribution of Working Group II to the Fourth Assessment Report of the Intergovernmental Panel on Climate Change*, Parry, M.L., Canziani, O.F., Palutikof, J.P., van der Linden, P.J., Hanson, C.E. (eds). Cambridge University Press, Cambridge, 653–685.

36 Jensen, Ø., 2008: Arctic shipping guidelines: towards a legal regime for navigation safety and environmental protection? *Polar Record* 44, 107–114.

37 Anisimov, O.A., et al. 2007.

38 Instanes, A., Anisimov, O., Brigham, L., Goering, D., Ladanyim B., Larsen, J.O., Khrustalev, L.N., 2005: Infrastructure: buildings, support systems, and industrial facilities. *Arctic Climate Impact Assessment, ACIA*. Symon, C., Arris, L., Heal, B. (eds), Cambridge University Press, Cambridge, 907–944.

39 Walsh, J.E., Anisimov, O., Hagen, J.O.M., Jakobsson, T., Oerlemans, J., Prowse, T.D., Romanovsky, V., Savelieva, N., Serreze, M., Shiklomanov, I., Solomon, S., 2005: Cryosphere and hydrology. *Arctic Climate Impacts Assessment, ACIA*. Symon, C., Arris, L., Heal, B. (eds), Cambridge University Press, Cambridge, 183–242.

40 Anisimov, O.A. et al., 2007.

41 ACIA, 2004.

42 Keay, K., Simmonds, I., 2006: Road accidents and rainfall in a large Australian city. *Accident Analysis and Prevention* 38, 445–454.

43 Arkell, B.P., Darch, G.J.C., 2006: *Impact of climate change on London's transport network*. Proceedings of the Institution of Civil Engineers Municipal Engineer 159, 231–237.

44 AEAT, 2003: *Railway Safety Implications of Weather, Climate and Climate Change. Final Report to the Railway Safety and Standards Board.*

www.railwaysarchive.co.uk/documents/RSSB_SafetyClimateChange2003.pdf
(retrieved February 2011).

45   Keay, K., Simmonds, I., 2006.

46   BMBF, 2007: *Time to Adapt – Climate Change and the European Water Dimension Discussion Paper: Inland Waterway Transport. Federal Ministry for the Environment, Nature Conservation and Nuclear Safety, Germany*, www.climate-water-adaptation-berlin2007.org/documents/transport.pdf (retrieved November 2008).

47   BMBF, 2007.

48   UN World Tourism Organization. 2008: *Historical perspective of world tourism*. http://unwto.org/facts/eng/historical.htm (retrieved November 2008).

49   European Commission, 2006: Tourism: a big business for small businesses. http://ec.europa.eu/enterprise/library/ee_online/art05_en.htm. (retrieved November 2008).

50   European Commission, 2006.

51   European Commission, 2006.

52   Elsasser, H. Burki, R., 2002: Climate change as a threat to tourism in the Alps. *Climate Research* 20, 253–257.

53   Hantel, M., Ehrendorfer, M., Haslinger, A., 2000: Climate sensitivity of snow cover duration in Austria. *International Journal of Climatology* 20, 615–640. Computed for the most sensitive elevations – 600 m in winter and 1400 m in spring.

54   Beniston, M., Keller, F., Goyette, S., 2003: Snow pack in the Swiss Alps under changing climatic conditions: an empirical approach for climate impact studies. *Theoretical and Applied Climatology* 74, 19–31.

55   Hamilton, L., Lyster, P., Otterstad, O., 2000: Social change, ecology and climate in 20th century Greenland. *Climatic Change* 47, 193–211.

56   Amelung, B., Viner, D., 2006: Mediterranean tourism: Exploring the future with the tourism climatic index. *Journal of Sustainable Tourism* 14, 349–366.

57   Hanson, C.E, Palutikof, J.P., Dlugolecki, A., Giannakopoulos, C., 2006: Bridging the gap between science and the stakeholder: the case of climate change research. *Climate Research* 13, 121–133.

58   Amelung, B., Viner, D., 2006.

59   Maddison, D., 2001: In search of warmer climates? The impact of climate change on flows of British tourists. *Climatic Change* 49, 193–208.

60   Amelung, B., Viner, D., 2006.

61   Nicholls, R., Wong, P., Burkett, V., Codignotto, J., Hay, J., McLean, R., Ragoonaden, S., Woodroffe, C., 2007: Coastal systems and low-lying areas. In: *Climate Change 2007: Impacts, Adaptation and Vulnerability. Contribution of Working Group II to the Fourth Assessment Report of the Intergovernmental Panel on Climate Change*, Parry, M.L., Canziani, O., Palutikof, J., van der Linden, P., Hanson, C. (eds). Cambridge University Press. Cambridge. UK. 315–356.

62   Eurosion, 2004: *Living with coastal erosion in Europe: Sediment and space for sustainability*. www.safecoast.org/editor/databank/File/Eurosion%20findings.pdf. Retrieved February 2011.

63   EC, 2007: Green paper: *Adapting to climate change in Europe – Options for EU action*. European Commission, Brussels, 29.6.2007. COM(2007) 354 final. SEC(2007) 849.

64   Feenstra, J., Burton, I., Smith, J., Tol, R. (eds). 1998: *Handbook on Methods for Climate Change Impact Assessment and Adaptation Strategies*. United Nations Environment Programme and Free University of Amsterdam. www.cordelim.net/extra/cd%20forestal/Adaptaci%F3n%20al%20CC%20y%20MNR/Literatura/APF/HMC.pdf. Retrieved November 2008.

65   BMBF, 2007.

66   Fenger, S., 2000: Implications of Accelerated Sea-Level Rise (ASLR) for Denmark: Proceeding of SURVAS Expert Workshop on European Vulnerability and Adaptation to impacts of Accelerated Sea-Level Rise (ASLR), Hamburg, Germany (19–21 June)

67   Arkell, B.P., Darch, G.J.C., 2006.

68   Parry, M.L, 2000.

69   Sievanen, T., Tervo, K., Neuvonen, M., Pouta, E., Saarinen, J., Peltonen, A., 2005: Nature-based tourism, outdoor recreation and adaptation to climate change. FINADAPT working paper 11, *Finnish Environment Institute Mimeographs* 341, Helsinki, 52 pp.

70   Amelung, B., Viner, D., 2006.

71   EEA, 2008: *Impacts of Europe's changing climate – 2008 indicator-based assessment*. EEA Report No. 4/2008.

72   Fukushima, T., Kureha, M., Ozaki, N., Fujimori, Y., Harasawa, H., 2002: Influences of air temperature change on leisure industries: case study of ski activities. *Mitigation and Adaptation Strategies for Global Change* 7, 173–189.

73   Hanson, C.E. et al., 2006.

74   Foresight Programme, 2004: *Foresight Future Flooding, Flood and Coastal Defence project of the Foresight programme*. Office of Science and Technology, London, UK.

75   Swiss Re, 2000: *Storm over Europe*. Swiss Re, Zurich, 27 pp.

76   Klawa, M., Ulbrich, U., 2003: A model for the estimation of storm losses and the identification of severe winter storms in Germany. *Natural Hazards in Earth System Science* 3, 725–732.

77   Dlugolecki, A. (ed.), 2001: *Climate Change and Insurance*. Chartered Insurance Institute, London, UK.

78   Dlugolecki, A., Keykhah, M., 2002: Climate change and the insurance sector: its role in adaptation and mitigation. *Greener Management International* 39, 83–98.

79   Munich Re, 1999: *Naturkatastrophen in Deutschland: Schadenerfahrungen und Schadenpotentiale*, Publication of the Munich Re, Order Number 2798–E-d www.munichre.com.

80   Klawa, M., Ulbrich, U., 2003.

9 Summing up impacts and adaptation

Climate impacts: everywhere, but different

Until now we have seen that the impacts of current and future climate change in Europe are pervasive, wide-ranging and diverse. In this chapter we depart from our detailed examination of impacts, and instead step back and look at the big

Life in Europe Under Climate Change, First Edition. Joseph Alcamo and Jørgen E. Olesen.
© 2012 Joseph Alcamo and Jørgen E. Olesen. Published 2012 by John Wiley & Sons, Ltd.

Figure 9.1 Key vulnerabilities of European systems and sectors to climate change during the 21st century for the main biogeographic regions of Europe: TU: Tundra, pale turquoise. BO: Boreal, dark blue. AT: Atlantic, light blue. CE: Central, green; includes the Pannonian Region. MT: Montane, purple. ME: Mediterranean, orange; includes the Black Sea region. ST: Steppe, cream. SLR: sea-level rise. NAO: North Atlantic Oscillation. Source: Alcamo et al.[1] Source of map of biogeographic regions: European Environment Agency, copyright EEA, Copenhagen. www.eea.europa.eu. Reprinted with permission from European Environmental Agency, Copenhagen. (See color plate.)

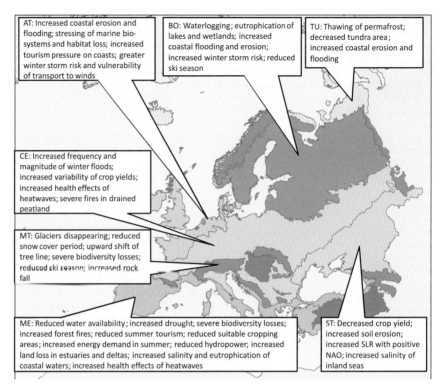

picture of how climate change will play out on the pan-European scale. We also review some of the key issues that have to be dealt with in order to adapt to a changing climate.

Earlier chapters have shown that the impacts we now observe are only a preview of how global warming will play out in different parts of Europe. As we have said many times, and stress again in Chapter 10, *these future impacts are not all inevitable*, and it is still possible for us, as members of society, to slow down changing climate to a relatively safe level.

Although it is expected that impacts will occur everywhere in Europe, they will not have the same character or intensity. On the contrary, as shown in Figure 9.1, we expect that different regions of Europe will experience climate impacts in a particular way. Along the Atlantic coast, sea-level rise and an increasing frequency of storms will accelerate coastal erosion and flooding, and cause the loss of wetlands and other habitats for flora and fauna. These losses of coastal wetlands will be even larger along coasts with little or no tidal influence, such as along the coast of the Baltic Sea and the Black Sea. In the boreal region of Europe (characterized by evergreen trees) increasing precipitation will lead to waterlogging of its forests, while warmer temperatures will contribute to increased eutrophication of its lakes and wetlands. The coastline in this region will also experience an increasing risk of winter storms accompanied by a greater threat of coastal flooding and erosion.

In the tundra region of Europe and elsewhere, permafrost areas will melt and sharply contract. Trees and shrubs, but even more so annual plants, insects and higher fauna, will move steadily into this area from the south, changing the character of the landscape and natural biodiversity. In central Europe, winter precipitation will increase and with it the frequency and intensity of winter floods. Crop yields here will become more variable and heatwaves will pose a threat to the population. In the montane parts of Europe glaciers will mostly disappear, the tree line will move up and plants and animals will be under threat of extinction. Increasing rock fall due to newly exposed rocks will make travel more dangerous and the ski industry will be endangered for lack of snow. Meanwhile, in the southeast of the continent, the salinity of the Black Sea will increase as sea level rises and freshwater outflow diminishes.

What are the most vulnerable areas of Europe?

Among the pervasive climate changes happening in Europe, some will pose a greater threat to society than others. What would we say are the most vulnerable areas? Of course, a precise answer is not possible because of the uncertainty of future climate change and our lack of understanding of how different impacts will play out. Scientists do not even agree on the definition of "vulnerability" (see Boxes 9.1 and 9.2). With these provisos in mind, here is a tentative list of the most vulnerable European areas up to the end of the 21st century:[1]

- the particular stretches of Europe's coastline (discussed in Chapter 6) facing a growing risk of flooding and erosion owing to increasing storminess and sea-level rise; particularly vulnerable areas existing along the south coast of the North Sea, in the Baltic Sea region and along low-lying coastal stretches of the Mediterranean, Adriatic and Black Sea;
- the Mediterranean region, which is threatened by a combination of warmer and drier conditions leading to longer and more frequent droughts, aggravated water scarcity, declining crop productivity, and higher fire risk;
- mountainous regions subject to stress due to increasing temperatures, a shift in the form of precipitation from snowfall to rainfall, reduced snow cover, melting permafrost, accelerated glacier melting, changing river hydrology, destabilized slopes and more frequent rock falls;
- the cities of Europe, particularly in its central and southern parts, which will be increasingly exposed to warmer average summer temperatures and short-term heatwaves as in 2003;
- agriculture in southeastern Europe, where higher summer temperatures and more frequent droughts will greatly reduce crop yields, but where winters remain too cold to allow for crop growth;
- lower-lying, densely populated areas along rivers throughout much of Europe, which are threatened by more frequent flooding due to the increasing volume and intensity of winter precipitation;

> **Box 9.1 Understanding "vulnerability" to climate change**
>
> Although the word "vulnerable" is often used in conjunction with climate impacts, it is often assigned a different meaning by different authors. One school of thought (promoted by the authors of this book) divide vulnerability into two components: "exposure" and "susceptibility". "Exposure" is the severity of climate change or the degree of pressure put on a human population or some other part of nature by climate change. The changes in climate could be any alteration of temperature, precipitation, frequency of storms, sea-level rise or other variables directly attributable to climate change. Most experts consider exposure to be "high" if the degree of change is high relative to some base period. As an example, a 50% decrease in precipitation up to the 2080s relative to the period 1990–2000 might be considered "high" whereas a decline of 5% is likely to be considered a mild change. The second component, "susceptibility", is an attribute of the population exposed to climate change and a function of its "adaptive capacity" (sometimes termed "coping capacity"). The scientific community does not yet agree about what determines this susceptibility or adaptive capacity. Indeed, different scientific disciplines can have quite contrasting opinions (see Box 9.2). But one thing is clear. To better assess climate impacts and select successful options for adaptation it is necessary to better understand the nature of vulnerability of a population or an ecosystem (managed or natural), and this should be a priority for scientific research.

- the Arctic and sub-Arctic regions, where large areas of permafrost will disappear and infrastructure and housing will need to be rebuilt, and current ecosystems are greatly threatened.

Sharpening north–south differences

Another finding of climate impacts studies is the striking geographic difference between north and south. In Chapter 2 it was reported that air temperature will increase everywhere in Europe, but not to the same extent, nor during the same seasons. The largest temperature increases are bound to occur in winter in the north, but in summer in southern and central parts of Europe. Equally important are the projections that average annual precipitation will increase north of the Alps and decrease in the south.

Differences in changing climate will also become apparent in the different types of climate impacts. In Chapter 4 it was shown that hotter and drier conditions will steadily reduce the average productivity of crops in the south. Meanwhile, in

Box 9.2 The many factors affecting susceptibility of a population to climate extremes from different disciplinary viewpoints – The case of exposure to droughts in Southern Portugal

(a) political science   (b) socio-economic   (c) environmental psychology

Different factors will affect the susceptibility or "adaptive capacity" of a human population or some other part of nature. The above "spider diagrams" are taken from the scientific literature and depict the factors influencing susceptibility of the population in southern Portugal to impacts of drought. The longer the "leg of the spider" on an axis, the more important the factor. For example, in the left-most diagram, the factor "lack of state will" is more important than the other factors. Each diagram depicts a different disciplinary viewpoint. These diagrams make two important points: First, a wide variety of factors affect susceptibility (see original publications for an explanation of factors).[2] Second, different disciplinary viewpoints come to different conclusions about the importance of different factors. Source: Alcamo et al.[3] Reprinted with permission from Springer Science + Business Media.

the north, wetter and warmer conditions will work in the opposite direction and boost crop productivity in the short run. But these gains in productivity are likely to occur only during the first phase of climate change, up to the point where annual average temperature (very roughly) increases by 2°C above temperatures in the mid-20th century. After that, warmer conditions will shorten crop growth duration and stimulate so much water loss in plants and surrounding soils that the negative effects of the temperature rise will outweigh its positive influence in many crops. Of course, farmers will adjust to a future climate, as they have to the year-to-year variation in current climate; they will experiment with new types and varieties of crops, modify their cropping practices, and irrigate where crops need additional water and where water is available. Yet, as explained in Chapter 4, not all farmers will be capable of changing their current practices.

The same conditions impacting crop production in Europe will also affect its forests. Warmer temperatures will encourage the growth of vegetation and it is expected that the area densely covered by forest will expand northwards in Europe;

windswept areas of open tundra near to forests will eventually become thick with bushes and trees. By contrast, in the south, increasing aridity will amplify tree mortality, and the climate will become too dry in many areas to support trees altogether. Wooded terrain will become thinner and some areas will eventually be transformed into open shrub- and grassland. In the first phase of climate change, forests in the north will increase their productivity while forest productivity will decline in the south.

The wetter climate in the north will, over the long run, tend to increase run-off into rivers.[4] Occasional droughts may cause temporary water shortages here, but in general water availability north of the Alps is likely to increase. By contrast, drying conditions in the south will prolong drought periods and lead to long-term decreases in the water available to society and the rest of nature.

Endangered nature and biodiversity

While some climate impacts will play out differently in the north and south, all parts of Europe will experience the effects of global warming on their nature and biodiversity. Changing temperatures and precipitation will disrupt the life cycles of individual species, the communities of plant and animals, and the ecosystems made up of many communities. One of the many threats to ecosystems will be climate-related sea-level rise which in this century will inundate up to 20% of existing coastal wetlands[5] – assuming they are not filled in beforehand by new beach-side communities, harbors or recreation areas. Diminishing wetlands implies disappearing habitat for many species that breed or forage in low-lying coastal areas. Many of these coastal wetlands are protected areas under the EU Natura 2000 network (Chapter 7).

Likewise, habitat for some species will be reduced by shrinking permafrost areas and encroaching forested areas in the Arctic, and disappearing ephemeral streams in the Mediterranean region. High in the Alps, warmer temperatures and changed habitat of plants could cause 60% of all species to disappear under most severe climate change scenarios.[6] What are the chances for these species to adapt? While coastal authorities will respond to an increasing number storm surges by building thicker dikes, cities respond to more frequent heatwaves with early warning systems and emergency plans, and water managers respond to more numerous river floods with floodways, plants and animals do not have similar response options. In Chapter 7 it was explained that ecosystems in mountainous and sub-Arctic regions have only a limited number of adaptation alternatives, and tundra and alpine vegetation even fewer. Normally, plants would be able to disperse their seeds to adjacent areas that have better climate conditions, but this does not hold for plants and animals adjusted to the harsher climates of mountainous and Arctic regions. Hence, their habitat is disappearing because of warmer temperatures, and there are no new areas to which they can migrate. It may be possible to preserve mountain plants in managed gardens at high elevation, but of course this is not the best option.[7] The reality is that plants throughout the continent are likely to experience difficulty in adapting to climate change, with the consequence that by the end of the century a large percentage of European flora is likely to become vulnerable, endangered, or committed to extinction.[8] The same risks of extinctions apply to the many insects and other animals whose lives depend on these species. A few options for enhancing the survival chances of these species are discussed in Chapter 7 and in the section on adaptation below.

Climate change threatens Europe's economy

The weather is already a costly factor in the economic life of Europe. A telling example is the list of economic losses incurred from the ten costliest winter storms since 1980 (Table 9.1). The two worst storms caused insured losses of US$5.9 billion and US$5.8 billion, respectively; overall losses were US$10 billion and US$11.5 billion. Changes in climate will further increase the risk of financial loss, in two ways. First, the growing frequency of extreme climate events in particular regions (e.g. flooding along the coastline and rivers) will threaten infrastructure

Table 9.1 Economic impacts of the weather: Ten costliest winter storms in Europe 1980–2007 ordered by insured losses. Source: Munich Reinsurance Group[9]

Date	Winter storm	Region	Overall losses* (m US$)	Insured losses* (m US$)	Fatalities
26.12.1999	Lothar	esp. France, Germany	11,500	5,900	110
18–20.1.2007	Kyril	esp. Britain Germany	10,000	5,800	49
25–26.1.1990	Daria	Western, Northern, Eastern Europe	6,900	5,100	94
15–16.10.1987	87J	esp. Northern Europe	3,900	3,100	18
7–9.1.2005	Erwin	esp. Northern Europe	5,800	2,600	30
27–28.12.1999	Martin	France, Spain, Switzerland	4,100	2,500	30
3–4.12.1999	Anatol	esp. Denmark	3,000	2,400	20
25–27.2.1990	Vivian	Europe	3,200	2,100	52
26–30.10.2002	Jeanett	esp. Western Europe	2,600	1,700	37
28.2–1.3.1990	Wiebke	Western, Southern Europe	2,300	1,300	64

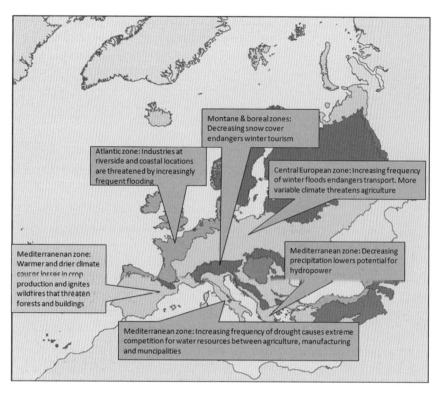

Figure 9.2 Some expected negative economic impacts of climate change according to biogeographic regions of Europe. See caption of Figure 9.1 for explanation of biogeographic regions. For details, see text, especially Chapter 8. Source of map of biogeographic regions: European Environment Agency, copyright EEA, Copenhagen. www.eea.europa.eu (See color plate.)

and buildings. Second, the more gradual changes in temperature, precipitation and sea level will have a disruptive effect on many of Europe's economic sectors (Figure 9.2). In the north, more frequent coastal and river flooding will force some industries to move away from the their waterside locations, either voluntarily because of ongoing danger of dike failure, or because of pressure from local authorities to move economic activity away from flood plains. More frequent flooding will also threaten the movement of goods via railways and highways. In the south, a side effect of climate change will be a growing demand for irrigation water because higher temperatures and decreased precipitation will decrease the natural moisture available to crops. Areas with rainfed agriculture may have to be irrigated, and already irrigated areas may have to be more intensively watered. Manufacturers needing large quantities of water for their industrial processes will be confronted with declining water supplies and increased competition from irrigated agriculture and municipal water suppliers. Some factories may move away from the south in search of more reliable water supplies. In addition, the movement of goods will be curtailed by higher temperature extremes that will damage rail and roadway surfaces throughout Europe.

High temperatures are likely to reduce summer tourism around the Mediterranean and perhaps increase tourism in spring and autumn; alternatively, tourists may decide to visit cooler, northern parts of Europe instead. Meanwhile, shrinking snow cover will reduce or eliminate winter tourism in the Alps and other mountainous regions.

More frequent and severe heatwaves will cause significant increases in mortality in Europe (Chapter 3). Increases in the body core temperature can lead to heat illness, or death from heatstroke, heart failure and a range of other ills. Most heatwave fatalities (heart attack and stroke) occur in people with pre-existing cardiovascular disease or chronic respiratory diseases. There are also considerable health risks associated with other climatic extreme events such as winds, storms and floods, and some new diseases that are transmitted by insect vectors requiring higher temperatures may become a problem in Europe. Overall, however, most European countries have health sectors that can deal with these issues – once they are recognised as being important.

Some good news is that by the 2050s, warmer temperatures will markedly reduce heating demands throughout Europe, in the Mediterranean perhaps by 2–3 weeks each year. On the other hand higher summer temperatures will require an additional 2–5 weeks of cooling along the Mediterranean.[10] Cooling requirements will also go up for much of central Europe – Towards the end of the 21st century, central European countries will, on average, have the same number of hot days as southern Europe now experiences.[11] The need to power air conditioners is likely to cause major peaks in electricity demands during summer, which will have to be met with new peak power generation or load transfers.

The electricity industry will also have to confront other challenges to the energy supply. As reported in Chapter 8, increasing run-off will increase the potential hydropower by 15–30% until the 2070s in northern and eastern parts of Europe, but lower run-off will decrease hydropower potential by 20–50% around the Mediterranean.[12] Less water will also be available for producing steam and for cooling turbines in thermal power plants (coal-fired, oil-fired, or nuclear facilities). Furthermore, warmer river temperatures will reduce the effectiveness of this water supply as a coolant for turbines. On the other hand, increasingly sunny conditions in central and southern Europe will increase the potential for solar power.

Interacting impacts

Although many different individual impacts of climate change have been described in this book, it is important to bear in mind that many of these impacts will interact with one other. Indeed, Europe is so well integrated economically, politically and physically that changes occurring on one part of the continent or in one sector are bound to influence changes in another. As global warming forces more farmers in the south to irrigate their crops, they will increase their withdrawal of surface water; this will further reduce the water available for aquatic ecosystems already under pressure from declining run-off. Competition for remaining water resources will increase between farmers, power plant operators and municipal officials responsible for public water supply. Decreases in agricultural production in the south will, of course, encourage higher production in the north, while the north has to confront a higher threat of winter flooding of agricultural fields. While anticipating life in Europe under climate change, it is crucial to bear in mind the comprehensive and interlocking nature of its impacts.

Impacts of a different scale: exceeding thresholds in the earth system

Most of the impacts discussed in this book will emerge gradually over years and decades rather than from one year to the next. But there are other types of impacts, more sudden and comprehensive, that occur once a critical threshold is exceeded. Such a threshold was discussed in Chapter 6 when we described how coastlines compensate for the scouring effects of storms by redepositing sediments. However, coastlines can keep pace with erosion only up to a point; eventually a point will be reached where the coastal system's self-capacity to replenish its beaches and wetlands will be overwhelmed by rising sea levels and an increasing frequency of storms. Beyond this threshold, the rate of erosion will exceed the rate of replenishing sediments, and the coastline will deteriorate rather rapidly.

Scientists have identified special cases in which going beyond a threshold might have particularly threatening implications; these cases are called "dangerous climate change" or "tipping points", because these changes could mean that parts of the earth system changes into a new state with little chance of returning to the original state. An important task of the scientific community is to identify when this dangerous climate change may occur. Unfortunately, the limits of the earth system are not very well understood, and where limits have been identified, it is difficult to predict if and when they will be exceeded. Nevertheless, research has advanced far enough so that some of these potentially dangerous changes have been described. One such case affecting Europe has to do with a slowing of the ocean circulation in the North Atlantic arm of the Gulf Stream. The ocean currents here are particularly important because they deliver warmth from the tropics to the North Atlantic, keeping the temperature of northwestern Europe much higher than it would otherwise be. Oceanographers tell us that the ocean currents that deliver this warmth are driven not only by persistent southerly winds, but also by differences in the density of ocean water in the North Atlantic, caused by vertical differences in the temperatures and salinity of seawater at this location. By the time the warmer surface current has arrived in the North Atlantic it has become saltier and colder and therefore more dense than underlying ocean layers. As a result, the surface layer sinks, or "downwells" in oceanographic parlance. The sinking effect is called "North Atlantic deep water formation". Because this circulation is partly driven by the saltier upper layer of the North Atlantic, it is also called "thermohaline circulation".

A particularly important region of downwelling occurs off the coast of Norway, as mentioned in Chapter 2. After these masses of water sink, they begin their return journey to the tropics, completing this part of the ocean's large-scale circulation. Oceanographers estimate that this circulation pattern in the ocean has persisted for centuries, if not longer. But climate change has the potential to alter this pattern. Warmer temperatures accompanying climate change are causing the large-scale melting of glaciers in Norway, Greenland and elsewhere in the north, emptying new masses of freshwater into the North Atlantic. This freshwater input is steadily making the upper layer of the North Atlantic fresher and this will eventually slow down the sinking tendency of the upper layers of ocean water; slowing down the sinking tendency means that less warm water will be pulled from the south. Consequently, less warm water will arrive from the tropics, causing a cooling tendency in northwestern Europe. However, since greenhouse gases will also be increasing during this time, the cooling tendency will be compensated by a parallel warming trend. In the end, the net effect may be that northwestern Europe warms, but not to the degree it would have if the ocean circulation had not slowed down. If the North Atlantic deep water formation is significantly slowed down, then a temporary cooling, rather than warming, could be observed over this part of Europe with negative consequences on crop production, heating demand and run-off. The important point is that once a threshold is reached, a major slowing of the northern branch of the Gulf Stream could occur within one or two decades, according to recent estimates.[13] On the other hand, model experiments indicate that the onset of this event is unlikely to happen before the end of the 21st century;[14] but the risk increases as the effects of climate change add up over many decades.

Another event, in the category of "dangerous" climate, is the accelerated melting of the Arctic ice cap. In Chapter 8 it was shown that the area of ice coverage in late summer might be sharply reduced by the end of the 21st century, clearing the way for new shipping lanes in the high North. But there is a negative side to the melting of the ice cap. As anyone blinded by a snowy landscape on a sunny day can confirm, snow and ice are very effective reflectors of sunlight. It follows that the Arctic ice cap now reflects incoming solar radiation very effectively (i.e., when the sun is above the horizon). In stark contrast, the ocean underlying the ice, especially when turbulent, *absorbs* solar radiation very efficiently. This means that the shrinking of the ice cap during the summer season (as discussed in Chapter 2) will expose the underlying ocean to sunlight for a much longer period of time each year. Hence, much more heat from the sun will be absorbed by the northern ocean than previously, and this will disrupt the long-term heat balance of the Arctic. What will follow is unclear, but it seems likely that global climate patterns will be disrupted.

Another large-scale change falling into the category of tipping points is the melting of permafrost at high latitudes in Europe and elsewhere. The melting of these areas may result in the inadvertent release of large quantities of methane gas which are now "locked up" in permafrost ice.[15] Since methane is a potent greenhouse gas, the release of this gas would have a "positive feedback" (positively re-enforcing effect) on regional and global temperatures; as the soil becomes warmer, more methane will be released which will add to the methane concentration in the atmosphere, which will stimulate further global warming, which will cause more permafrost to be melted, which will release more methane, and so on.

How can "dangerous climate change" be avoided? Some scientists believe that preventing global temperature from rising more than 2°C (relative to its pre-industrial level) would lower the risk of these and other threats. As a result, a group of countries agreed to a 2°C limit to global warming at the Copenhagen climate summit in 2009 and nearly all countries agreed to this target a year later at the climate summit in Cancun. Chapter 10 returns to this crucial policy issue and its impact on climate protection in Europe.

Why adaptation should begin now

The impacts described above and in previous chapters make up a good case for acting on climate change in Europe. The only question is, how should we proceed? Two main options are at our disposal: The first is to mitigate climate change by reducing the concentration of greenhouse gases in the atmosphere. This approach is discussed in Chapter 10. The other option is to adapt to climate change, which is dealt with here. Earlier chapters presented some of the alternatives available for adapting to particular climate impacts, and now we address the overriding issues that have to do with selecting and implementing adaptation measures.

Before discussing how to select adaptation measures we should step back and review the arguments for immediately planning and beginning adaptation actions. The first is the simple fact that climate change already has a negative impact on Europe and the rest of the world, and these impacts merit a response. Most impacts are attributed to a warming trend in temperature, and have affected the growing characteristics of both wild plants and managed crops, leading to earlier growth, a longer growing season and a shift in species. It has shrunken the extent of permafrost and glacier area. An increase in temperature extremes has been linked with the great European heatwave of 2003, which led to thousands of deaths, severely affected major ecosystems and had many other detrimental impacts.

The second is that climate change is expected to intensify, bringing with it more serious impacts, even if we immediately roll back the level of greenhouse gases in the atmosphere. We expect further climate change because of the inertia of the climate system, which leads to a long time lag between the emissions of greenhouse gases and the system's response (Box 9.3). One estimate derived from climate model experiments is that an additional increase of 0.6°C global average temperature "is in the pipeline", in other words, will be difficult to avoid.[16]

But there are other important time lags, and these provide the third incentive for adaptation. This other kind of delay is caused by the slowness of society to respond to climate change. This slowness comes from the sluggishness of the policy process but also from the phenomenon named "technological lock-in" by which society invests in large-scale infrastructure which has both a long lifetime

Box 9.3 Why some climate change is inevitable: Lag times in the climate system

The climate change happening today has more to do with the cumulative effect of greenhouse gases injected into the atmosphere over the past several decades than with the emissions over the past year. Likewise, greenhouse gases emitted this year into the atmosphere will affect climate for many decades to come. This time lag has to do with the inertia of the climate system, which leads to a long delay between emissions, changes to the climate system and climate impacts. The response of the atmosphere to emissions is fairly rapid since most of the important greenhouse gases mix within a few months to years in the atmosphere. The response of the atmosphere to these gases is also fairly rapid. But the response of the full ocean to atmospheric heating is very slow by human standards. The long delay arises, in particular, because the ocean is a huge sink of heat from the atmosphere, and only very slowly comes into equilibrium with the atmosphere because of the sea's enormous mass and slow vertical mixing.[17] Because of this time lag, it was recently estimated that at current concentration levels of greenhouse gases, we should expect an additional 0.6°C of global average temperature increase,[18] which would translate into even larger surface temperature increases and further precipitation changes in Europe.

(power plants, buildings) and produces substantial emissions of greenhouse gases. An example is a city that invests in a new downtown development made up of energy-wasting buildings that produce more emissions per square meter of office space than state-of-the-art energy-efficient ones. Once these buildings are constructed it is said that their emissions are "locked-in" for the lifetime of the structures and can be reduced only at great expense. Of course, in reality, some cost-effective measures can be taken to reduce emissions from virtually every source. In the case of the office buildings, actions can be taken to reduce after-hours office illumination and air conditioning so that the buildings consume less energy and produce lower emissions. Individuals can also be slow to change their behavior, both when it comes to transportation, where car and air traffic causes large emissions, and when it comes to food preferences where meat (in particular beef) is associated with large emissions.

A fourth motivation is that the greater the impacts of climate change, the more expensive the measures to adapt to it. Chapter 8 discussed some examples of how the costs of damage might grow much more rapidly than the intensity of a natural hazard. We related, for example, that the property damage caused by wind storms increases roughly as the cube of maximum wind speed during a storm. As climate impacts intensify and accumulate, society will have a more difficult and expensive time in keeping up with these impacts. This, of course, is also a clear motivation for slowing the tempo of climate change by reducing greenhouse gas emissions, as discussed in Chapter 10.

It can be argued that the international community has already accepted the call for adapting to climate change. Article 4 of the Framework Convention of Climate Change, which already entered into force in 1994, stipulates that all Parties to the Convention should:

- Formulate, implement, publish and . . . update national and . . . regional programmes containing . . . measures to facilitate adequate adaptation to climate change (Section 1b).
- Cooperate in preparing for adaptation to the impacts of climate change . . . (Section 1c).[19]

More recently, at the climate summit on Bali in December, 2007, the international community agreed to set up a fund of several hundred million US dollars for adaptation projects in developing countries.[20] This was confirmed at the climate summit in Copenhagen in 2009, where a Green Climate Fund was established to assist developing countries in mitigating and adapting to climate change. Subsequent climate summits in Cancun and Durban have further developed the plans for this fund. But action has not been limited to the international level. In fact, many communities, businesses and local governments on the continent are also concerning themselves with adaptation actions or planning, as described in previous chapters.

Types of adaptation

Earlier chapters have shown that measures for adaptation cover a wide palette of different types and scales. The measures used at a particular place will depend on the intensity of climate change, and the sensitivity and susceptibility of the human population or ecosystems faced with changing climate.

With this wide variety of options, how can adaptation measures be classified in some orderly way? Table 9.2 gives a few possibilities. Often a distinction is made between autonomous and planned adaptation. With autonomous adaptation

Table 9.2 Classification of adaptation measures (see text for definition)

Autonomous	↔	Planned
Anticipatory	↔	Reactive
Centralized	↔	Decentralized
Gradual	↔	Abrupt

there is no governing body influencing the adaptation practices. Examples are range shifts of plants and animals in response to warming. Here species and individuals change their geographical location in response to the changing climatic conditions, unless migration is obstructed by rapid climate change or a fragmented landscape. Other examples of autonomous adaptation in response to observed climate change are the changes in planting times and crop choices made by farmers. Planned adaption by contrast typically involves some sort of anticipation of future climate change to reduce costs of future climate change or take advantage of benefits. Planned adaptation is particularly important when investments or measures have long-term effects (e.g. building a new dam for hydropower or rebuilding coastal defenses).

Another alternative is to divide them into classes of "anticipatory" (i.e. measures that cope with anticipated climate impacts), or "reactive" (i.e. measures taken in response to an already experienced climate impact).[21] The current heightening of coastal dikes in Germany in anticipation of sea-level rise is an example of an anticipatory measure, whereas the recent strengthening of the river dikes along the Rhine can be thought of as a reactive action since it follows the near breaching of existing dikes in 1994.

Another possible delineation is between "centralized" and "decentralized" adaptation. A centralized action is steered by a central political authority or other institution (for example, a parliament adopts a national law requiring new climate-sensitive building standards); a decentralized measure is implemented at the initiative of local organizations (for example, a business carries out a water conservation program to reduce its exposure to declining water availability).

Finally, adaptation measures can be distinguished between "gradual" and "abrupt". An example of a gradual adaptation would be the actions of forest managers to gradually replace tree species with new species in response to slowly changing climate conditions. An abrupt adaptation might be the introduction of snow-making machines to a ski resort to counteract deteriorating snow conditions, or the closing down of a water-intensive industry because of repeated droughts and its relocation to a water-rich area. Of course, the boundaries between gradual and abrupt are fuzzy and depend very much on the situation being considered.

Yet another useful way of classifying and organizing adaptation options is given by Burton et al.[22] Here options are sorted according to the goals of the adaptation (with examples added by authors):

- *Preventing loss* – Action is taken to reduce susceptibility to climate change (e.g. building new facilities to store water, or making buildings more flood-resistant).
- *Tolerating loss* – Accepting adverse impacts under the assumption that they will not hinder the well-being of society or the rest of nature (e.g. projecting only a small decrease in water availability in a river basin due to climate change, and accepting this reduction because decreasing population will lead to lower water demands in any case). This, of course, is the equivalent of the "no action" alternative, rather than an adaptation measure.
- *Spreading or sharing loss* – Action is taken to spread out the costs and other burdens of the expected impact (e.g. increasing governmental compensation to flood and drought victims). This also covers insurance against climate related losses.

- *Changing use or activity* – Modifying the activities of a community in the light of climate change (e.g. introducing irrigated agriculture where rainfed agriculture is no longer viable).
- *Changing location* – Action is taken to move an activity to another location or time, where climate conditions are more suitable (e.g. shifting tourist seasons to spring and autumn, when summers become too hot).
- *Restoration* – Rehabilitating climate-related damage (e.g. rebuilding a town after it experiences flood damage).

Assessing adaptation options

Although there has been considerable research into how people and the rest of nature adapt to climate variations or climate hazards, there has been relatively little work on how they can adapt to climate change.[23] In the face of the potential impacts we have described throughout this book, it is clear that now is the time for assessments of adaptation to climate change. But where to begin?

Box 9.4 shows an early effort by the IPCC to sketch out a procedure for climate impact and adaptation assessment.[24] An elaboration of this approach is given in

---

Box 9.4 Procedures for adaptation assessment

The IPCC approach:[28]

1 define objectives;
2 specify important climatic impacts;
3 identify adaptation options;
4 examine constraints;
5 quantify measures / formulate alternative strategies;
6 weight objectives / evaluate trade-offs;
7 recommend adaptation measures.

The stages in the UK-CIP framework for adaptation assessment.[29,30]

1 identify problem and objectives;
2 establish decision-making criteria;
3 assess risk;
4 identify options;
5 appraise options;
6 make decision;
7 implement decision;
8 monitor, evaluate and review.

a handbook from the United Nations Environment Programme.[25] More recently, the Climate Impact Programme of the United Kingdom (UK-CIP) introduced a further refinement of these earlier approaches, which they called a "decision making framework".[26,27] Putting "decision making" in the forefront is sensible because the ultimate aim of an adaptation assessment is to take decisions on how to cope with climate change. The UK-CIP approach is given in Appendix 9.1 and the eight steps of the framework are summarized in Box 9.4.

Limits and barriers to adaptation

One of the key things to keep in mind during an adaptation assessment is that many barriers, some obvious, some more subtle, will stand in the way of a particular course of action. First of all, the feasibility of an adaptation measure will depend on the magnitude and tempo of climate change; the stronger and faster the rate of climate change, the larger the number of options that fall away as undesirable, ineffective, or infeasible.[31] Reinforcing this point, the IPCC made it clear that "adaptation is more feasible when climate change is moderate and gradual than when it is massive and/or abrupt".[32]

Adaptation options can also be limited by physical factors, financial costs, politics and/or organizational considerations:

- *Physical or biological limits.* Adaptation measures may have a physical or practical limit. It may not be possible, for example, to reduce water demands without threatening the health of water consumers or economic activity.[33] Likewise, it may not be feasible to find an adequate replacement for crops when their productivity declines under future climate conditions. It is likely that critical thresholds exist, beyond which humans or other parts of nature simply cannot easily adapt. IPCC points out that these thresholds could exist for natural systems as diverse as kelp beds, coral reefs, rangelands and lakes.[34] In nature, the loss of "keystone species", those plants or animals that play a central role in the functioning of an ecological system, could lead to an irreversible decline of an ecosystem or to regime shifts, where new forms of ecosystems replace existing ones.
- *Financial limits.* Adaptation measures may not be economically feasible. Some communities may simply not be able to afford the costs of thicker, higher sea walls or other expensive coastal protection structures. Another example is the adaptation of ski resorts to warmer temperatures by artificial snow-making. It may happen that the costs of producing artificial snow at a ski resort will become prohibitively expensive once winter warming reaches a particular threshold. Such a threshold was found in a theoretical study of Canadian ski resorts.[35]
- *Political limits.* Political or social considerations may limit the implementation of an adaptation measure. For example, political opposition to the use of genetically modified (GM) crops in Europe is likely to be a barrier to using GM technology for developing drought-resistant crops that could sustain the expected increased frequency of droughts in southern and other parts of Europe. On the other hand, there are often ways to address such barriers. For crop modification some

molecular techniques that exploit existing genetic variation in crops can be almost as effective as genetic modification in developing drought tolerant crops.

- *Organizational limits.* In some circumstances existing organizations and institutions do not have the capacity or will to take on the additional duties required to implement adaptation. Municipal and provincial agencies in some parts of the continent may already be overloaded with current responsibilities and unable to take on new tasks for an adaptation program.

Helping nature to adapt

We have seen above that many factors can limit the success of measures aiming to help society cope with climate change. Yet the reader will recall from Chapter 7 that the alternatives available to natural flora and fauna are even more limited. Their options are usually two: to remain at their current location and adjust to new climate conditions, or migrate to a more climate-conducive location. We have seen that many organisms and ecosystems have built-in mechanisms for adapting to climate change in Europe. However, some of the more sensitive species and ecosystems are extremely vulnerable and their natural tendency is to pursue the second option, namely, to move to an area where the climate is more favorable. Plants, for example, will naturally seek out these places by dispersing their seeds over a wide geographic area. In fact, some climate-related migration of flora and fauna has already taken place in Europe, as mentioned in Chapters 1 and 7; two examples are the increase in the elevation of the tree line in montane areas, and changes in the geographic range of certain butterflies.

But the high population density of Europe presents a practical barrier to the migration of plant and animal communities: These communities are hemmed in by settlements, roads, industrial areas, and in mountain environments by the lack of room at the top. Clearly new thinking is needed to raises the chances of Europe's sensitive ecosystems.[36] One promising concept is the "ecological corridor", a strip of territory extending from one protected area to another which allows plants and animals to eventually move in response to changing climate (Figure 9.3). But corridors themselves are insufficient; what is needed is a concept that provides

Figure 9.3 Different types of ecological corridors. Such corridors can support the migration of ecosystems as they seek more climate-suitable areas in the face of climate change. Source: van der Sluis et al.[40] Reprinted with permission from Alterra, Wageningen University Research.

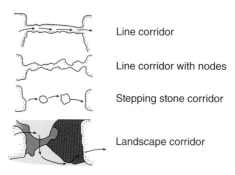

Line corridor

Line corridor with nodes

Stepping stone corridor

Landscape corridor

substantial territory for new habitat. In Europe this is termed "an ecological network", defined as a "system of areas which are connected via ecological links or physical links".[37] Such networks contain a multiplicity of different geographic units including "core areas", corridors, buffer zones and, in some cases, nature development or restoration areas.[38] Among the many ideas for preserving ecosystems and the diversity of nature in Europe, the concept of "ecological network" is among the best thought out and elaborated.[39]

But what of the many plant and animal communities who cannot migrate? Their chances for adjusting to climate change can be enhanced by increasing their overall resilience to change. This resilience can be boosted by lessening the other pressures placed on ecosystems such as habitat fragmentation and destruction, overexploitation, eutrophication, desertification and acidification.[41] In short, the key to adaptation of natural ecosystems in Europe is two-fold: reducing the other sources of stress on these ecosystems, and making more room available to them to migrate and re-establish themselves at more climate-favorable sites.

Mainstreaming adaptation

Developing a good adaptation strategy is a worthwhile enterprise, but adaptation measures still need to be brought to fruition. This raises the basic issue: How can public and political support be won for the financing and implementation of adaptation actions? One alternative is to consolidate all adaptation measures into a separate, stand-alone "national adaptation program". This is a good option but has the disadvantage that it has to directly compete with other public policy priorities such as expanding education, developing new products and markets, or maintaining or strengthening the military. "Adaptation", as a political theme, is often (but not always)[42] unattractive compared to these other issues, because it addresses impacts that will largely occur in the future.

An alternative to national programs, and perhaps a more realistic way to proceed, is to integrate adaptation plans and measures into mainstream public policy actions, and avoid the competition problem. This is called "mainstreaming". The slogan of "mainstreaming climate policies" has already been popularized in the water sector by the network "Cooperative Programme on Water and Climate".[43] The objective of mainstreaming would be to fine-tune existing planning efforts in order to make society more resilient to growing risks of climate change. Previous chapters have given many examples of such mainstreamed policies: several countries are now raising their current coastal levees to take into account future sea-level rise and/or an increased frequency of coastal storms. Some communities factor in climate change when planning new water storage facilities for their growing service population.

Another popular concept related to mainstreaming is called "climate-proofing". This means making facilities in the planning stage (such as dams, reservoirs, buildings, bridges, transportation corridors, and suchlike), robust to the potential

impacts of climate change.[44] Although mainstreaming and climate-proofing policies will lessen the need for a large centralized adaptation bureaucracy, it is likely
that some new adaptation experts will be needed at key agencies (such as those
responsible for coastal protection or electricity production) to carry out adaptation policies.[45]

A very effective way of "mainstreaming adaptation" would be to embed it in an
even larger planning concept for a region or part of a country. In this way, adaptation goals could be better harmonized with other visions for the region. Two such
planning concepts exist in Europe. The first is Integrated Coastal Zone Management (ICZM) (discussed in Chapter 6) which is a broad set of principles for
managing change in the coastal zone. ICZM, as it is called, advocates planning
from a long-term perspective, and provides an ideal platform for introducing
measures for coping with future climate change into local planning. Since ICZM
has been adopted as a cornerstone of EU policy, it will be an important vehicle
for implementing adaptation in Europe. The second case is Integrated Water
Resource Management (IWRM) (see Chapter 5) which also advocates a longer-
term view, in this case, for river basin management. An important element of
IWRM is the dialogue it fosters among different stakeholders in a river basin
(municipal water managers, irrigation district representatives, leaders of conservation organizations) with the aim of achieving consensus on a future plan for
the river basin. It is easy to imagine how the topic of adaptation to climate change
could be integrated into this dialogue, and how adaptation measures could eventually become part of an overall strategy for the river basin.

While modern coastal management and river basin management provide
opportunities to mainstream adaptation measures at particular geographic locations, another opportunity exists that cuts across all geographic areas, namely, to
introduce adaptation plans into existing disaster management programs. The
advantage here is that nearly all European communities, regions and countries
have institutions (emergency and rescue agencies, emergency plans, early warning
systems) for coping with natural disasters.[46] Hence, these institutions could be
used as a platform for implementing adaptation measures. For example, existing
early warning systems for river flooding could be extended to include an increased
frequency of heavy precipitation due to climate change. Similarly, the portfolio of
rescue agencies could be expanded to include public assistance during more frequent heatwaves.

It can be argued that the idea of adaptation management is very compatible
with disaster relief planning since both are based on the concept of "risk assessment and management". In fact, the adaptation assessment procedure outlined in
Appendix 9.1 closely follows the thinking of risk assessment.

Adaptation, but where and how?

Up to now we have seen the many justifications and possibilities for adaptation,
but have not yet discussed where these adaptations should be planned or carried

out. What is the most appropriate level? Local, regional, or global? Considering the diverse character of adaptation measures, it is evident that no single level of organization, neither the individual nor the international level, nor any level in between, can adequately cover all of these measures. The question can be better answered if we divide it into two parts: Where should adaptation be planned, and where should it be implemented?

There are good reasons to plan adaptation at the highest sensible geographic level in order to minimize duplication and to increase efficiency. The duplication problem arises, for example, when both local and national governments pass regulations about requiring shading elements for new buildings as a measure to adapt to warmer temperatures. If these regulations are not coordinated nationally, consumers may be confronted with a patchwork of contradictory regulations. Coordination at the highest geographic level should be able to minimize these contradictions. The efficiency argument arises, for example, in planning future water storage as a hedge against more frequent droughts. It is obvious that upstream communities cannot impound and store a river water supply unilaterally without interfering with the rights of downstream users. National or river basin coordination is needed to prevent inefficiencies of water resource planning and injustice in the distribution of water resources.

Many European countries are already proceeding with assessing and planning adaptation measures on the national level. One of the first examples is a comprehensive Finnish study called "FINADAPT" which examines many different aspects of climate impacts and adaptation options for a wide range of sectors including forestry, the built environment, transport, water resources, and health.[47] A pan-European approach has been advanced by the Commission of the European Community (CEC) in its Green Paper *Adapting to climate change in Europe – options for EU action*.[48] The CEC argues for a European-level approach for three main reasons. First, it believes that climate change is a challenge facing all of its members, and is therefore an EU concern. Second, the CEC argues that some areas affected by climate change transcend national boundaries (e.g. some large river basins) and therefore must be dealt with internationally. Finally, the Commission argues that the key economic sectors to be impacted by climate change (agriculture, fisheries, energy) are highly integrated at the European level and have to be dealt with on this level. Assuming that its member states accept these arguments, the Commission proposes "four pillars" for pan-European action on adaptation (Box 9.5).

The First Pillar is "early action in the EU" involving (i) integrating the aims of adaptation in EU legislation and policies, for example, having to do with European agricultural and rural development; (ii) integrating adaptation goals into EU funding programmes for infrastructure such as bridges and motorways, and (iii) "developing new policy responses" such as developing novel building codes and climate-resilient crops.

The Second Pillar is "integrating adaptation into EU external actions", which means adding the goals of adaptation to EU foreign policy. This involves taking political and financial steps to support adaptation planning outside of the EU.

Box 9.5 The "four pillars" of EU policy for adapting to climate change[49]

The first pillar: *Early action in the European Union* – Build adaptation policies into forthcoming legislation and policies, and into European funding programs. Develop new policy responses to climate change (e.g. modifying building codes and methods, and developing and promoting climate-resilient crops).

The second pillar: *Integrating adaptation into EU external actions* – Promote adaptation measures in developing countries through various foreign policy actions; promote adaptation actions in neighboring countries through activities under the European Neighborhood Policy. Exchange information on adaptation practices with industrialized countries.

The third pillar: *Reducing uncertainty by expanding the knowledge base through integrated climate research* – Develop comprehensive scientific approaches to assessing impacts, vulnerability and adaptation to climate change. Identify more precisely regional impacts in Europe.

The fourth pillar: *Involving European society, business and public sector in the preparation of coordinated and comprehensive adaptation strategies* – Intensify discussions about climate change adaptation on the pan-European scale. Include the private and public sector, and the European Commission in these discussions.

The Third Pillar is "reducing uncertainty by expanding the knowledge base through integrated climate research", which has to do with expanding European funds for research related to climate impacts and adaptation. It also involves support for further developing and using different early warning systems in Europe for anticipating crop failures, forest fires and river flooding.

The Fourth Pillar is to involve ". . . European society, business and public sector in the preparation of coordinated and comprehensive adaptation strategies". The basic idea here is that some European economic sectors, such as agriculture, forestry, renewable energy, water fisheries and tourism, are particularly weather-dependent and might require "significant restructuring" to adapt to climate change. As part of its program, the CEC proposes to stimulate a European-wide discussion about adaptation actions among "the parties and civil society" involved in these sectors.

While the EU seems ready to take a pan-European approach to adaptation, they are to some extent held back by the European principle of "subsidiarity", which says that governance should be exerted at the lowest possible geographic level. Hence, as stated in the EU Green Paper, the most reasonable way to plan adaptation in Europe might be through partnerships between different levels of government and society,[50] for example, shared responsibility between the European Commission and national governments, or between national governments and regional governments within a country. Under some circumstances, as in the case of adaptation of agriculture, it may even be reasonable to forge a coalition among European, national, regional and local authorities, as well as other stakeholders.

As to the question, where should adaptation be implemented, the answer is not the same as for planning. Many adaptation measures, practically speaking, take place from the very local level (families taking decisions on insurance and where to live) up to the regional level (cities or farmers banding together to cooperate on adaptation). But sometimes it is sensible to implement adaptation nationally, as in the case of a nationwide program to reinforce coastal levees, or internationally, as in the case of climate-proofing transportation links in Europe. In general, the scale for implementing particular adaptation measures depends very much on the situation. Important factors will be the kinds of climate risks the measure deals with, the geographic scale at which decisions are taken (local, regional, national or EU scale), and the source of financing for the measure.

Adaptation impacts

An important point to keep in mind when planning adaptation measures is that the measures themselves could have negative consequences.[51] Adaptation actions could have a direct impact on society or the rest of nature, or they might have an indirect impact by hindering mitigation of climate change or by taking funds away from other public expenditures.

An example of a direct impact of an adaptation measure is the construction of river levees as a hedge against river flooding; these embankments will sever the ties between ecosystems in the river and ecosystems established along its banks and flood plains. Freshwater wetlands often depend on replenishment by river flooding and this source of replenishment will be cut off by river levees. Levees also decrease the amount of water stored in a flood plain and thereby can increase the severity of floods downstream.

To understand how adaptation could hinder the mitigation of climate change, consider the case of using air conditioning as a way of coping with summer heat-waves; the additional energy costs and emissions of running air conditioners will make it more difficult to reduce overall greenhouse gas emissions.

A third example of the impact of adaptation has to do with the costs of paying for the measures. If adaptation requires the construction of new thick sea walls, it is possible that the high costs of these dikes will take away from other planned public expenditures. This is a strong argument for giving high priority to "no regret" adaptation measures, as mentioned above, which not only help society cope with climate change but have other clear benefits to the community as well.

In short, analysts should not overlook the possibility that adaptation actions may have unintended, negative side effects. An analysis of these side impacts should always be part of an adaptation assessment.

Appendix 9.1 Procedure for adaptation assessment

This appendix presents a procedure for assessing adaptation options laid out by the United Kingdom Climate Impact Programme (UK-CIP).[52] Also included is information from the assessment approaches of the Intergovernmental Panel on Climate Change (IPCC)[53] and the United Nations Environment Programme (UNEP).[54] The authors of the UK-CIP approach divide adaptation assessment into six main "stages" or phases, as explained below.

Stage 1 Identify problem and objectives

At the start of an adaptation assessment the analyst defines the problem and articulates the objectives of the assessment and eventual actions. Since impact and adaptation are such wide-ranging and vague topics, it is crucial at the outset to be clear about the goals of the assessment. To what impacts is adaptation needed? Where is the adaptation targeted? At the community, regional, or national level? Who is affected by the impacts and the adaptation?

A useful way of identifying goals of the adaptation measures is to first identify some general goals and to break them down further into very specific objectives. Table 9.3 depicts two very general goals for the adaptation actions (sustainable development and reducing vulnerability), and assigns specific objectives to each of these. These objectives should be specific enough to guide the assessment.

Table 9.3 Example of setting goals for an adaptation assessment. From Carter et al.[55]

Overall goal	Specific objective	Evaluation criteria
Sustainable development	Regional economic development	Income, employment
	Environmental protection	Biodiversity, habitat areas, wetland types
	Equity	Distribution of employment, minority opportunities
Reduce vulnerability	Minimize risk	Population at risk, frequency of event
	Economic issues	Personal losses, insured losses, public losses
	Increase institutional response	Warning time, evacuation time

In this part of the assessment the analyst identifies the "climate-sensitive" decisions that are at the center of the assessment. "Climate-sensitive" decisions are those decisions that have to take into account climate change as a factor. Falling under this category are decisions to do with planning large infrastructure, such as reservoirs and power plants or new nature conservation areas.

The UK-CIP suggests the following questions (modified by the present authors) to help structure the assessment and identify its objectives:

- Where does the need to make the (adaptation) decision come from? What are the main drivers behind the decision? What beneficial objectives are intended?
- Is the problem explicitly one of managing present-day climate or adapting to future climate change (i.e., is the problem perceived to be a climate adaptation decision problem)?
- Who or what will benefit or suffer as a consequence of the (climate impact) problem being addressed? Who are the key stakeholders representing these interests?
- Have timescales been established for making and/or implementing a(n) (adaptation) decision?
- Is the decision expected to provide benefits in the long-term (> 10 years) or have other long-term consequences?

Stage 2 Establish decision-making criteria

To ultimately decide on an adaptation measure or measures, the analyst or decision maker needs a benchmark for ranking options. Criteria are needed to help the analyst or decision maker decide whether one option is better than another, or if they are altogether acceptable. These criteria are identified in this step. Examples of possible criteria are:[56]

- ease of implementation;
- risk that the measure will not be successful;
- cost;
- equity;
- public approval;
- public acceptability.

Another important point is that the criteria must be matched to the objectives from Stage 1. Table 9.3 gives an example of matching criteria to goals.

During this stage of the assessment, the UK-CIP also recommends that the "receptor at risk" be specified in detail. This is another name for the subject of the assessment, whether it be an organization, social group, community, city, or ecosystem. Another parameter to be decided upon is the "exposure unit", meaning the organizational level that is covered by the study. For example, in an assessment of the adaptation options available to municipal water supply authorities to climate change within a river basin, the "receptors at risk" are the water supply agencies within the river basin, and the "exposure unit" is the water supply system or sum of water resources within the river basin.

Important questions at this stage are (again paraphrasing from UK-CIP):[57]

- What are the legislative requirements or constraints influencing various adaptation options?
- What are the "rules" for making a decision? For example do national guidelines already exist for carrying out an assessment?
- How will the decisions on adaptation affect other policy decisions?
- Who will finally take decisions on adaptation?

Stage 3 Assess climate impact risk

One more step remains before the analyst can actually begin to assess adaptation options, and that is to identify the risks of climate impacts to the subject and area of interest. Adaptation measures will be matched to these risks, so it is important to articulate them as clearly as possible. Different climate impact assessment methods are available for this task, as discussed in Chapter 1. Impact assessment involves projecting the magnitude and spatial extent of impact, its frequency, duration, speed of onset and seasonality. The impacts must also be compared to a future reference case in which no climate change occurs. As an example, the assessment of climate impacts on water supply agencies within a particular river basin could involve the following steps:

1 Specify temporal and spatial dimensions of the assessment (e.g. time horizon: 2050s, spatial coverage: entire river basin).
2 Articulate a reference scenario of no climate change. Such a scenario needs to describe the impact on water resources of different socio-economic developments. Included in these developments are changes within the river basin (for instance, increasing population, higher income levels, changes in energy production) and outside the basin (e.g. international or national demand for agricultural products from cultivated land within the basin).
3 Assemble climate scenarios for the basin.
4 Articulate a set of scenarios that take into account both climate changes and socio-economic developments.
5 Compare the scenarios under climate change with those without, and identify the first order impacts on water resources due to climate change (e.g, changes in annual water availability, changes in frequency of floods and droughts).
6 Based on the impacts in step 5 compute the secondary impacts of climate change on water management within the basin (e.g, effects on municipal water supply on withdrawals for power plants and manufacturing facilities), on availability of water for irrigated agriculture.

A climate risk assessment can be carried out with various levels of detail, depending on the requirements of the assessment. Here are three alternative levels of detail.

Option 1 Risk screening. A "screening" or preliminary assessment of the risk of climate change is appropriate if little is known in advance about the impacts of climate change on the study area or how it could affect decision making in the area.

Option 2 Qualitative risk assessment. This option is appropriate if preliminary knowledge about climate risks is available and if the goal of the assessment is to prioritize and rank risks qualitatively rather than depict risks in numerical terms.

Option 3 Quantitative risk assessment. To convince decision makers or the general public about new large investments in adaptation measures, it is likely that the analyst must put the risks of climate change in quantitative terms. Under many circumstances the risks will be expressed as a quantitative estimate of the probability of different levels of damages due to climate change.

Stage 4 Identify options

Once the risks of climate change have been articulated, we can proceed to identify the options available for adapting to these risks. The UK-CIP suggests the following questions (paraphrased by the authors) to help the analyst identify options:[58]

- What type of options should be considered?
- What are the consequences of the "do nothing" option?
- Can "no regret" or "low regret" options be identified? These are adaptation actions that have clear benefits in addition to those related to climate change. An example is household water conservation, which is an adaptation option because it lessens the impact of droughts on households, but it has the additional benefits of reducing consumer expenses for water services and avoiding impacts of new water storage facilities on ecosystems.
- Can options be defined that are flexible enough to handle the uncertainty of future climate impacts? Since climate impacts are inherently uncertain, a good characteristic of an adaptation measure is that it can be changed in the future if new knowledge shows that the climate impact is more or less severe than anticipated. For example, supplementary irrigation might be a flexible way of dealing with changed precipitation patterns. If long-term precipitation turns out to be different than projected, the commitment to irrigation can be adjusted fairly flexibly.

Depending on the goals of the assessment, the information about options should include:[59]

- approaches currently or previously used or those that have not been tried but are feasible in the future;
- purpose, cost, and effectiveness of different options;
- measures that have to do with changes in infrastructure or buildings;
- legal and legislative changes required for adaptation;
- regulatory measures;
- administrative measures;
- public and institutional education;

- financial incentives and disincentives, such as direct support, taxes, user fees, and tariffs;
- measures having to do with research and development;
- market mechanisms, such as water pricing so that users pay for amount of water consumed;
- technological changes.

Since the number of options is so large (see previous chapters) it is helpful to use a particular scheme for assembling adaptation options. For example, the analyst could use the different categories of adaptation strategies described in Table 9.2.

Stage 5 Appraise options

The aim of this stage is to provide a robust foundation for selecting an adaptation measure in the next step. Here the options assembled in Stage 4 are compared and evaluated against the criteria from Stage 2. The costs and benefits of each option are articulated. Benefits in this case are the climate change damages avoided (e.g. flood damage avoided) or the positive effects exploited, whereas costs are the financial and other expenses for carrying out an adaptation option (e.g, financial costs of building new levees to avoid flood damage, ecological costs of new levees isolating aquatic and riparian ecosystems, and social costs of relocating people to new homes where the sea is allowed to overflow existing land). The individual measures can be ranked according to their effectiveness, robustness and resilience vis-à-vis the objectives.

It is likely that some of the options compiled in the previous step are more feasible than others. Each option will have its own set of constraints, whether they are limited by current laws, unacceptable because of local or national norms, too expensive, or simply physically limited. A careful analysis of these constraints is needed for the next step.

As in Stage 3, the evaluation of adaptation options here can be conducted with different levels of detail:[60]

Option 1 Systematic qualitative analysis. Under this option, the analyst ranks, in qualitative terms, the relative size of the risks, costs and benefits of different options.

Option 2 Semi-quantitative analysis. Here the options are ranked with a mix of qualitative and quantitative information, depending on the uncertainty of the different options.

Option 3 Fully quantitative analysis. Under this option, quantitative information is provided about risks, costs and benefits. Where possible, this information is given in monetary terms.

In selecting the appropriate method for evaluating measures, Feenstra et al. suggested the following criteria:[61]

- How well does the method address the goals of the assessment (e.g. needed precision for decision making, contribution to building consensus)?
- How well can the method address uncertainties (e.g. related to timing and magnitude of impacts)?
- What is the availability of inputs needed for the method (e.g. socio-economic data)?
- Are adequate resources available for using the method (e.g. expertise, financial, time)?

A range of methodological options is given in Table 9.4.

Stage 6 Make a decision

The objective of this stage is to decide on a "preferred option". The basic question is, has a clearly preferred option emerged up to this point? If the answer is yes,

Table 9.4 Summary of some available techniques for assessment of adaptation measures. Source: Feenstra et al.[62] Reprinted with permission from Free University of Amsterdam.

Method	Level of precision	Ability to address uncertainties	Inputs needed	Resource requirements
Forecasting by analogy	Low	No formal provisions	Detailed historical/ contemporary account	Moderate
Screening	Low	No formal provisions	Expertise to make a yes/no evaluation against selected criteria	Low
Tool for environmental assessment and management	Moderate	Assumptions can easily be modified to test sensitivity	Expertise to establish a relative ranking against elected criteria; computer and software	Moderate
Adaptation decision matrix	Moderate	Assumptions can easily be modified to test sensitivity	Expertise to establish a relative ranking against selected criteria, ability to estimate costs of measures	Moderate
Benefit–cost analysis	High	Variety of approaches	Data intensive	High
Cost–effectiveness analysis	Useful when either benefits or costs fixed	Variety of approaches	Moderately data intensive	High
Implementation analysis	Useful when benefits can be assumed to be positive and comparable among measures	No formal provisions	Expertise to estimate requirements for overcoming barriers	Low

then the analyst moves on to the next step. If the answer is no, then two further questions should be addressed:

1 Has the problem been well enough defined in Stage 1, or should it be further elaborated?
2 Are the criteria set out under Stage 2 adequate for taking a decision or should they be further extended?

If the answer to either of these questions is no, then Stages 1–5 may have to be repeated entirely or in part.

Stage 7 Implement decision

Here the preferred option from Stage 6 is implemented. As with any other political decision or action, it is important that the decision maker or decision-making agency communicate to the general public and specific stakeholders not only the benefits, but also the costs and uncertainties associated with the recommended adaptation policy.

Stage 8 Monitor, evaluate and review

A successful adaptation action does not end with its implementation in Stage 7. It is essential that the decision be followed up and evaluated. One possible outcome of this evaluation is that a mid-course correction is needed for the adaptation decision. Furthermore, an evaluation of adaptation actions provides invaluable input to future decisions about adapting to climate risks.

Notes

1 Alcamo, J., Moreno, J.M., Novaky, B., Bindi, M., Corobov, R., Devoy, R.J.N., Giannakopoulos, C., Martin, E., Olesen, J.E., Shvidenko, A., 2007: Europe. Chapter 12 in: *Climate Change 2007: Impacts, Adaptation and Vulnerability. Contribution of Working Group II to the Fourth Assessment Report of the Intergovernmental Panel on Climate Change (IPCC)*, Parry, M.L., Canziani, O.F., Palutikof, J.P., van der Linden, P.J., Hanson, C.E. (eds), Cambridge University Press, Cambridge, UK.
2 Alcamo, J., Alcosta-Michlik, L., Carius, A., Eierdanz, F., Klein, R., Krömker, D., Tänzler, D., 2008: A new approach to quantifying and comparing vulnerability to drought. *Regional Environmental Change* 8, 137–149.
3 Alcamo, J. et al., 2008.
4 Warmer temperatures will tend to increase evapotranspiration and reduce water availability, but this effect will not be as strong as the effect of increasing precipitation which will tend to augment water availability
5 Alcamo, J., et al., 2007.

6   Thuiller, W., Lavorel, S., Araújo, M.B., Sykes, M.T., Prentice, I.C., 2005: Climate change threats plant diversity in Europe. *Proceedings of the National Academy of Sciences* 102, 8245–8250.

7   Guisan, A., Theurillat, J.-P., 2005: Appropriate monitoring networks are required for testing model-based scenarios of climate change impact on mountain plant distribution. In: *Global change in mountain regions*, Huber, U. M., Bugmann, H., Reasoner, M. A. (eds). Kluwer, 467–476.

8   Alcamo, J. et al., 2007.

9   Munich Re Group. www.munichre.com/en/service/search.aspx (retrieved November 2008).

10   Giannakopoulos, C., Bindi, M., Moriondo, M., LeSager, P., Tin, T., 2005: Climate change impacts in the Mediterranean resulting from a 2oC global temperature rise. WWF report, Gland Switzerland. Accessed 01.10.2006 at http://assets.panda.org/downloads/medreportfinal8july05.pdf

11   Beniston, M., Stephenson, D.B., Christensen, O.B., Ferro, C.A.T., Frei, C., Goyette, S., Halsnaes, K., Holt, T., Jylhä, K. Koffi, B., Palutikof, J., Schöll, R., Semmler, T. Woth, K., 2007: Future extreme events in European climate: an exploration of regional climate model projections. *Climatic Change* 81, S71–S95.

12   Lehner, B., Czisch, G., Vassolo, S., 2005: The impact of global change on the hydropower potential of Europe: a model-based analysis. *Energy Policy* 33, 839–855.

13   Wood, R., Collins. M., Gregory, J., Harris, G., Vellinga, M., 2006: Towards a risk assessment for shutdown of the Atlantic thermohaline circulation. In: Schellnhuber, J., Cramer, W., Nakicenovic, N., Wigley, T., Yohe, G. (eds). *Avoiding dangerous climate change.* Cambridge University Press.

14   Solomon, S., Quin, D., Manning, M. Chen, Z., Marquis, M., Averyt, K., Tignor, M., Miller, H. (eds). 2007: *Climate Change 2007: The Physical Science Basis. Contributions of Working Group I to the Fourth Assessment Report of the Intergovernmental Panel on Climate Change.* Cambridge University Press. Cambridge. UK.

15   For an overview of potential events caused by exceeding thresholds in the earth system, see Schellnhuber, J., Cramer, W., Nakicenovic, N., Wigley, T., Yohe, G. (eds). 2006: *Avoiding dangerous climate change.* Cambridge University Press.

16   Hansen, J., Nazarenko, L., Ruedy, R., Sato, M., Willis, J., del Genio, A., Koch, D., Lacis, A., Lo, K., Menon, S., Novakov, T., Perlwitz, J., Russell, G., Schmidt, G., Tausnev, N., 2005: Earth's energy imbalance: confirmation and implications. *Science* 308, 1431–1435.

17   Hansen, J. et al., 2005.

18   Hansen, J. et al., 2005.

19   *UN Framework Convention on Climate Change.* http://unfccc.int/essential_background/convention/items/2627.php (retrieved November, 2008).

20   UNFCCC, 2007: Bali Action Plan. Decision 1/CP. 13. Report of the Conference of the parties on its thirteenth session, held in Bali from 3 to 15

December 2007. http://unfccc.int/resource/docs/2007/cop13/eng/06a01.pdf#page=3. Accessed February 2010.

21 Wilibanks, T., Lankao, R., Bao, M., Berkhout, F., Cairncross, S., Ceron, J.-P., Kapshe, M., Muir-Wood, R., Zapata-Marti, R., 2007: Industry, settlement, and society. Chapter 7 in: *Climate Change 2007: Impacts, Adaptation and Vulnerability. Contribution of Working Group II to the Fourth Assessment Report of the Intergovernmental Panel on Climate Change*, Parry, M., Canziani, O., Palutikof, J., van der Linden, P., Hanson, C. (eds). Cambridge University Press. Cambridge. UK, 357–390.

22 Burton, I., Kates, R., White, G., 1993: *The environment as hazard*. Guilford Press, NY.

23 Wilibanks, T. et al., 2007.

24 Carter, T.R., Parry, M.L., Harasawa, H., Nishioka, S., 1994: IPCC Technical Guidelines for Assessing Climate Change Impacts and Adaptations. Geneva. Intergovernmental Panel on Climate Change.

25 Feenstra, J., Burton, I., Smith, J., Tol, R. (eds), 1998: *Handbook on Methods for Climate Change Impact Assessment and Adaptation Strategies*. United Nations Environment Programme and Free University of Amsterdam.

26 UK-CIP (United Kingdom Climate Impact Programme), 2008: *Identifying adaptation options*. 34 pp. www.ukcip.org.uk/images/stories/Tools_pdfs/ID_Adapt_options.pdf. Retrieved November, 2008.

27 Willows, R., Connell, R. (eds), 2003: *Climate adaptation: Risk, uncertainty and decision-making*. UKCIP Technical Report. UKCIP, Oxford. 154 pp.

28 Carter, T. et al., 1994.

29 UK-CIP, 2007.

30 Willows, R., Connell, R., 2003.

31 While this may be true in general, it also happens that an extreme climate event serves as a catalyst for adaptation. For example, after the European heat wave of 2003 (see Chapter 3), several Southern European cities quickly implemented city-wide heat wave warning systems.

32 Wilibanks, T. et al., 2007.

33 Arnell, N., Delaney, E., 2006. Adapting to climate change: Public water supply in England and Wales. *Climatic Change* 78, 227–255.

34 Adger, W.N., Agrawala, S., Qader, M.M., Conde, C., O'Brien, K., Pulhin, J., Pulwarty, R., Smit, B., Takahashi, K., 2007: Assessment of adaptation practices, options, constraints and capacities. Chapter 17 in: *Climate Change 2007: Impacts, Adaptation and Vulnerability. Contribution of Working Group II to the Fourth Assessment Report of the Intergovernmental Panel on Climate Change (IPCC)*, Parry, M.L., Canziani, O.F., Palutikof, J.P., van der Linden, P.J., Hanson, C.E. (eds). Cambridge University Press, Cambridge, UK, 717–743.

35 Scott, D., McBoyle, G., Mills, B., 2003: Climate Change and the skiing industry in southern Ontario (Canada): exploring the importance of snowmaking as a technical adaptation, *Climate Research* 23, 171–181.

36   Alcamo et al., 2007.

37   van der Sluis, T., Bloemmen, M., Bouwma, I.M., 2004: *European corridors: Strategies for corridor development for target species ECNC*, Tilburg, the Netherlands & Alterra P.O. Box 90154 NL-5037 AA Tilburg. The Netherlands. 32 pp.

38   van der Sluis, T. et al., 2004.

39   van der Sluis, T. et al., 2004.

40   van der Sluis, T. et al., 2004.

41   Fischlin, A., Midgley, G.F., Price, J.T., Leemans, R., Gopal, B., Turley, C., Rounsevell, M.D.A., Dube, O.P., Tarazona, J., Velichko, A.A., 2007: Ecosystems, their properties, goods, and services. *Climate Change 2007: Impacts, Adaptation and Vulnerability. Contribution of Working Group II to the Fourth Assessment Report of the Intergovernmental Panel on Climate Change*, Parry, M.L., Canziani, O.F., Palutikof, J.P., van der Linden, P.J., Hansson, C.E. (eds), Cambridge University Press, Cambridge, UK, 211–272.

42   Climate impacts are not always just associated with future events; some extreme weather events have already been attributed to climate change and have led to public support for immediate adaptation measures. For example, after the great European heatwave of 2003 (see Chapter 3), the risk of high temperature events was interpreted as an immediate threat to society and the public was ready to support, if not demand, that measures be taken to protect them from similar events in the future. As a result, many cities in France, Spain, Portugal and elsewhere almost immediately introduced early warning systems for heatwaves.

43   Kabat, P., van Schaik, H., 2003: *Climate changes the water rules. The Dialogue on Water and Climate*. www.waterandclimate.org/UserFiles/File/changes.pdf

44   Biemans, H., Bresser, T., van Schaik, H., Kabat, K., 2006: *Water and climate risks: a plea for climate proofing of water development strategies and measures*. Co-operative Programme on Water and Climate. www.waterandclimate.org/UserFiles/File/manifest.pdf

45   Wilibanks et al., 2007.

46   Yohe, G.W., Lasco, R.D., Ahmad, Q.K., Arnell, N.W., Cohen, S.J., Hope, C., Janetos, A.C., Perez, R.T., 2007: Perspectives on climate change and sustainability. *Climate Change 2007: Impacts, Adaptation and Vulnerability. Contribution of Working Group II to the Fourth Assessment Report of the Intergovernmental Panel on Climate Change*, Parry, M.L., Canziani, O.F., Palutikof, J.P., van der Linden, P.J., Hansson, C.E. (eds), Cambridge University Press, Cambridge, UK, 811–841.

47   Finnish Environment Institute. Finadapt website: www.environment.fi/default.asp?node=13867&lan=en

48   Commission of the European Communities. Green Paper: *Adapting to climate change in Europe – options for EU action*. SEC(2007) 849. COM (2007) 354.

49   European Commission, 2007.

50   European Commission, 2007.

51   Wilibanks et al., 2007.

52   UK-CIP. Identifying adaptation options.
53   Carter, T. et al., 1994.
54   Feenstra, J. et al., 1998.
55   Carter, T. et al., 1994.
56   UK-CIP, 2007.
57   UK-CIP, 2007.
58   UK-CIP, 2007.
59   Carter, T. et al., 1994.
60   UK-CIP, 2007.
61   Feenstra, J. et al., 1998.
62   Feenstra, J. et al., 1998.

10 Climate protection in Europe

Life in Europe Under Climate Change, First Edition. Joseph Alcamo and Jørgen E. Olesen.
© 2012 Joseph Alcamo and Jørgen E. Olesen. Published 2012 by John Wiley & Sons, Ltd.

Introduction

Up to this point we have drawn on the latest findings of science to draw a picture of what life would be like in Europe under climate change. We have seen how its far-reaching consequences will bring changes to air temperature, patterns of rainfall and snowfall, the supply of water available for municipalities and industries, the frequency of flooding, the types of vegetation, and the shape of the coastline. Throughout this book we have often used the expression "expected" impacts, but we have never posed the basic question, is climate change immutable? Can we avoid this future? This is a big question, so it is not surprising that it has at least two answers. One is that the scientific community believes that some further global warming is inevitable, even if we were to take strong action to reduce greenhouse gas emissions. In Chapter 9 we explained that this is because of the very long time lag between the emissions of greenhouse gases and the response of the oceanic part of the global climate system (Box 9.3). Indeed, we saw in Chapter 1 that climate impacts are not only inevitable but already with us in Europe, not to mention other parts of the world. As we have argued in earlier chapters, if we accept the unavoidability of some impacts of climate change, we come to the inevitable conclusion that it is already time to begin to adapt.

We can also give a second answer to the question of whether climate change is inevitable, and this one is more optimistic: While some global warming is inevitable, the intensity and tempo of climate change is not. Indeed, society can take actions that both lessen its intensity and slow its pace.

Among the most effective actions is to reduce emissions. In fact, there are at least four good reasons for decreasing their size. The first is that the lower the emissions of greenhouse gases into the atmosphere, the lower the intensity and tempo of climate change. Figure 2.5 in Chapter 2 shows computer modeling calculations of how surface air temperature could increase over the course of the 21st century under various assumptions regarding emissions of greenhouse gases. The lower the level of greenhouse gas emissions, the lower the curve of temperature. Similar tendencies are noted for changing precipitation and other climate variables. From this and other evidence it is obvious that limiting greenhouse gas emissions will slow the rate of climate change and its ultimate intensity. We return later to the question of how much and how soon emissions should be reduced.

Second, as the rate and intensity of climate change are reduced, so too will its impacts on the water cycle, crop production, growth of vegetation, and the frequency of heatwaves. Think of a pot of water being heated on a stove; the lower the heat the longer it takes to boil. If the heat is turned out after a few minutes, then the water temperature will stop rising. In the same way, if we slow and eventually end the warming of the atmosphere and ocean, we will also slow the rate and intensity of climate change impacts.

Third, lowering climate impacts has an additional pay-off, namely, the lower the impacts, the lower the costs of adaptation. If sea level is projected to increase up to 10 cm by the end of the century rather than 30 cm, then coastal levees do not have to be made as high, and not as much property will have to be moved inland. If annual precipitation decreases by 10% rather than 30%, then fewer new water storage facilities will be needed.

Dangerous climate change

The fourth argument for slowing the tempo of climate change is to lessen the risk that humanity pushes up against thresholds in the global climate system. Some of these limits are discussed in Chapter 9, which explains that if climate change happens rapidly, there is a risk that the climate system could undergo rapid and fundamental changes causing high risk of damage to society. One example is the possible disruption of the northern Atlantic branch of the Gulf Stream, which could be set off quickly by a chain of processes beginning with the warming and melting of ice masses at high latitudes leading to a flood of freshwater into the North Atlantic and an alteration of the salt concentration in the ocean.

Another "non-linear" threshold could be reached if global warming rapidly defrosts permafrost regions at high latitudes. This melting could lead to the relatively rapid release of large volumes of methane stored in the permafrost. Since methane is a greenhouse gas, its rapid release could further accelerate warming at high latitudes and elsewhere, which would lead to further melting of permafrost and a further release of methane, and so on, in a kind of snowballing climate effect.

A global target for protecting climate

How, then, can we protect society and the rest of nature from the negative impacts of climate change? Or put another way, how much climate change is acceptable? Considering the awesome complexity of the climate system, it is nearly impossible to identify a limit that is "acceptable". Such a limit would have to take into account thresholds which, if exceeded, would lead to dangerous climate change, but also thresholds at which gradually occurring impacts such as shifts in the ranges of flora and fauna finally become unacceptable. Another consideration is that different thresholds are likely to coexist for different aspects of human society and nature. For example different ecosystems, different coastal zones, and different water systems all have their particular sensitivities and thresholds. Under such a situation it is very difficult for science to provide a clear picture of the limits of the climate system.

Nevertheless, the story does not end here, because one of the pillars of European policy is the "precautionary principle" which provides the political justification for setting goals for climate policy even though the level of scientific uncertainty is high. Perhaps the best known formulation of this principle is found in the Rio Declaration from the Earth Summit of 1992:

> In order to protect the environment, the precautionary approach shall be widely applied by States according to their capabilities. Where there are threats of serious or irreversible damage, lack of full scientific certainty shall not be used as a reason for postponing cost-effective measures to prevent environmental degradation.

Based on this principle, the EU went ahead in the face of significant scientific uncertainty and proposed to limit the increase of global average surface temperature to 2°C above its pre-industrial level. This was proposed by the EU in 1996 during the Kyoto Protocol negotiations and was reaffirmed in 2005 by the European Council. Outside Europe, this limit was also agreed to by numerous countries as part of the Copenhagen Accord in 2009 and by an even larger group in the Cancun Agreement of 2010.

Since the 2°C target was adopted by the EU in 1996 the scientific community has assembled much more knowledge about current and potential climate impacts, as shown for Europe in earlier chapters. An interesting question is whether this knowledge supports or undermines the temperature goal. One way this knowledge has been summed up is in the form of "burning ember" diagrams shown in Figures 10.1 and 10.2. They get their name from their appearance – to some readers – as burning logs. A redder glow in these diagrams indicates that the negative impacts or risks are becoming more widespread and/or more intense. Although no clear threshold emerges from these diagrams, they do suggest that the world

Figure 10.1 "Burning embers" diagram showing vulnerability of different parts of the European and global environment to global warming. The "X" axis uses the increase in average global surface temperature (°C) (relative to pre-industrial climate) as a proxy for the specific changes in climate occurring at the regional and global levels. The redder the color, the more widespread and or greater magnitude of negative impacts or risks. Note that the impacts of climate change on potential food production may be positive in the first phase of climate change because of warmer and moister conditions, but they eventually turn negative because of the impact of higher temperatures. (See Chapter 4). Sources: MNP,[1] redrawn by EEA.[2] Reprinted with permission from Netherlands Environmental Assessment Agency. (See color plate.)

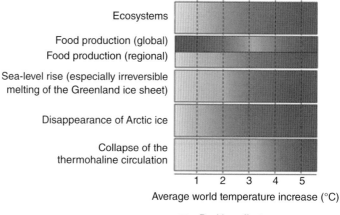

Figure 10.2 "Burning embers" diagram showing increasing threat to ecosystems as climate changes. Change in global surface temperature relative to pre-industrial conditions is used as a proxy for specific changes in climate occurring at the regional and global levels. The redder the color, the more widespread and or greater magnitude of negative impacts or risks. Source: IPCC.[3] Reprinted with permission from Intergovernmental Panel on Climate Change, Geneva. (See color plate.)

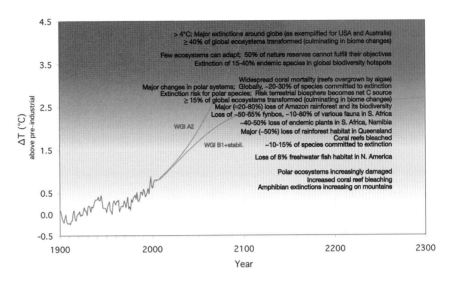

enters deeper into the "red zone" of serious impacts and risks when temperature rises 1.5–2.5°C above its historic level.

The next big question is, what does the 2°C target imply about emission reductions? This is another difficult question because the relationship between future emissions and future temperature change is uncertain as shown by the fuzziness of the curves in Figure 2.5. But computer models take into account some of this uncertainty and have been used to make estimates of future emissions. Several modeling groups came together under the umbrella of the United Nations Environment Programme and estimated that there would be a "likely" chance of staying within the 2°C target if global greenhouse gas emissions peaked before 2020 and then dropped to around 50 to 60 percent of their 1990 levels by 2050.[4] Another way of looking at this problem is illustrated in Figure 10.3, which shows that each 10-year delay in action to curb emissions increases the global mean temperature at its peak by about 0.5°C.[5]

Considering that the trend in global emissions is still generally upwards, achieving the 2°C target will clearly require concerted action by Europe and the rest of the world. We return to this topic later when we discuss global action for climate protection.

How should we act to protect climate?

Later we will see that the EU and the rest of the international community has indeed decided that the threat of climate change is real enough to justify climate

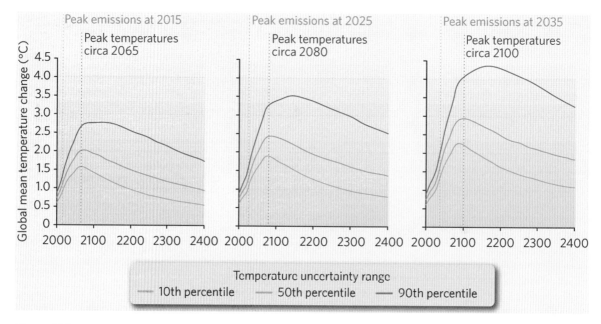

Figure 10.3 Increase in average global surface temperature for scenarios of peak greenhouse gas emissions in either 2015, 2025 or 2035 with 3%-per-year reductions in greenhouse gas emissions. Note: the later the peak in global emissions, the later and higher the peak in temperature. Source: Parry et al.[6] Reprinted with permission from Macmillian Publishers Limited: Nature. (See color plate.)

protection. But it is one thing to decide to protect climate, and yet another to act on this decision. What are the possibilities for dealing with climate change? In Chapter 9 we mentioned that the options fall into two categories:

- *Adaptation,* which is the set of actions to reduce vulnerability to expected or experienced climate change. The IPCC defines adaptation as "the adjustment in natural or human systems in response to actual or expected climatic stimuli or their effects, which moderates harm or exploits beneficial opportunities".[7]
- *Mitigation,* which are actions to lessen the intensity of climate change, or, according to the IPCC, an "anthropogenic intervention to reduce the anthropogenic forcing of the climate system, [including] strategies to reduce greenhouse gas sources and emissions and enhancing greenhouse gas sinks".[8]

In previous chapters we saw many examples of adapting to climate change and these were also summarized in Chapter 9. Whereas "adaptation" is reactive against a perceived threat, "mitigation" is proactive by avoiding or minimizing the cause of the threat. There are three main ways to mitigate greenhouse gas emissions, the root cause of climate change:

1 *Reduce* emissions by switching to low carbon fuels such as natural gas, by increasing the efficiency of energy production, or by modifying farming and other land use practices.

Figure 10.4 Share of 2006 greenhouse gas emissions in EU-27 by gas, weighted according to their global warming potentials. The numbers show the percentage contribution of the individual gas. Source: EEA.[9] Redrawn with permission from European Environmental Agency, Copenhagen.

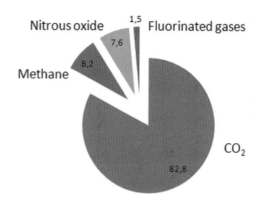

2 *Avoid* emissions by replacing fossil fuels with renewable energy sources or by simply stopping the activity that leads to emissions.

3 *Offset* emissions by stimulating uptake of carbon dioxide by vegetation or through "carbon capture and storage".

Each of these options has its advantages and disadvantages as shown in the following paragraphs.

Which emissions need to be mitigated? We explained in Chapter 2 that many different gases contribute to warming of the atmosphere. Of these, six in particular are controlled under the Kyoto Protocol because they are either important or becoming important. These include three gases occurring in nature – carbon dioxide (CO_2), methane (CH_4), and nitrous oxide (N_2O), as well as two classes of man-made gases – hydrofluorocarbons (HFCs) and perfluorocarbons (PFCs), and an individual man-made gas – sulfur hexafluoride (SF_6). The magnitude of their emissions depends on the intensity and type of economic activities in a region. Emissions in Europe are dominated by CO_2, which accounts for nearly 83% of the total, followed by CH_4 and N_2O, which have roughly the same magnitude (Figure 10.4). The man-made fluorinated gases make up the remainder of emissions.

Reducing emissions

Fuel substitution

The notion of "reducing emissions" implies that current practices remain basically the same, and we make small adjustments in order to reduce emissions. One common example is continuing to use the same types of power-generating turbines and internal combustion engines, but changing the types of fuels used to power these machines. This is called "fuel substitution", which aims to replace fuels that produce high levels of emissions with ones that have lower emissions. The simplest case is to replace high-carbon fuels such as coal (high in the sense that

Figure 10.5 Greenhouse gas emissions from different fuels and energy systems. Emissions are given in units of tons equivalent CO_2 emissions, per gigawatt – hour of energy produced, meaning that non-CO_2 gases such as methane and nitrous oxide were converted to CO_2 equivalents according to their global warming potential. The bars indicate the uncertainty range, depending on technologies applied. Source: WEC.[10] Redrawn with permission from World Energy Council.

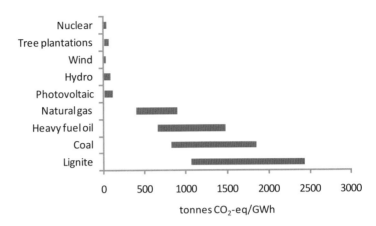

their carbon content is large relative to their heat value) with natural gas and other lower-carbon fuels. In general, the lower the carbon content of the fuel, the lower the emissions of carbon dioxide per unit energy produced (Figure 10.5). A current example is that some European countries plan to replace coal in power plants with natural gas because gas produces much less carbon dioxide per unit energy generated (as shown in Figure 10.5).

But switching fuels sometimes raises other problems, including questions of cost, availability and, in some cases, security. The growing reliance of western Europe on Russia for natural gas has kindled a spirited discussion about energy security. This discussion reached a new peak in the particularly cold winter of 2008–09 when Russia interrupted the flow of natural gas to Ukraine because of a contract dispute. This also disrupted a main source of fuel for western Europe's heating systems.

Another example of fuel substitution policy is the European goal to replace a portion of the gasoline used in motor vehicles with biofuels. We see below that this can be an effective way to reduce emissions, but it raises questions about the side effects of energy crop cultivation and biofuel production.

Improving energy efficiency

It has been shown that fuel substitution can lower emissions without having to make major changes in the structure of the energy system. Another approach along these lines is to improve the efficiency of existing energy systems, in other words, to get the same energy service (lighting, heating, etc.) for less input of energy. This is not a new idea; governments have sponsored large energy conservation programs since at least the 1970s when OPEC temporarily cut back oil deliveries to the West.

But there remains tremendous potential for further energy savings without having to sacrifice the benefits of energy services. The European Commission recently estimated that one-fifth of current (2005) primary energy use in the EU can be saved by 2020 with reasonable financial investments.[11] Out of this one-fifth, the largest potential savings lie in residential buildings (27%) and commercial buildings (30%). Indeed, there are opportunities to boost the energy efficiency of nearly the entire housing stock in Europe. As an example, the EU-funded SOLANOVA project retrofitted a Communist-era public housing project in Hungary with energy saving and renewable energy equipment and achieved a reduction of 84% fossil fuel use. Before, the retrofit residents paid out 24% of their monthly income for energy costs, afterwards only 4%.[12]

The options for realizing energy savings in buildings are too numerous to list here, but some alternatives under consideration in the European Union are: (1) developing pan-European legislation for regulating the energy performance of new buildings; (2) promoting the construction of "zero-energy" houses that utilize super efficient insulation, shading and other energy efficiency features; and (3) establishing a European-wide eco-label to reward companies that show particular proficiency in saving energy.

Although we have concentrated on the building sector, other parts of the economy have similarly high potentials for saving energy. A few of the many possible measures are:

- labeling the energy performance of televisions, washing machines and other appliances so that consumers can seek out the most energy efficient items;
- improving the efficiency of energy conversion in power plants, industrial processes, and other installations;
- making transport more efficient by modifying vehicle driving practices, improving the efficiency of motors, encouraging the use of public rather than auto transport, and encouraging telecommuting;
- enforcing minimum energy efficiency standards in energy provision sectors and for final energy use in transport, housing, appliances, and so on.

We return to the topic of energy saving measures when we discuss European climate policy.

Improving agricultural management

We mentioned in Chapter 2 that three major greenhouse gases – carbon dioxide, methane and nitrous oxide – originate not only from energy and industry, but also from farming and other land use activities. Sources include nitrous oxide emitted from the soil when an excess of nitrogen fertilizer is applied to fields; methane produced by the digestion of livestock, in particular from ruminant animals (cattle, sheep and goats); and methane released when organic wastes in landfills and sewage treatment plants decompose in the absence of oxygen. In fact, agriculture and other land use activities are among the most important worldwide sources of methane and nitrous oxide emissions. The IPCC gives a short list of the many options available for reducing these sources of emissions. The ones judged to have the highest potential in terms of cost–effectiveness are:[13]

- restoration of cultivated organic soils;
- improved cropland management (including nutrient management, and tillage/residue management) and water management (including irrigation and drainage) and set-aside/agro-forestry;
- improved grazing land management (including adjustment of grazing intensity, increased productivity, nutrient management, fire management and species introduction);
- restoration of degraded lands (using erosion control, organic amendments and nutrient amendments).

Other actions with a lower, but still substantial potential for reducing emissions, are:

- management of rice fields; and
- livestock management (including improved feeding practices, dietary additives, breeding and other structural changes, improved manure management, and improved storage and handling of wastes).

The advantage of these measures is that under some circumstances they can be a cheaper way to reduce the emissions of carbon dioxide, methane and nitrous oxide than intervening in the energy system. An additional advantage is that some of these actions will reduce more than one greenhouse gas. For example, better fertilizer management of fields will tend to increase the carbon content of soils, thereby removing some carbon dioxide from the atmosphere, and at the same time reduce the nitrous oxide coming from excess fertilizer usage.

But these measures have their share of drawbacks. Later, we point out some of the disadvantages of modifying land use practices for offsetting emissions, and several of these arguments also apply here. A key point is that it is difficult to monitor or control agricultural and land use practices indefinitely; as these practices inevitably change, the release of greenhouse gases will also change, and not necessarily in a favorable direction. Second, some thorny questions appear once land use practices become part of an international agreement to reduce emissions. What will the international community do if the agreement is not complied with? Is it expected that the international community will interfere with how land is used within a sovereign country?

Avoiding emissions: renewable energy

Although substituting fuels can be an effective tactic under some circumstances, its potential to lower emissions is limited. An essential part of any mitigation strategy, therefore, is to move away from energy production based on fossil fuel use and towards renewable energy sources such as wind, solar and biomass, which produce small amounts of greenhouse gases compared to fossil fuels (Figure 10.5). These energy sources are "renewable" in the sense that they are continuously renewed by nature and not drawn down as are the reserves of coal, oil and uranium. But producing usable energy from renewable energy sources is also not 100% emission-free because energy input is required to manufacture, dispose of, and maintain renewable energy installations. Hence, when summed over their entire life time, these technologies also produce small amounts of greenhouse gas emissions (Figure 10.5 and Box 10.1).

The share of all renewable energy sources in the world's primary energy supply was 15.4% in 2005. Most of this (9.4%) was made up of wood, wastes, and other combustible material used mainly in developing countries as fuel (Table 10.1). Hydroelectricity made up the next largest fraction (5.3%) and solar, wind and other renewables accounted for only about 0.7% of the world's primary energy usage.[14] But the global use of modern renewable energy (hydro, wind, solar, geo-thermal, modern biomass and wave energy) is projected to triple between 2008 and 2035. In the EU, renewable energies in 2006 made up 7% of the total energy

---

Box 10.1 Why renewable energy sources are not emissions-free

While renewable energy carriers are low emitters of greenhouse gases they are not completely free of emissions. This is apparent from "life cycle analyses" conducted on renewable energy facilities such as wind parks and solar photovoltaic arrays. These studies carefully note the energy used to produce, run and dispose of the renewable energy equipment. For example, a wind generator is made up of a motor and various other components and fuel is required to manufacture them. Since most energy used by industries in Europe comes from fossil fuels this means that building wind generators has to burn up fossil fuels. Likewise, further fossil fuels are expended by delivering raw materials to the turbine manufacturing plant, by delivering the wind generator to its outdoor site, for disposing of the generator and other scrap metal once its useful life is finished, and so on.

The point is that a wind turbine will require energy input during various phases of its life cycle, and this energy will come mostly from fossil fuels, and when these fuels are used they emit greenhouse gases. As might be imagined, the amount of emissions depends very much on how the facility is built, managed and disposed of. A representative example of emissions produced over the life cycle of different energy systems is shown in Figure 10.5. From this diagram we can see that coal and other fossil fuel energy systems produce emissions in the range of 400–1000 tons of CO_2-equivalents greenhouse gases per gigawatt-hour of generated electricity. By comparison, almost all renewable energy systems produce substantially less than 100 tons of CO_2-equivalents, even after taking into account their life cycle emissions. One could argue that eventually the energy used to manufacture a wind generator might come from nearly 100% renewable energies, and thereby become virtually emission-free. But that is something for the future.

Table 10.1 Contribution of renewable energy sources to energy consumption or production in the EU and worldwide. Sources: EC[15] and Sims, IPCC[16]

Renewable energy source	Europe: Percentage share of gross inland energy consumption of the European Union-25 in 2006 (%)[17]	World: Percentage share of total world primary energy production in 2005 (%)[18]
Hydro (>10 MW)	–	5.1
Hydro (<10 MW)	–	0.2
Hydro (total)	1.4	5.3
Wind	0.4	0.2
Biomass (modern)	–	1.8
Biomass (traditional)	–	7.6
Biomass (total)	4.8	9.4
Geothermal	0.3	0.4
Solar PV	–	<0.1
Concentrating solar	–	0.1
Solar (total)	0.06	–
Ocean (all sources)	–	0

Note: Dash indicates no data available from cited source.

consumed within its borders (Table 10.1), with the largest fraction coming from biomass (4.8%), followed by hydroelectricity (1.4%), wind energy (0.4%), geothermal (0.3%) and solar (0.06%). Hence, although solar and wind are now highly promoted, they still account for only a small share of European energy supply. Nevertheless, the use of renewable energy is on the rise in Europe and elsewhere, as we see in the following paragraphs.

Wind energy

The basic idea of exploiting the wind is to convert its kinetic energy into electricity by having the wind's force turn the long blades of a turbine. The turbine is usually installed on a tall pole 50–90 meters above the ground in order to capture the stronger winds that prevail at higher elevations. A typical wind installation in Europe has a three-blade generator with a capacity around 1.5 MW. However, the size and the capacity of wind turbines are steadily increasing, and wind turbines installed in 2010 typically had capacities of 2–3 MW.

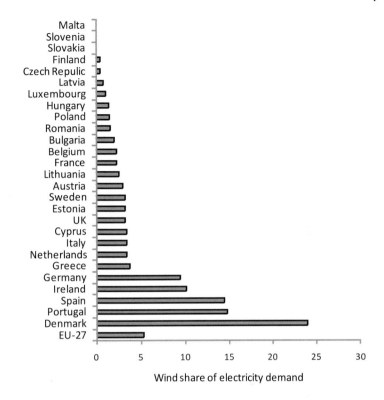

Figure 10.6 Share of wind power in electricity production up to 2010. European Union countries. Source: EWEA[20] Redrawn with permission from European Wind Energy Association.

Wind energy is one of the fastest growing energy technologies in the world, with Europe leading the world with 69 percent of total capacity.[19] Wind energy now satisfies about 5% of the EU's total electricity demand (up to the end of 2010). (Its contribution to total energy use is smaller, as noted above.) It accounts for over 5% of electricity usage in five countries (Germany, Ireland, Portugal, Spain, and Denmark,) (Figure 10.6). In terms of total capacity, the leaders in the EU are Germany with over 22,000 MW and Spain with more than 15,000 MW installed capacity. Indeed, in these countries the sights and sounds of large wind turbines have become a familiar part of the landscape.

In the mid-1990s the average cost of wind-generated electricity was about 0.07 euros per kilowatt-hour, but with mass production and public subsidies wind devices have become competitive with fossil fuel energy.[21] On good, unobstructed, windy sites, wind power can be generated at a cost of around 0.025–0.035 euros per kilowatt-hour.[22]

But wind generation has not been expanding without opposition. The problem is that wind itself is a thinly-distributed energy source and so a substantial amount of land is required to harvest its potential. Roughly speaking, around 20 hectares of land is needed per megawatt of installed capacity.[23] Hence, producing the electricity of a typical 1000 MW conventional power plant requires about 20,000

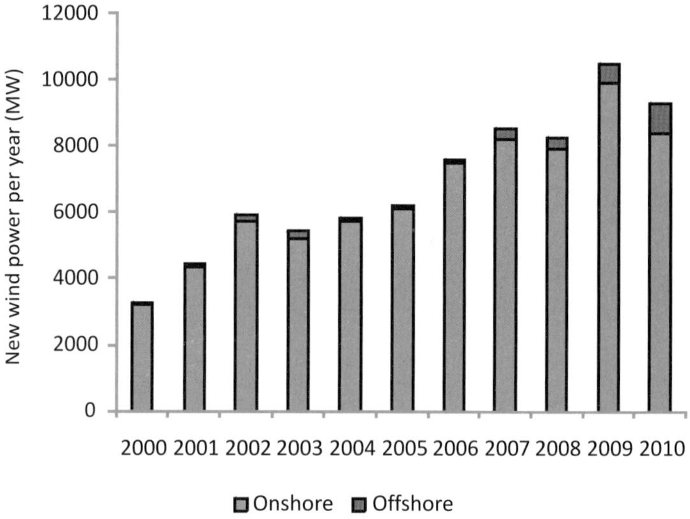

Figure 10.7 Newly installed capacity of onshore and offshore wind power in the European Union. Source: EWEA.[24] Redrawn with permission from European Wind Energy Association. (See color plate.)

hectares or $200\,km^2$ of land surface. This hunger for land is the main reason for citizen opposition, as large numbers of wind generators sprout up on previously undisturbed natural and agricultural landscapes in Denmark, Germany, and elsewhere. There have also been reports of generators disturbing wildlife as in case of birds being killed by the blades of wind turbines in large wind parks (Chapter 7). On the other hand, recent experience shows that giving the community a say in where wind turbines are located can lead to less disturbance of the landscape, and ultimately greater public acceptance of this low-emission energy source.

Some governments are trying to lessen the disturbance to the countryside by putting large-scale wind parks offshore the coasts of Denmark, Germany, and the Netherlands. These are clusters of dozens of wind generators located within 2–10 km of the shoreline. The annual installed new capacity of offshore wind electricity in Europe has quickly risen to around 1,000 MW (Figure 10.7) with substantially more capacity in the planning stage.

Solar electricity and heating

Concentrating solar power plants

There are two main strategies for generating electricity from sunlight. The first is through "concentrating solar power plants" (CSP), also known as "solar thermal power plants". These are facilities that use a large array of mirrors to concentrate sunlight onto a central focal point. One variant is to focus the sunlight onto a central tower ("solar tower"). A special fluid then transports the heat from the tower to a conventional steam turbine. An alternative design uses a row of parabolic mirrors to concentrate sunlight onto a pipe which contains fluid that collects the heat and delivers it to a turbine ("solar trough"). Sometimes these facilities are equipped with heat storage tanks where heat is stored in the form of molten salt or other substance until it is used for night-time power generation. Because of the machinery and materials involved in constructing a CSP, they are found to be most cost-effective if they are built fairly large and located where sunlight is plentiful, usually in low latitude sunny climates.

Until recently the main CSP facilities were located outside Europe, but now the Continent is investing heavily in this technology. The first commercial CSP plant in Europe opened up in 2007 near Seville, using 624 movable mirrors, each with a 120 m^2 surface for focusing solar radiation on a 115-meter tall central tower. The facility now has an 11 MW capacity, but will be expanded to 300 MW by 2013. Many other installations are in the planning or construction phase; by the end of 2007, more than 50 CSP plants with a total capacity of around 2150 MW had been registered with the Spanish Ministry of Industry.[25]

As noted, one of the keys to the success of a CSP system is to locate them where solar radiation is intense. With that in mind, very ambitious schemes are under discussion to build a huge network of CSP plants in the Sahara and Middle East which would provide not only the energy required for countries in these regions but also enough for export to Europe. One such concept is "DESERTEC" which envisions a huge capacity of concentrated solar plants feeding electricity to Europe through high-voltage direct current lines.[26]

Solar photovoltaic

Besides large facilities for concentrating solar energy with mirrors, the second way to utilize the sun is on a smaller scale with "photovoltaic" (PV) or "solar" cells. These cells are made up of semiconductor materials (typically silicon) that convert radiant light energy directly into electricity. Several cells are commonly assembled into a module and oriented on rooftops in a way to best exploit the sun. These modules go by different names, including "photovoltaic arrays" or "solar panels".

As with wind generators, PV arrays provide small amounts of electricity on-site or can be fed into an electrical utility grid. They are already widely used throughout the developing world in small-scale rural applications for water pumping, illumination, and refrigeration. In industrialized countries they are used as a decentralized source of power for traffic lights, water pumps, telecommunications (for example, emergency call boxes near interstate highways), security and lighting systems, resource monitoring, and electric load management.[27] But over the past few years they have become a common sight on rooftops in many European countries. Globally, about half the existing PV capacity is grid-connected, and half is used off the grid.[28] The situation is different in Europe where most PV arrays are used to feed the electrical grid. As with wind power, photovoltaic technology is growing at a tremendous rate; grid-connected panels are increasing at an annual rate of 50–60%, primarily in Germany, Japan and California, and off-grid panels at 15–20% per annum.[29] Total installed capacity of photovoltaic panels in the EU (up to the end of 2007) was 1,541 MW at peak performance, led by Germany (1,103), Spain (341) and Italy (50).[30]

In high sunshine areas, PVs can generate electricity at a cost of around 0.18 euros per kilowatt-hour.[31] But costs are declining for two reasons. First, the efficiency by which cells convert solar radiation to electricity is steadily improving. Hopes are centering on the further development of "thin-film" PV cells which now achieve a conversion efficiency of around 5% but are expected to reach 15% soon.[32] The second reason is that manufacturers are quickly gaining experience in producing PV cells and are using this experience to manufacture cells more cheaply.

But both wind energy and photovoltaic technology are difficult to integrate into the electric grid because they are "intermittent" sources of energy in that they only provide power when the wind is blowing or solar radiation is strong enough. Hence their production is subject to the vagaries of weather. With this in mind, engineers estimate that the total share contributed by wind and solar to the electrical grid is limited. Despite this pessimism, some new developments may make it possible to provide more than 50% of a grid's electricity with intermittent sources such as wind and PV modules.[33]

- Wind and solar technologies can be better integrated into the grid with new demand-management techniques. Price signals might be used to compensate for the variable nature of this electricity. Consumers would be informed about the availability and price of this electricity and they could plan their electrical use accordingly. Perhaps "intelligent" appliances could be designed that switch on when signaled by the grid that electricity is available.
- Energy suppliers could compensate by transferring electricity from different parts of the electricity grid. Management of interconnected grids could be optimized to take best advantage of these intermittent sources in the same way that the contribution of base load plants are now optimized.
- To compensate from the supply side, wind and solar could be used in conjunction with hydroelectric plants and power plants that can be operated economically at a low capacity and have a fast response time such as gas turbine and combined cycle plants.
- Storage of electricity can also compensate somewhat for the intermittent nature of solar and wind energy sources. For example, compressed-air storage has been shown to be economically promising when used with large wind energy farms. In some cases, existing hydropower plants can be modified to store energy by lifting water from a lower lake to an upper lake. Eventually, storage facilities will probably be integrated into a utility system to dampen out the irregularities of wind and PV power production.[34]
- There is currently a rapid development in battery technology that is likely to lead to a major increase in the number of electric cars within a few years. This will increase the demand for electricity in the transport sector, but could contribute to a net reduction in CO_2 emissions per kilometer of travel if less fossil fuels are used overall for transport. However, the real benefit of battery-operated cars comes from their ability to efficiently use the intermittent supplies of electricity generated by wind and solar. With a smart electrical grid, it should be possible to charge the

car batteries during periods of excess supply of renewable electricity and conversely to let the car battery buffer the supply on the grid during periods of insufficient supply from the renewable sources.

Solar heating

One of the oldest and most widely used solar technologies are roof-top thermal solar collectors which use the sun's energy to directly heat water in homes, building spaces, and swimming pools. Thermal collectors are also used to provide heat input to industrial processes, district heating, and crop drying. A thermal collector consists of a box oriented towards the sun, containing a black metal plate for absorbing solar energy; heat is collected by a fluid running in a piping system winding over the plate; the heat is transported to a heat storage tank from which it is distributed to end users. The box is covered with glass or other glazing to reduce heat loss. In many parts of the world, as diverse as India and southern Europe, roof-top solar collectors are a common feature of the urban and rural landscape. The capacity of thermal solar collectors in the EU (2006) is 14,289 MW_{th}, led by Germany (2,301), Austria (1,987) and France (812).[35] New capacity of thermal solar collectors is being added on at the same tremendous pace as wind turbines and photovoltaic cells. Current global growth rates are around 20% per annum.[36]

Bioenergy

About 10% of total global primary energy comes from "bioenergy", the utilization of different forms of biomass with a range of conversion techniques to provide usable energy. About two-thirds of the global total involves the age-old use of wood, dung and other traditional forms of biomass used as a cooking or heating fuel in developing country households.[37] But when we speak of biomass in connection with climate protection we usually mean "modern biomass" in the form of sugar cane, sugar beets or other feedstocks cultivated or collected especially for energy conversion. "Modern biomass" could also mean using a traditional fuel, such as wood, in a modern installation such as a power plant fired by wood chips.

Modern bioenergy technology refers to any number of different energetic processes by which the energy content of harvested biomass is extracted for heat, electricity or transport fuel. The range of technologies already in use is very wide, covering such disparate facilities as anaerobic tanks that produce methane gas from municipal waste and manure, power plants that produce electricity from wood pellets, and refineries that produce ethanol from sugar cane. Biomass can be converted into gaseous fuels (biogas, hydrogen, synthesis gas), liquid fuels (methanol, ethanol, biodiesel), or solid fuels (wood pellets and chips, briquettes, logs), or can be directly converted into heat or electricity. Feedstocks for conversion processes include dedicated energy crops such as sugar cane, rape seed, miscanthus plants, and sugar beets; or short rotation plantations of willow, poplar and eucalyptus or other woody biomass; or the organic part of municipal solid waste; or residues from crops, livestock raising (manure), and forestry.

Bioenergy is different from the other renewable energy sources in that large quantities of greenhouse gases are produced when biomass is converted to usable energy. Why then is it considered a low carbon dioxide source? The point is that growing the next crop of biomass is expected to absorb about the same amount of carbon dioxide from the atmosphere as was emitted when the last crop was burned or fermented. From this perspective, over a few years the cycle of growing, collecting, burning and then regrowing biomass should add up to close to zero net carbon emissions.

In practice, biomass conversion has some characteristics that do not make it as environmentally friendly as most wind and solar energy applications. First of all, biomass conversion may be a minimal net source of carbon dioxide, but it does release other air pollutants including volatile organic compounds and carbon monoxide. Second, the requirements for growing biomass are similar to any other crop, and so growing biomass can have the same impacts as any other modern cultivated crop (release of pesticide and fertilizer residues to the environment, reduction of diversity of vegetation under some circumstances, relatively high energy and water requirements). Using fertilizer to grow rapeseed and other energy crops can also release large amounts of the greenhouse gas, nitrous oxide, to the atmosphere.[38]

Third, biomass requires space and encroaches on natural land, and sometimes competes with food production. The classic case occurred in 2007 and 2008 when corn was subsidized in the United States for ethanol production and this diverted some tonnage of corn from its normal food export markets to Mexico and elsewhere leading to higher CO_2 emissions from biofuel use than from petrol use.[39] The diversion of corn for fuel raised its price as a food commodity and made it more difficult for poor people to purchase food.

A fourth issue, and one that has received scant attention, is the question of the reliability and security of the bioenergy supply. Energy crops, as any other crops, are subject to large swings in annual production, depending on weather conditions where they are grown. A poor harvest can, in principle, mean a cutback in energy supply for a region dependent on energy crops. Of course, this risk can be tempered by stocking up on energy crops or by being ready to substitute bioenergy with other fuels.

But just as modern agriculture can be made more environmentally-friendly, so too biomass can be exploited in a way to minimize environmental impacts. What is needed is a conscious effort to grow and convert biomass to usable energy in a "sustainable" way. The German Environment Agency has proposed some criteria for judging the sustainability of bioenergy projects. They would be more sustainable if they:[40]

1 make a significant contribution to greenhouse gas reduction;
2 minimize negative consequences of indirect land use changes and compensate for competing land use;

3 avoid loss of ecological areas with high natural value;
4 avoid loss of biodiversity;
5 minimize negative effects on soil, water and air;
6 avoid causing the local population to suffer any disadvantages;
7 follow internationally recognized standards for working conditions.

Going a step further, the European Environment Agency (EEA) showed that the potential for environmentally-friendly biomass in Europe is in fact much greater than the 69 MtOE (million ton of oil equivalents), or 4% of the EU's primary energy consumption, that biomass now provides.[41] By 2030, the potential annual production of "environmentally compatible" biomass could be 40 MtOE from lumber and forest residues, 142 MtOE from energy crops, and 96 MtOE from the waste sector (mostly agricultural residues and municipal waste).[42] This amounts to 15–16% of Europe's projected primary energy production in 2030.[43]

Meanwhile, costs for producing electricity with biomass are becoming competitive with other energy carriers, although costs vary greatly (between 0.04 and 0.09 €/kWh[44]) depending on the kind of feedstock and conversion technology.

Hydroelectric

One of the oldest and most heavily exploited sources of renewable energy, besides wood and other traditional biomass, has been the potential energy of falling water. Hydroelectric plants use the force of water falling downhill to push the blades of a turbine which then generates electricity. The contribution of hydroelectric facilities to overall electricity production is substantial, providing 17% of global electricity[45] and 17.9% of electricity in the EU.[46]

Hydroelectric plants fall into two size categories with greatly different social and environmental impacts. The larger facilities (greater than 10 MW) are responsible for most of the hydroelectricity production in the world. To achieve large outputs, these facilities first have to store water behind large dams and then gradually let this water flow through a series of turbines. The back-up of water behind large dams often means that large numbers of people are displaced from their homes, and terrestrial ecosystems destroyed. Some of the other drawbacks of large dams, as summarized by the World Commission on Dams, are given in Box 10.2.

Smaller facilities (less than 10 MW) are more environmentally-friendly because they do not require large dams and reservoirs to store water. Instead, turbines are set right amid the main currents of a stream, which then push the blades of the turbine and generate electricity. Because river flow is not obstructed by dams or reservoirs, run-of-the-river facilities generally have much lower ecological and social impacts than larger facilities. Nevertheless, in-stream turbines can interfere with the migration of fish and have other smaller ecological impacts.

Box 10.2 Some drawbacks to large dams as identified by the World Commission on Dams[47]

- Large dams display a high degree of variability in delivering predicted water and electricity services – and related social benefits – with a considerable portion falling short of physical and economic targets, while others continue generating benefits after 30–40 years.
- Large dams have demonstrated a marked tendency towards schedule delays and significant cost overruns.
- Large dams designed to deliver irrigation services have typically fallen short of physical targets, did not recover their costs and have been less profitable in economic terms than expected.
- Large hydropower dams tend to perform closer to, but still below, targets for power generation, generally meet their financial targets but demonstrate variable economic performance relative to targets, with a number of notable under-and over-performers.
- Large dams generally have a range of extensive impacts on rivers, watersheds and aquatic ecosystems – these impacts are more negative than positive and, in many cases, have led to irreversible loss of species and ecosystems.
- Efforts to date to counter the ecosystem impacts of large dams have met with limited success owing to the lack of attention to anticipating and avoiding impacts, the poor quality and uncertainty of predictions, the difficulty of coping with all impacts, and the only partial implementation and success of mitigation measures.
- Pervasive and systematic failure to assess the range of potential negative impacts and implement adequate mitigation, resettlement and development programmes for the displaced, and the failure to account for the consequences of large dams for downstream livelihoods have led to the impoverishment and suffering of millions, giving rise to growing opposition to dams by affected communities worldwide.
- Since the environmental and social costs of large dams have been poorly accounted for in economic terms, the true profitability of these schemes remains elusive.

In thinking about hydroelectricity as part of future European climate policy, some special factors have to be considered. First of all, the potential for hydropower may substantially decrease south of the Alps because of declining precipitation (see Chapter 5). Second, we cannot count on building many new large-scale hydropower facilities in Europe because the public is unlikely to accept the flooding of valleys that are populated or are under nature protection. Europe's strong web of environmental laws and regulations is also likely to stand in the way of building new large facilities.

The situation is different for small hydropower. Although these facilities will also be affected by declining run-off in southern Europe, they do not have the negative image of their larger cousins. There are now about 17,200 small hydro-electric plants (less than 10 MW each) operating in the EU with a total installed capacity of 11 GW.[48] This represents only about 65% of the potential of small hydropower.[49] Hence, replacing fossil fuel power plants with small-scale hydropower is another option for reducing emissions.

Geothermal and other renewable sources of energy

Renewable energies are not limited to wind, sun or falling water, but also include the heat of the earth and the movement of oceans. While these other sources can make a contribution to reducing CO_2 emissions, their long-term global potential for producing electricity is not as great as for other renewable sources.[50]

Tapping the natural warmth of the earth is the basic idea behind "geothermal energy". Heat energy originates deep in the earth from its molten interior and from radioactive decay. It is slowly brought to the surface by the upward movement of molten magma, by the deep circulation of groundwater, and by crustal plate movements. In the end, this geothermal energy tends to be relatively diffuse, indeed too diffuse to exploit, except for the fact that tectonic activity creates convenient pockets of high temperature at easily accessible depths.[51] Geothermal energy contributed about 0.3% of primary energy consumption in the EU in 2006.

The history of this energy source is closely associated with Europe since the first geothermal power plant began production near Larderello, Italy, in 1911. Geothermal energy is used in three main ways. First, hot springs and reservoirs close to the surface, or as deep as one kilometer or more, are used as a direct source of heat for buildings and district heating systems through a network of heat exchangers. It is estimated that the entire city of Copenhagen could obtain half of its heating from such a geothermal source, and that this energy supply could last for more than a thousand years.[52] Second, higher temperature sources (above 250°C), normally located deeper in the earth, are used as a heat source for running turbines in a power plant. Third, geothermal "heat pumps" take the constant temperatures maintained in soils a few meters below the earth's surface and pump it into buildings as a source of heat during cool months, and conversely, as a source of cooling during warmer months. But this heat source is not really geothermal in the sense of the first two cases. Much of this heat is actually derived from solar radiation at the surface, and therefore it deserves to be called "solar" as much as "geothermal".

A completely different kind of renewable energy source is the extraction of energy embodied in the wave and tidal movements of the ocean. The basic idea is to situate power turbines in the path of tidal flows or steady ocean currents so that the movement of water turns the blades of the turbine and generates power. Although the potential for generating power is huge the economic cost of building and sustaining a power plant in the harsh environment of the ocean is also huge.[53] Hence, very few large facilities have been constructed up to now. One of the three tidal barrages that have been built, the plant on the Rance estuary in France, has been in operation since the 1960s.

Offsetting emissions: enhancing CO₂ uptake from the atmosphere

We have seen how emissions can be reduced or avoided, and now we examine a third option, namely, how they can be offset by enhancing carbon dioxide uptake from the atmosphere. The first example of offsetting emissions is by increasing the size of the stock of carbon in forests. This sometimes called "carbon sequestration in forests". Carbon is stored in copious amounts in woodlands, both in the form of wood and other biomass, and in the organic matter of its soils. One option for increasing the carbon stock is to increase the area or density of tree coverage through "afforestation" (planting trees where there have not been any for at least several decades) or "reforestation" (restocking forests that have been recently thinned out or cleared). Another approach is to store more carbon by altering forest management, for example, by shortening the time between tree planting and harvesting. This works because younger, faster growing trees take up more carbon dioxide from the atmosphere than older, slower growing trees. But this net benefit will be lost if the harvested wood immediately begins to decompose and return carbon to the atmosphere.

Forests have a bigger role to play in climate protection than just carbon uptake. In particular, taking steps to slow down deforestation and soil degradation would also reduce the emissions of carbon dioxide that originate from these processes. Also, substituting wood for fossil fuels, and using lumber products as a substitute for more energy-intensive materials, would eventually lessen the amount of carbon dioxide emissions released into the atmosphere.

But trying to increase the uptake of carbon dioxide by forests can be more complicated than reducing emissions through energy conservation or using renewable energy. The difficulties are summarized by Block et al.:[54]

- *Non-permanence or reversibility.* Although tree growing and harvesting practices can be changed so that forests have a net uptake of CO_2 from the atmosphere, this uptake can easily change into emissions if the harvested wood is burned or quickly decays.
- *Small anthropogenic changes relative to large natural turnover.* The uptake and emissions from land use is only a very small component of a very large natural turnover. Since the biosphere stores and exchanges huge amounts of carbon with the atmosphere, small changes in the biosphere's stock of carbon could make large changes in the exchange of CO_2 to/from the atmosphere.
- *Uncertainty of fluxes.* It is usually much more difficult to estimate the uptake and emissions of forestry and other land use practices than it is to estimate emissions from energy and industrial sources. Emissions and uptake from land use depend on many uncertain variables that are difficult to measure over large areas.

- *Impact on global carbon cycle.* While the terrestrial biosphere is, as a whole, now acting as a sink of CO_2 from the atmosphere, this natural uptake is expected to decline over time because of climate change and other factors. This effect could interfere with policies that are meant to encourage the uptake of CO_2 or reduce the emissions of CO_2, methane, or other greenhouse gases.
- *Delay between action and emission/removal.* Actions taken now to reduce forestry emissions or increase their uptake of CO_2 from the atmosphere are likely to have an effect many years after the initial human intervention (because of the slow growth of forests) whereas taking action to reduce emissions from fossil fuels has an almost immediate effect on emissions.

Indeed the difficulty in establishing land use policies has led to a very complicated accounting system for sinks of greenhouse gases as compared to other sources.[55] But all is not lost because various "guidelines" are now available to explain to analysts how to assess the capacities of forests to take up carbon dioxide from the atmosphere.[56]

A more technological approach for offsetting emissions is through "carbon capture and storage" (CCS). This is a set of technologies used to "capture" carbon dioxide gas from industrial sources. After being captured, the gas is transported by pipeline to permanent storage in a geological formation. Carbon capture and storage is now being promoted within the EU and elsewhere as a major option for mitigating climate change.

While CCS has been used on the small scale, large-scale applications are at the experimental stage. Three alternative technologies are being considered for application to large power stations. The first is "post-combustion capture" whereby carbon dioxide is scrubbed from flue gases after combustion, the advantage being that this equipment can be retrofitted onto existing power stations. The second technology is "pre-combustion capture" whereby carbon dioxide is removed or separated before the fuel in the power plant is combusted. Fossil fuels are converted into a gas made up of hydrogen and carbon dioxide, and then the carbon dioxide is removed and the hydrogen is combusted to produce electricity. This process has a higher energy efficiency than post-combustion capture. The third approach is "oxyfuel technology" which works by burning fossil fuels at the power plant with nearly pure oxygen. This produces a flue gas made up of only carbon dioxide and steam. These two are then separated by cooling, which condenses the steam and leaves behind almost pure carbon dioxide.

Once carbon dioxide is captured by the above technologies, it typically will be compressed into a liquid state and transported to geological formations for permanent storage. Planners in the UK, for example, are considering storing CO_2 in various formations under the North Sea, including deep saline aquifers or depleted oil and gas fields.[57]

But planning of CCS facilities has slowed because the public is not convinced of the "permanence" of the CO_2 storage. The debate centers around the possibility of large- or small-scale leakages of the gas from its storage facilities. While carbon dioxide is not toxic in itself, a sudden release and exposure to concentrations greater than 7–10% of carbon dioxide by volume in air poses a serious danger to human health.[58] Views vary as to the size of the risk, but the Intergovernmental Panel on Climate Change in an assessment of CCS stated that "with appropriate site selection", and other measures, "the local health, safety and environment risks of geological storage would be comparable to the risks of current activities such as natural gas storage".[59]

Despite some public skepticism, the development and application of CCS has become part of the EU's climate protection policies, as shown below.

Economic costs and benefits of mitigation

Reducing greenhouse gas emissions carries costs compared with business as usual. However, a business as usual baseline may also be difficult to define, since increasing scarcity of oil could also mean higher costs. For example, in the case of Denmark, costs of using fossil fuels are expected to grow significantly over coming decades, and this will set in motion many economic adjustments including continuing efforts to save on energy use. Against this background, phasing out fossil fuels altogether was estimated to entail a modest additional cost on the order of 0.5% of Denmark's gross domestic product.[60] As to the feasibility of a complete phase-out, a study of the country's energy and transport system showed that it is technically possible to become completely independent of fossil fuels by 2050 without using nuclear energy.[61] And this would reduce emissions by about 80% relative to 1990.

The costs of reducing greenhouse gas emissions vary greatly depending on the economic sector and which measures are taken to reduce emissions. From the point of view of society, the cost of reducing each ton of CO_2 should be kept as low as possible. However, since greenhouse gas emissions carry costs in terms of damages associated with climate change, society may be willing to pay a price for individuals and companies to reduce their emissions, or society could require firms and individuals to do so. The EU Emissions Trading Scheme puts a price on CO_2 emitted from the EU energy sector, and this encourages the introduction of emission-savings procedures that are less costly than the price of CO_2. Similar trading schemes are not available to reduce emissions from other sectors such as transport and agriculture. Here, instead, other incentives such as taxation or efficiency requirements are put in place to curb emissions.

Mitigation measures also provide economic benefits by reducing the impacts of climate change, and the costs associated with them. In addition, they can bring economic benefits by reducing local air pollution and energy resource depletion. The economic activities stimulated by climate mitigation also make a real and substantial contribution to the emerging "Green Economy". The United Nations Environment Programme noted that investments in wind generators, solar power plants, and other low-carbon energy technologies amounted to about US$180–200 billion in 2010, with much of this coming from non-OECD countries.[62] Worldwide, more than 2.3 million people were already employed by the renewable energy sector in 2006.[63] Within Europe, it is estimated that over 1.5 million people are producing wind generators, photovoltaic arrays, and otherwise employed as part of the renewable energy industry. By 2020 this figure could increase by 3 million.[64]

Global action for climate protection

In earlier chapters we reviewed the many measures available for adapting to climate change. To complete the picture we have now had a brief survey of the options available for mitigating climate impacts including ways to reduce, avoid and offset the emissions that cause climate change. How then can we bring these options for climate protection into practice? A key starting point on the global level is the United Nations Framework Convention on Climate Change signed in 1992 and ratified by over 175 states. This was a landmark advance because it was the first clear statement by the international community that action is needed to cope with climate change. It is called a "Framework" convention because it is a legal structure onto which various concrete regulations and actions can be hung, rather than specifying these actions itself. Article 2 of the Convention states that the goal of international climate protection is "to stabilize greenhouse gas concentrations in the atmosphere at a level that would prevent dangerous anthropogenic interference with the climate system". Although the meaning of "dangerous interference" has never been agreed upon, policy makers have been urged by scientists, environmental organizations, and citizens to take preventive action against the risks of climate change.

In 1995, justified by the "Berlin Mandate", the world community began negotiating binding actions to mitigate climate change. Two years of negotiations culminated in the Kyoto Protocol to the Climate Convention, signed in 1997. This new law stipulated emission limits for industrialized countries up to a compliance period running from 2008 to 2012. Developing countries were not required to control their emissions.

Under the protocol, industrialized countries aimed to reduce their emissions by at least 5% up to the compliance period, compared to 1990. The EU was allotted an "umbrella" reduction of 8% and allowed to determine itself how its 15 Member States at that time would divide up this commitment. These states later agreed to the "burden sharing" of emission limits. This burden sharing was set to reflect both the current per capita emissions in the respective countries (Figure 10.8) and their economic strength. When 10 new Member States joined the EU in 2004 they retained the emission reduction objectives they had individually agreed to in 1992.

After long debate, emission targets were only set for industrialized nations since developing countries had argued that most global emissions of greenhouse gases up to that point had been produced as a side effect of the economic growth of richer countries. It is only fair, they insisted, that poorer countries also be given

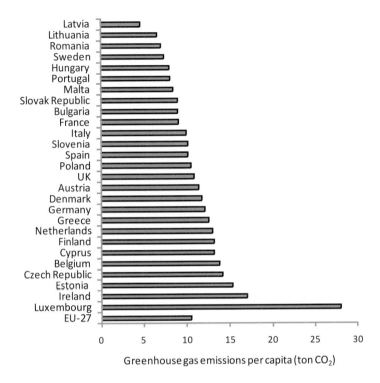

Figure 10.8 Greenhouse gas emissions per capita in 2005 for members of the EU-27. Source: EEA.[65] Redrawn with permission from European Environmental Agency, Copenhagen.

a chance to expand their economies using relatively cheap and available fossil fuels without being hindered by emission limits. While there is a certain justice to this argument, the reality is that climate change cannot be slowed without stabilizing and then reducing global emissions, and this cannot be achieved unless all major emitting countries agree to limits on their release of greenhouse gases.

Since the Kyoto Protocol runs out in 2012 a great effort is underway to develop a new international climate agreement. The new treaty has to address the growing scientific consensus that significantly larger reductions in emissions are necessary to protect global climate (see below). It must also address the debilitating draw-backs of the Kyoto Protocol such as the lack of participation by the United States, and the lack of emission restrictions on China, India and other rapidly indus-trializing countries, which account for much of the increasing trend in global emissions.

These issues were discussed up to, and during, the COP-15 meeting ("Confer-ence of Parties") of the United Nations Framework Convention on Climate Change in Copenhagen in December 2009. The aim was to agree on a global climate treaty as a follow-up to the Kyoto Protocol. However, owing to the great complexity of the issues and reluctance of some of the major global economies to commit to drastic emission reductions, the ambitious targets for COP-15 were not fulfilled. Instead the "Copenhagen Accord" was agreed to by a large, but still limited, set of countries. The Accord recognizes the need for deep emission cuts to keep global warming below a target of 2°C, and that both developed and developing countries should contribute to this. But the Accord only calls for voluntary pledges to do so. It does, however, set up institutions for reducing emissions from deforestation, for adaptation measures in developing countries, and for accelerating the development and diffusion of low-carbon technologies.

A key issue is whether the emissions reductions called for in the Accord will be adequate for meeting its own 2°C target. On one hand, global annual emissions should be around 44 gigatonnes equivalent-CO_2 or lower in 2020 to have a "likely" chance of staying within the 2°C target.[66] On the other hand, expected emissions under business-as-usual conditions may reach 56 gigatonnes by that year.[67] (These are only medium range estimates, and the band of uncertainty is wide.) That leaves a substantial "emissions gap" of 12 gigatonnes in 2020. How much does the Copenhagen Accord help in closing the gap? Adding up all the voluntary pledges under the Accord would lower expected emissions from 56 to around 49 gigatonnes in 2020, assuming "stringent" implementation of these pledges.[68] While this would leave a significant 5 gigatonne emissions gap, on the positive side these pledges would manage to narrow the gap by 60% as compared to business-as-usual. Hence, the voluntary pledges can be seen as moving in the right direction for global climate protection. Now the challenge is to find cost-effective and feasible measures to close the remaining gap.

The follow-on from the Copenhagen conference was held in Cancun, Mexico in 2010 and is referred to as COP-16. At this meeting many more countries signed on to the 2°C target and it has now become an official reference point of international climate negotiations. The so-called "Cancun Agreement" confirms the establishment of a Green Climate Fund, which will assist poorer countries in financing emissions reductions and promoting adaptation. It was further agreed to set up a technology mechanism to facilitate networks that can promote development of cheap and efficient green technologies. Indeed, the focus of negotiations is increasingly on how to promote new technologies that can solve the climate change challenges while, at the same time, helping to build a "green economy" as described above.

Another important aspect of the Cancun Agreement is its strong provisions supporting land use management in developing countries as a way to reduce greenhouse gas emissions and enhance the uptake of carbon dioxide from the atmosphere.[69] This was adopted despite the difficulties mentioned above regarding accounting and other aspects of land use policies.

Climate policy in Europe

The 20 20 policy

While the Kyoto Protocol was a step in the right direction, public support for climate protection in Europe has pushed the EU even further down the road of climate protection. In 2007 the Member States agreed to a "20 20 by 2020" reduction policy calling for stronger emission reductions within Europe. This includes:[70]

- a decrease of at least 20% in carbon dioxide emissions in 2020 relative to 1990 (the EU has also pledged an even deeper emissions cut of 30% "provided that other developed countries commit themselves to comparable emission reductions");
- a 20% share of renewable energies in EU primary energy supply by 2020;
- an "indicative" increase of 20% of energy efficiency.

The costs of reducing emissions by 30% in 2020 instead of 20% has been estimated at 33 billion euros, and this deeper reduction goal would make it easier to meet a long-term reduction target of 80–95% in the EU. Some EU countries have adopted similarly strong targets. Germany, for example, has pledged to reduce greenhouse gas emissions (relative to 1990) up to 21% by 2012, 40% by 2020, and 80% by 2050.[71]

The climate and energy package

The "20 20 by 2020" policy has ambitious goals, and the question is, how does the EU plan to meet them? The answer from policy makers is found in the "Climate and Energy Package" of legislation agreed by the European Parliament and Council in December 2008.[72] This is a series of actions on many different fronts and is a combination of extending old policies and initiating some new actions. If a single word has to be applied to this diverse collection of policies and measures, the word "pluralistic" might fit best. It has obviously evolved from a tug of war between different political and economic interests across the wide landscape of European politics. The package is made up of the following four main elements:

1 *Emissions trading* The current European Emissions Trading System will be revised and strengthened by setting a single EU-wide cap on emission allowances after 2013, by expanding its coverage to include more sectors and greenhouse gases, and by auctioning emission allowances. The basic idea of emissions trading, and two other related schemes – "clean development mechanism" and "joint implementation" – is to enable emission reductions to take place where they are cheapest and thereby lower the overall costs of mitigation. These schemes involve financial arrangements between European countries and between industrialized and developing countries (Box 10.3).

Box 10.3 Emissions trading and other "mechanisms" for reducing emissions

Apart from the question of the technologies and practices needed to reduce emissions, there is still the issue of how to organize the financing required to mitigate climate change. Many of these efforts are organized under three so-called "flexible mechanisms" mentioned in the Kyoto Protocol: "joint implementation", "clean development mechanisms", and "emissions trading".

"Joint implementation" (JI) and "clean development mechanisms" (CDM) are arrangements by which industrialized countries can invest in approved projects in other countries that lead to emission reductions or increased sequestration of carbon. For example, Finland has financed the replacement of an oil-fired power plant in Estonia with a facility fired by wood chips. Credits gained from these investments can be used by the investing country to meet their targets under the Kyoto Protocol. For instance, by financing the wood chip power plant in Estonia, Finland received a credit for the resulting emission reductions. JI covers projects in which an industrialized country invests in another industrialized country (most JI projects take place in Eastern European countries and other nations with economies in transition), CDM projects, by comparison, have to do with an industrialized country investing in a developing country.

A third mechanism permitted under the Kyoto Protocol is "international emissions trading". This is a financial scheme that allows countries with emissions below their targets in the Kyoto Protocol to sell their "excess" reductions to countries looking for low-cost ways to achieve their targets.

An approach related to international emissions trading, called the "EU Emissions Trading Scheme" (EU ETS), has become a focal point of reducing emissions within the EU. This is an arrangement allowing the trading of emission permits between major greenhouse gas emission sources in EU countries such as power plants, iron and steel manufacturers, cement manufacturers, and oil refineries. The basic idea, as with the other mechanisms above, is to enable emission reductions to take place where it is cheapest, and thus to lower the overall costs of reducing emissions. The scheme works by national governments allocating each year a certain amount of emissions to each of these installations. These allotments are based on "National Allocation Plans" which set caps on emissions and must be approved by the European Commission. This allocation has now been taken out of the hands of national governments and placed with the European Commission. Installations receiving allowances have to monitor and report their emissions and have to surrender allowances to account for their actual emissions. If their emissions are above their allowance, they are allowed to buy additional allowances from other installations to cover the gap. If their emissions are below their allowance, they can sell this "unused" part of their allowance to another installation. In principle, an installation that can cheaply reduce its emissions will do so and sell its unused allocation on the ETS market, thereby making a profit from its emission reductions. In this way the give-and-take of trading under ETS should lead to overall lower costs of reducing emissions in Europe. As of 2008, the EU ETS covered more than 40% of total greenhouse gas emissions in the EU.[74]

2   *Reducing emissions not covered by emissions trading*   A so-called "Effort Sharing Decision" sets out to control emissions from sectors not covered by the European Emissions Trading System, such as transport, housing, agriculture and waste. Under this law EU Member States agree to binding national targets to limit emissions from these sectors. The national targets have to result in an overall reduction of 10% of the emissions from these sectors in 2020 relative to 2005.

3   *Renewable energy*   Here Members States are required to agree on binding national targets for the percentage of renewable energy used in their economies. When aggregated, the national targets have to increase to the percentage of renewable energy used in the EU economy by 2020.

4   *Carbon capture and storage*   Included in the package of measures is a legal framework to promote "carbon capture and storage" (CCS). As described above, this is a set of technologies that capture the carbon dioxide emissions of concentrated industrial sources and store them underground in geological formations. Various CCS facilities are in planning stage in Europe. For example, a facility is planned for a gas-fired power plant on Norway's west coast, which will capture carbon dioxide and deposit it in an underground saline formation.[73] The EU has a policy of encouraging this approach to mitigating climate change with the proviso that it can be done "safely", that is, there must be a very low risk that the gases will later escape and endanger public safety or return to the atmosphere and add to global warming.

Apart from the Climate and Energy Package, the EU has other programs which have an important effect on reducing emissions. These are summed up in Table 10.2 and described in the following paragraphs.

The transport sector

Because transportation accounts for one-fifth of Europe's greenhouse gas emissions it is has earned special attention in European policy. One pillar of the transportation strategy aims to reduce carbon dioxide emissions from new automobiles from an average of 186 g of CO_2 per km in 1995 to 120 g/km. This should be accomplished by voluntary commitments by Japanese, Korean and European car manufacturers, by providing information to car purchasers about CO_2 emissions, and by basing car taxation rates on the CO_2 emissions of automobiles.

Another pillar of the transportation strategy is to shift some share of freight transport from road to energy-saving rail and water transport. Co-financing is provided for start-up of non-road freight services and for exchanging know-how on ways to make freight transport more energy efficient.

A third part of the transportation strategy concerns the incorporation of environmental costs in tolls and other charges paid by heavy freight vehicles. Taxation policies have also been changed to encourage more efficient use of coal, natural gas, and electricity where they are used as transport and heating fuels. Meanwhile, as part of this strategy, the fluorinated greenhouse gas (HFC-134a) used in auto

Table 10.2 Main policies and measures that make up the EU regulations on climate change. Source: EEA[75] and EC[76]

Sector	Common coordinated policies and measures
Cross-cutting	EU emission trading scheme (Directive 2003/87/EC)
	Directive linking the EU CO_2 emission trading scheme with the Kyoto mechanisms (COM(2003) 403)
	Effort Sharing Decision on reducing emissions by 10% in 2020 relative to 2005 for non-ETS sectors (406/2009/EC)
Energy supply and use	Directive on the promotion of electricity from renewable energy sources (2001/77/EC)
	Directive on Combined Heat and Power to promote high efficiency cogeneration (2004/8/EC)
	Directive on the Energy Performance of Buildings (2002/91/EC)
	Directive restructuring the Community framework for taxation of energy products and electricity (2003/96/EC)
	Directive on establishing a framework for setting of ecodesign requirements for energy-using products (2005/32)
	Motor Challenge Programme, voluntary programme from 2003 for improving energy efficiency of motor driven systems
	Directive on energy end use efficiency and energy services (2006/32/EC)
	Renewable Energy Directive (2009/28/EC)
Industry	Regulation on certain fluorinated greenhouse gases (EC 842/2006)
Transport	Reduction in average CO_2 emissions of new passenger cars (voluntary 1998/98)
	Directive on use of biofuels in transport (2003/30/EC)
	Directive on emissions from air-conditioning systems in motor vehicles (2006/40/EC)
Agriculture	Support for certain practices under Common Agricultural Policy (Regulation 1782/2003)
Waste management	Recovery of methane from biodegradable waste in landfills (1999/31/EC)

air conditioning is being phased out. A program called "STEER" supports fuel diversification, biofuel use and energy efficiency in transport systems.

The industry sector

Industry is a special focus of EU climate policy. A main goal is to prevent the fluorinated gases used for refrigeration, air conditioning, fire-fighting equipment and various industrial processes to escape into the atmosphere. To do so, financing is provided for better containment and monitoring of these gases. Under a directive, "Prevention of emissions of greenhouse gases from industrial and agricultural installations" national authorities can impose greenhouse gas emission limits on these sectors. Under the Landfill Directive the amount of biodegradable wastes delivered to landfills has to be reduced to 35% of their 1995 level by 2016. This will reduce the emissions of methane, which stem from the degradation of these wastes. Methane emissions will be further reduced by other waste-related legislation. which promotes recycling and waste prevention to reduce the amount of biodegradable wastes.

The agriculture and forestry sectors

Co-financing is available from the EU's Common Agriculture Policy (CAP) to promote land management including afforestation and reforestation to enhance the uptake of carbon dioxide from the atmosphere. A regulation under CAP provides financial subsidies to energy crop growers. The Nitrates Directive aims to promote lower-fertilizer farming to reduce excess nitrogen build-up in soils, which leads to many environmental impacts including emissions of nitrous oxide from soils (a potent greenhouse gas), and the implementation of the Water Framework Directive will in many regions further reduce nitrogen fertilization thus reducing nitrous oxide emissions.

Other policies

An assortment of other policies also lead in the direction of lower greenhouse gas emissions. An "Energy Performance of Buildings" Directive specifies a high standard of energy savings in EU structures, while another directive requires labelling the energy requirements of appliances. A comprehensive directive on "End use efficiency and energy services" requires energy companies to offer energy conservation services.

The Green Public Procurement Handbook aims to inform procurers of goods and services in Europe how they can reduce energy use through the products they procure. Meanwhile, the Sustainable Energy Europe Campaign financially supports activities to popularize sustainable energy use in Europe, while a related Climate Change Awareness Campaign raises public consciousness about the need and possibilities of climate protection in Europe.

Research and development

The EU Climate Research Program supports research to better identify sources of emissions and ways of reducing these emissions. It also makes funds available for the development and improvement of climate-friendly technologies. The LIFE Environment program provides co-financing for innovative environmental demonstration projects, including those aimed at reducing greenhouse gas emissions.

Another EU program, "Structural and Cohesion Funds" provides financing for projects that should "eliminate economic and social disparities" within the EU. Many of the priority areas of this program concern climate-friendly actions such as "sustainable transport", "forest and nature protection", "sustainable urban centers", "environmental technologies in industry", "renewable energies", and "sustainable waste management".

Effectiveness of EU policies

Since some of the above initiatives have been running for a few years, it is reasonable to take stock of which have been most effective in reducing emissions. By "effective" we mean the degree to which these policies have led to lower emissions. When asked this question, EU Member States responded as shown in Figure 10.9. At the top of the list is emissions trading, followed by the EU directive on promoting renewable energy sources. But it is also clear from this diagram that a wide range of different policies are making a contribution to emission savings.

Another conclusion of climate policy making up to this point is that policies at the EU-level also have a strong impact on national climate policies, even when they do not require action from Member States.[78] Governments reported that more than 57% of national policies and measures executed to reduce emissions were a response to "coordinated common policies and measures" agreed by the EU.[79]

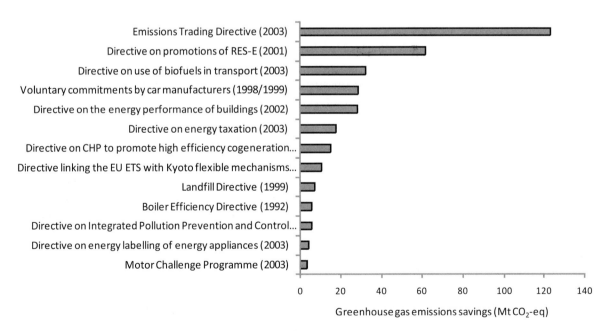

Figure 10.9 Estimating the effectiveness (in terms of total amount of emissions reduced) of different European policies in saving emissions by 2010. Based on country questionnaires. Opinions of Member States of EU-27. Source EEA.[77] Redrawn with permission from European Environmental Agency, Copenhagen.

As a final word regarding EU policy, the financial crisis that began in 2008 meant a downturn in the European economy, which reduced its energy consumption and thus also CO_2 emissions from fossil fuel use. Since the Emissions Trading System has a fixed cap on emissions, a reduction in the total emissions resulting from an economic downturn reduces the price of the CO_2 quota, and this is exactly what happened. The price per ton of CO_2 on the ETS market declined from about 25 euros in 2008 to 15 euros by the end of 2010. This low price unfortunately dampens the incentive to invest in renewable energy. On the other hand, Member States have also responded to the economic downturn with economic stimulus packages that include investments in renewable energy, which help keep this energy sector afloat.

What can cities do?

We have described in some detail the plans for climate protection at the European level, but we should not overlook what can be accomplished more locally. In some respects, the urban level is an ideal setting for carrying out climate policies because programs can be concentrated on a small geographic area and citizens have an opportunity to get directly involved in the programs. In fact European cities have been actively working together since at least 1993 as part of the "Cities for Climate Protection (CCP) – Europe" under the International Council for Local Environmental Initiatives. The CCP is a network of local authorities working to:[80]

1 strengthen local commitments to reducing greenhouse gas emissions;
2 develop and disseminate tools that increase local capacity and enhance strategies for energy efficiency;
3 promote best practices to reduce energy use in buildings and transport;
4 facilitate activities of local authorities in national and European climate strategies.

The CCP-Europe campaign has not limited itself to a single tactic for achieving its goals. Instead it has educated local officials about the principles of climate policy; organized itself as a municipal interest group and lobbied for local climate protection as part of the Conference of Parties to the UN Framework Convention on Climate Change; and even produced software for computing greenhouse gas emissions and provided user training. As a sign of the strong interest in its efforts, by 2011 more than 180 local authorities in Europe had joined the CCP-Europe network.

What can industry and corporate businesses do?

Industry accounted for 51% of the global delivered energy use in 2007,[81] and 37% of global primary energy use in 2004.[82] The large primary materials industries (i.e.

chemical, petrochemical, iron and steel, cement, and paper and pulp), account for more than two-thirds of this amount. It is estimated that up to 50% of emissions from industry can be reduced at relatively low costs using existing technologies. In the manufacturing sector emissions can be lowered by:[83]

- *measures to increase efficiencies* (e.g., installing more efficient motors, boilers and process heaters; and recycling waste materials);
- *process-specific options* (e.g., using wastes from the food and paper industries as a feedstock for bioenergy production; and recovering waste heat from manufacturing process streams);
- *Operating procedures* (e.g., minimizing steam and air leaks; making better use of insulation; and optimizing manufacturing processes to minimize energy and materials use).

The energy sector also has a large role to play in paving the way towards a society with much lower emissions of greenhouse gases. European industry is already a leader in many areas of renewable energy technologies, such as wind and solar power, and in technologies that enhance energy efficiency. But to be effective in reducing global emissions, such technologies must be used not only in Europe, but throughout the world. This provides an imperative for Europe to transfer clean technologies to less developed countries.

It is not only the manufacturing and energy sectors, but also the transportation industry with its shipping, airlines and land vehicle fleets that has a large role to play in reducing emissions. A range of technologies exist for reducing emissions per kilometer transport, and emissions can be saved by improving logistics (i.e. using good planning to decrease the transport distances of people and goods).

What can the individual do? The concept of "sustainable consumption"

Despite the complexity of climate change and climate policy, there are actually many concrete and effective steps that individuals can take to reduce their greenhouse gas emissions. All of these actions fall under the label "sustainable consumption" – that is, consuming resources and using products in a way that lessens their impact on the climate system and the environment in general. Of the many options available, here are a few examples from the "sustainable consumption and production" initiative of the United Nations Environment Programme[84] as well as energy tips for individuals provided by the EU:[85]

- Make sure your home is well insulated.
- Install efficient and appropriate thermostats.
- Lower your thermostat by 4 to 5 degrees Celsius during the night and when nobody's home.
- Avoid systematically switching on air conditioning.
- When possible, use renewable energies.

- Switch to an electricity provider that offers electricity derived from renewable sources.
- Switch off lights when they are not needed.
- Replace filament and halogen bulbs with energy-saving ones.
- Adapt lamps' wattage to actual needs.
- Do not leave appliances on standby; make sure they are completely switched off.
- Equip some appliances with a timer.
- Choose energy-saving appliances and keep them in good working order.
- Use alternatives to the automobile and airplane: ride a bicycle for shorter distances and consider a bus or train for longer distances.
- If possible, work at home instead of in an office building distant from home.
- Educate others. Tell people about the importance of an energy-efficient lifestyle.
- Avoid food wastage (don't buy more food than you can and will eat).
- Reduce your consumption of meat and other livestock-based food products.

See the EU energy portal[86] for more information about how to save energy in heating, laundry, lighting, cooking, refrigerating, cars, houses and offices.

Conclusion

Throughout this book we have seen the pervasive impact of climate change on life in Europe. We have examined how it will, or already does, affect the water cycle, agricultural system, and industries of the continent. We have also seen that we do not have to stand by passively in the face of these impacts. The first thing we can do is adapt. This is a sensible starting point since some climate impacts are already evident and are forerunners of more serious consequences of global warming.

But we are not limited to simply reacting to the climate challenge; it is also sensible and logical to be proactive, that is, to lower emissions in order to slow the tempo and ultimate intensity of climate change. The scientific community is confident that the lower the rate of emissions, the slower and less intense the change in temperature, precipitation and other climate variables and climate events. Furthermore, the slower the climate change the less likely we will surpass thresholds in the climate system and the easier it will be to adapt to the changes we cannot avoid. So, the more we do to mitigate climate change by reducing the level of emissions to the atmosphere, the more benefits we will accrue. And we have many options for mitigation: We can reduce greenhouse gases by making adjustments to our current way of doing things; we can avoid them by making essential changes in our energy and agricultural systems, and we can offset emissions by stimulating greater uptake of CO_2 in forests or possibly by carbon capture and storage.

We also have many options as to where we can act. Certainly, strong action is needed at the global level since emissions from all countries mix together in the atmosphere. On this level a new strong international regime is urgently needed for controlling emissions. Action can be, and is being, taken on the European level since the systems that cause emissions – energy, agriculture, transportation – are well integrated on the continent. Finally, action is needed and possible in our cities, communities, businesses, and homes where we can directly influence and participate in the actions for climate protection.

It is a complex and complicated matter, this business of coping with climate change. While the many changes, impacts and options may seem baffling, we know so much more now than we did just a few years ago, and the way ahead will become even clearer as we go along. While we are justified to be concerned, there is no need to panic about life under climate change in Europe.

Notes

1   MNP, 2005: *Limits to warming. In search of targets for global climate change.* Netherlands Environmental Assessment Agency. Report no. 2005999 MNP, Bilthoven, The Netherlands. 35 pp.

2   EEA, 2007: Chapter 3. Climate. In: *State of the Environment in Europe.* European Environment Agency. No. 1/2007.

3   Parry, M.L., O.F. Canziani, J.P. Palutikof, et al., 2007: Technical Summary. *Climate Change 2007: Impacts, Adaptation and Vulnerability. Contribution of Working Group II to the Fourth Assessment Report of the Intergovernmental Panel on Climate Change,* M.L. Parry, O.F. Canziani, J.P. Palutikof, P.J. van der Linden and C.E. Hanson, Eds., Cambridge University Press, Cambridge, UK, 23–78.

4   United Nations Environment Programme, 2010: *The emissions gap report. Are the Copenhagen accord pledges sufficient to limit global warming to 2° or 1.5° C?* United Nations Environment Programme, Nairobi, Kenya. www.unep.org/publications/ebooks/emissionsgapreport (retrieved April, 2011).

5   Parry, M.L., Lowe, J., Hanson, C., 2009: Overshoot, adapt and recover. *Nature* 458, 1102–1103.

6   Parry, M.L. et al., 2009.

7   IPCC, 2007: Appendix 1. Glossary. In: *Climate Change 2007: Impacts, Adaptation and Vulnerability. Contribution of Working Group II to the Fourth Assessment Report of the Intergovernmental Panel on Climate Change,* Parry, M., Canziani, O., Palutikof, J., van der Linden, P., Hanson, C. (eds). Cambridge University Press. Cambridge. U.K. p. 869–883.

8   IPCC, 2007.

9   EEA, 2008: *Greenhouse gas emission trends and projections in Europe. Tracking progress towards Kyoto target. EEA Report No 5/2008.* European Environment Agency, Copenhagen, Denmark.

10 WEC, 2004: *Comparison of energy systems using life cycle assessment – special report.* World Energy Council, July, London. www.worldenergy.org/documents/lca2.pdf (retrieved April, 2011).

11 These and other data presented in this section (unless otherwise noted) are taken from: *EC (European Commission), 2007: 2020 vision: Saving our energy*. European Commission, Directorate-General for Energy and Transport, BE-1049 Brussels. http://ec.europa.eu/dgs/energy_transport/index_en.html.

12 Hermelink, A., Hübner, H., 2006: Reality check: The example SOLANOVA, Hungary. In: *Proceedings of the European Conference and Cooperation Exchange 2006*. Sustainable Energy Systems for Buildings – Challenges and Chances (15–17 November). Vienna, Austria. www.solanova.org/resources/SOLANOVA_Paper_Pacific_Grove_08–2006_Hermelink_web.pdf.

13 Barker T., Bashmakov, I., Bernstein, L., Bogner, J.E., Bosch, P.R., Dave, R., Davidson, O.R., Fisher, B.S., Gupta, S., Halsnæs, K., Heij, G.J., Kahn Ribeiro, S., Kobayashi, S., Levine, M.D., Martino, D.L., Masera, O., Metz, B., Meyer, L.A., Nabuurs, G.-J., Najam, A., Nakicenovic, N., Rogner, H.-H., Roy, J., Sathaye, J., Schock, R., Shukla, P., Sims, R.E.H., Smith, P., Tirpak, D.A., Urge-Vorsatz, D., Zhou, D., 2007: Technical Summary. In: *Climate Change 2007: Mitigation. Contribution of Working Group III to the Fourth AssessmentReport of the Intergovernmental Panel on Climate Change*, Metz, B., Davidson, O.R., Bosch, P.R., Dave, R., Meyer, L.A. (eds), Cambridge University Press, Cambridge, United Kingdom and New York, NY, USA.

14 IEA, 2010: *World Energy Outlook 2010*. International Energy Agency.

15 EC (European Commission), 2008: *EU Energy in figures 2007/2008. Directorate-General for Energy and Transport*. European Commission. http://ec.europa.eu/dgs/energy_transport/figures/pocketbook/doc/2007/2007_energy_en.pdf. Retrieved January, 2009.

16 Sims, R.E.H. et al., 2007.

17 EC, 2008: EU Energy in figures 2007/2008. Directorate-General for Energy and Transport. http://ec.europa.eu/dgs/energy_transport/figures/pocketbook/doc/2007/2007_energy_en.pdf (retrieved January, 2009). Definition of "gross inland consumption" (from EC reference): "quantity of energy consumed within the borders of a country. It is calculated using the following formula: primary production + recovered products + imports + stock changes – exports – bunkers (i.e. quantities supplied to sea-going ships)."

18 Sims, R.E.H. et al., 2007.

19 IEA, 2007.

20 EWEA, 2011: *Wind in power. 2010 European statistics*. The European Wind Energy Association.

21 EEA, 2009: *Europe's onshore and offshore wind energy potential*. EEA Technical Report No 6/2009. European Environmental Agency, Copenhagen.

22 Sims, R.E.H., Schock, R.N., Adegbululgbe, A., Fenhann, J., Konstantinaviciute, I., Moomaw, W., Nimir, H.B., Schlamadinger, B., Torres-Martínez, J., Turner, C., Uchiyama, Y., Vuori, S.J.V., Wamukonya, N., Zhang, X., 2007: Energy supply. In *Climate Change 2007: Mitigation. Contribution of Working Group III to the Fourth Assessment Report of the Intergovernmental Panel on*

Climate Change, Metz, B., Davidson, O.R., Bosch, P.R., Dave, R., Meyer, L.A. (eds), Cambridge University Press, Cambridge, United Kingdom and New York, NY, USA.

23 FEMP, 2001: *Greening federal facilities*, Section 3.10.2 Wind energy. US Federal Energy Management Program. www.eren.doe.gove/femp/greenfed. 4pp.

24 EWEA, 2011.

25 Watson, C., 2007: Europe's first commercial CSP plant launched in Spain. *Energy Business Review*. www.energy-business-review.com/article_news.asp. (retrieved January, 2008).

26 DLR, 2005: *Concentrating solar power for the Mediterranean region*. German Aerospace Center. 285 pp. www.dlr.de/tt/Portaldata/41/Resources/dokumente/institut/system/publications/MED-CSP_complete_study-small.pdf

27 Photovoltaic Facts. Federal Energy Management Program. US Dept. of Energy, www.nrel.gov/ncpv/

28 Sims, R.E.H. et al., 2007.

29 Sims, R.E.H. et al., 2007.

30 Energy.eu – Europe's energy portal, 2009: www.energy.eu/#renewable (retrieved January, 2009).

31 Sims, R.E.H. et al., 2007.

32 Contreras, M.A., Egaas, B., Ramanathan, K., Hiltner, J., Swartzlander, A., Hasoon, F., Noufi, R., 1999: Progress toward 20% efficiency in Cu(In,Ga)Se2 polycrystalline thin-film solar cells. *Progress in Photovoltaics: Research and Applications* 7, 311–316.

33 Ishitani, H., Johansson, T.B. et al., 1996. Energy supply mitigation options. In: *IPCC Working Group II. Impacts, Adaptation and Mitigation Options*. Cambridge University Press, NY, USA. pp. 587–657.

34 Ishitani, H. et al., 1996.

35 Energy.eu – Europe's energy portal, 2009.

36 Sims, R.E.H. et al., 2007.

37 Sims, R.E.H. et al., 2007.

38 Crutzen, P.J., Mosier, A.R., Smith, K.A., Winiwarter, W., 2008: N2O release from agro-biofuel production negates global warming reduction by replacing fossil fuels. *Atmospheric Chemistry and Physics* 8, 389–395.

39 Searchinger, T., Heimlich, R., Houghton, R.A., Dong, F., Elobeid, A., Fabiosa, J., Tokgoz, S., Hayes, D., Yu, T.H., 2008: Use of U.S. croplands for biofuels increases greenhouse gases through emissions from land-use change. *Science* 319, 1238–1240.

40 Fehrenbach, H., Giegrich, J., Reinhardt, G., Sayer, U., Gretz, M., Lanje, K., Schmitz, J., 2008: Criteria for sustainable bioenergy use on a global level. IFEU (Institut für Energie- und Umweltforschung), Heidelberg. www.umweltdaten.de/publikationen/fpdf-k/k3514.pdf.

41 EEA, 2006: *How much bioenergy can Europe produce without harming the environment?* EEA Report No. 7/2006. European Environment Agency.

42 EEA, 2006.

43 EEA, 2006.

44 Sims, R.E.H. et al., 2007.

45 Data from 2003 from WEC (World Energy Council), 2004: *Survey of world energy resources.* Elsevier: Amsterdam. 446 pp.

46 Data for 2006 from EC (European Commission), 2008: *EU Energy in figures 2007/2008.* Directorate-General for Energy and Transport. http://ec.europa.eu/dgs/energy_transport/figures/pocketbook/doc/2007/2007_energy_en.pdf

47 WCD, 2000: Report of the World Commission on Dams. Executive Summary. World Commission on Dams. www.mindfully.org/Water/World-Commission-Dams-ExSumm.htm. Retrieved January, 2009.

48 European Renewable Energy Council. Small hydropower. Data from 2008.

49 This statistic only applies to the original EU-15 countries. European renewable energy council, 2009: Small hydropower. www.erec.org/renewableenergysources/small-hydropower.html. (retrieved January, 2009).

50 IEA, 2007.

51 IEA, 2007.

52 Danish Energy Agency, 2009: *Geothermal energy – heat from the inner of the earth. Status and opportunities in Denmark* (in Danish). Copenhagen.

53 Sims R.E.H. et al., 2007.

54 Blok, K., Höhne, N., Torvanger, A., Janzic, R., 2005: *Towards a Post-2012 Climate Change Regime.* Report to European Commission, DG Environment, Directorate C – Air Quality, Climate Change, Chemicals and Biotechnology.

55 Blok, K. et al., 2006.

56 Cienciala, E., Seufert, G., Blujdea, V., Grassi, G., Exnerová, Z., 2011: Harmonized methods for assessing carbon sequestration in European forests. Results of the Project "Study under EEC 2152/2003 Forest Focus regulation on developing harmonized methods for assessing carbon sequestration in European forests". 2010. Joint Research Center European Commission. 328pp. http://afoludata.jrc.ec.europa.eu/metaed/project_collection/JRC_54744_Mascaref_Full_Report_for_print.pdf

57 Carbon Capture and Storage Association, 2011: UK proposed CCS storage locations. www.ccsassociation.org.uk/ccs_projects/uk_projects.html. (retrieved May, 2011).

58 IPCC, 2005. *Carbon capture and storage. A Special Report of Working Group III of the Intergovernmental Panel on Climate Change. Summary for policy makers.* www.ipcc.ch/pdf/special-reports/.../srccs_summaryforpolicymakers.pdf

59 IPCC, 2005.

60 Richardson, K., Dahl-Jensen, D., Elmeskov, J., Hagem, C., Henningsen, J., Korstgård, J., Kristensen, N.B., Morthorst, P.W., Olesen, J.E., Wier, M., Nielsen, M., Karlsson, K., 2011: A Danish roadmap for breaking fossil fuel dependence. *The Solutions Journal* 2(4), 45–55.

61 Richardson, K. et al., 2011.

62 UNEP, 2011: Towards a green economy: Pathways to sustainable development and poverty eradication – a synthesis for policy makers, www.unep.org/greeneconomy (retrieved May, 2011).

63 UNEP, 2011.

64 EC, 2011: *Renewable energy: progressing towards the 2020 target*. European Commission. COM (2011) 31 final.

65 EEA, 2007: *Greenhouse gas emission trends and projections in Europe 2007*. EEA Report 5/2007. European Environmental Agency, Copenhagen.

66 UNEP, 2010: *The emissions gap report. Are Copenhagen Accord pledges sufficient to limit global warming to 2°C or 1.5°C? A preliminary assessment*. United Nations Environment Programme. www.unep.org/publications/ebooks/emissionsgapreport/

67 UNEP, 2010.

68 UNEP, 2010.

69 See Section C of the Cancun Agreement: "Policy approaches and positive incentives on issues relating to reducing emissions from deforestation and forest degradation in developing countries; and the role of conservation, sustainable management of forests and enhancement of forest carbon stocks in developing countries." http://unfccc.int/resource/docs/2010/cop16/eng/07a01.pdf#page=2 (retrieved April, 2011).

70 Agreed at meeting of European Council, 8 March 2007. See, e.g. UK government website: www.number10.gov.uk

71 Statement of German government, 5 December 2007, www.london.diplo.de/Vertretung/london/en/04/Science__and__technology/Climate__Change/Klima-energie-paket14–12–07,property=Daten.pdf

72 See, for example, the European Commission website: http://ec.europa.eu/clima/policies/package/index_en.htm (retrieved May, 2011).

73 http://sequestration.mit.edu/tools/projects/statoil_mongstad.html (retrieved May, 2011)

74 EEA, 2008.

75 EEA, 2008.

76 EC, 2011.

77 EEA, 2008.

78 EEA, 2008.

79 EEA, 2008.

80 ICLEI. Local governments for sustainability, website: www.iclei-europe.org/?ccpeurope and www.managenergy.net/download/nr101.pdf

81 US Energy Information Administration, 2010: International Energy Outlook 2010. wwww.eia.doe.gov/oiaf/ieo/world.html (retrieved April, 2011).

82 Bernstein, L., Roy, J., Delhotal, K.C., Harnisch, J., Matsuhashi, R., Price, L., Tanaka, K., Worrell, E., Yamba, F., Fenggi, Z., 2007: Industry. In: *Climate Change 2007: Mitigation. Contribution of Working Group III to the Fourth Assessment Report of the Intergovernmental Panel on Climate Change*, Metz, B., Davidson, O.R., Bosch, P.R., Dave, R., Meyer, L.A. (eds), Cambridge University Press, Cambridge, United Kingdom and New York, NY, USA.

83   Bernstein, L. et al., 2008.

84   UNEP, 2011: *Resource kit on sustainable consumption and production*. United Nations Environment Programme.

85   Energy.eu – Europe's energy portal, 2009: www.energy.eu/#saving (retrieved January, 2009).

86   Energy.eu, 2009.

Index

Life in Europe Under Climate Change, First Edition. Joseph Alcamo and Jørgen E. Olesen.
© 2012 Joseph Alcamo and Jørgen E. Olesen. Published 2012 by John Wiley & Sons, Ltd.